Technology

by
R. Thomas Wright
Professor, Industry and Technology
Ball State University
Muncie, Indiana

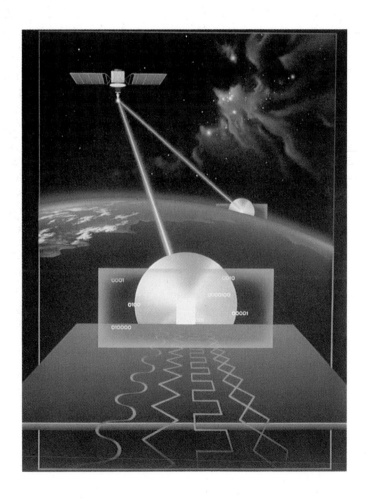

Publisher
The Goodheart-Willcox Company, Inc.
Tinley Park, Illinois

ABOUT THE AUTHOR

Dr. R. Thomas Wright is one of the leading figures in technology education curriculum development in the United States. He is the author or coauthor of many Goodheart-Willcox technology textbooks. Dr. Wright is the author of *Manufacturing Systems, Processes of Manufacturing,* and *Exploring Manufacturing.* He is the coauthor of *Exploring Production* with Richard M. Henak, and *Understanding Technology* with Howard Smith.

Dr. Wright's educational background includes a bachelor's degree from Stout State University, a master of science degree from Ball State University, and a doctoral degree from the University of Maryland. His teaching experience consists of three years as a junior high instructor in California and 25 years as a university instructor at Ball State. In addition, he has also been a visiting professor at Colorado State University, Oregon State University, and Edith Cowan University in Perth, Australia. Dr. Wright is currently a professor of Industry and Technology at Ball State University.

Cover: Copyright Bill Brooks, Masterfile Corporation
Title Page Photo: Courtesy of London Picture Service

Copyright 2000

by

THE GOODHEART-WILLCOX COMPANY, INC.
Previous editions copyright 1996, 1992

Library of Congress Catalog Card Number 98-53653
International Standard Book Number 1-56637-580-0

4 5 6 7 8 9 10   00   03 02 01

Library of Congress Catalog-in Publication Data

Wright, R. Thomas.
    Technology / by R. Thomas Wright.
       p. cm.
    Rev. ed. of: Technology systems. c1996.
    Includes index.
     ISBN 1-56637-580-0
     1. Technology. I. Wright, R. Thomas. Technology systems.
  II. Title.
T47.W74 2000   98-53653
600--dc21   CIP

INTRODUCTION

TECHNOLOGY will help you to understand:
- How people use technology to make our world work.
- Why technological systems work the way they do.
- In what ways technology affects both people and our planet.

It covers the four areas of technological activity describing how these areas work together and assist each other.
- Communication.
- Transportation.
- Construction.
- Manufacturing.

As you progress through TECHNOLOGY, you will learn that technology is a reaction to problems and opportunities, a human adaptive system. You will learn that technological systems are made up of many parts that require the use of tools. You will be introduced to the problem solving and design process, with special emphasis on the testing, evaluating, and communicating of design solutions.

Sections explore, in depth, the production of products and structures, communication, transportation, and the use of energy. Because every system must have direction, the management of technological systems is covered. The examination of societal and personal views of technology rounds out the book.

TECHNOLOGY is illustrated with photographs, drawings, diagrams, and original artwork to help explain the concepts in the text. Most of these illustrations are in color. This material has been carefully selected to make technology easy to understand. Each chapter begins with objectives to cue you in to what will be covered. Key words are in **bold** to help make you aware of them. A list of the key words is found at the end of each chapter. Review questions and activities will improve your understanding. The activities between sections provide you with valuable hands-on experience.

Impacts, both positive and negative, accompany the use of technology. The only way that people in the modern world can choose and apply technology responsibly is to understand how technology develops and how communication, manufacturing, transportation, and construction interact. The impacts of any technological activity must be carefully studied, and the negative impacts minimized.

A Student Activity Manual has activities and exercises that will give you important experience while fully enriching the concepts developed in the text.

A sound understanding of technology is vital if you are to make wise choices. As you study, you will begin to see the effects of your choices. These choices control how technology is used. Each person can make quite a difference to make sure that technology is used responsibly. With a solid understanding of technology, you can understand and take an active part in our human-built world.

R. Thomas Wright

CONTENTS

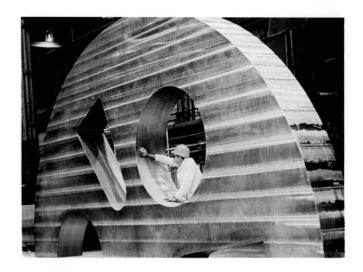

Section Five:
Applying Technology: Producing Products and Structures

Section Six:
Applying Technology: Communicating Information and Ideas

Section Seven:
Applying Technology: Transporting People and Cargo

Section Eight:
Applying Technology: Using Energy

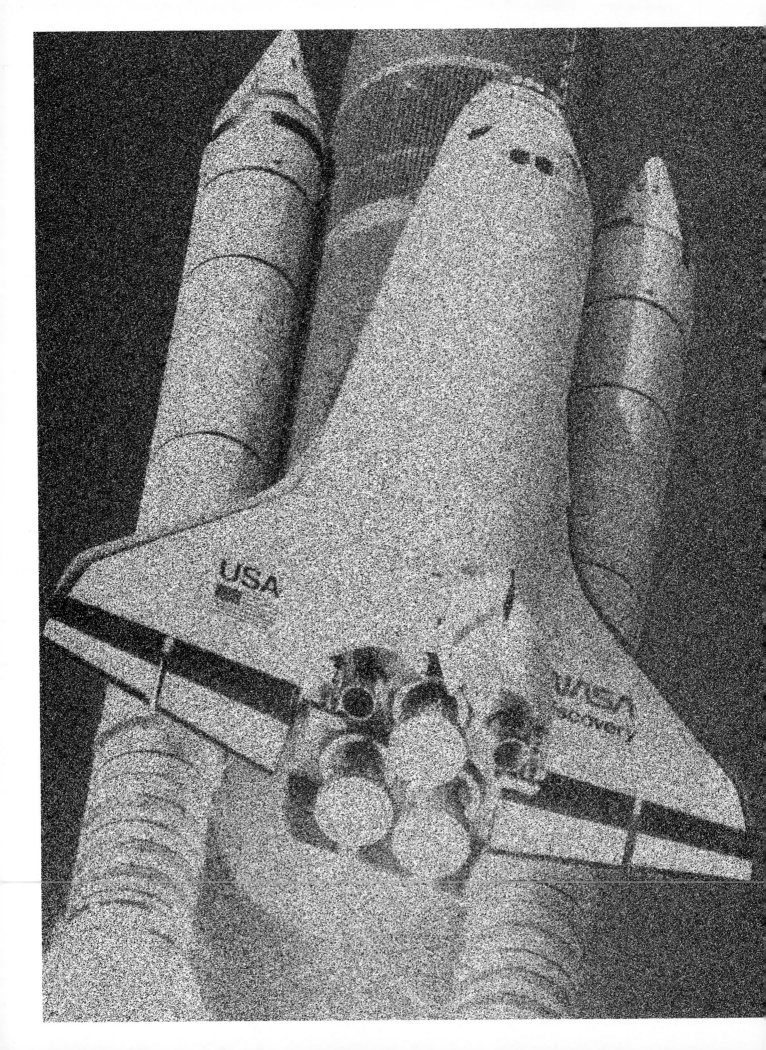

Technology

NASA

People use technology to extend their abilities. People are the most important part of any technological ststem.

TECHNOLOGY: A HUMAN ADAPTIVE SYSTEM

After studying this chapter you will be able to:
- *Define technology.*
- *Tell the difference between science and technology.*
- *Describe technology as an evolving, dynamic area of knowledge.*
- *Describe technology in the industrial and the information age.*

WHAT IS TECHNOLOGY?

Do you know what technology is? Is it mainframe and microcomputers, industrial robots, laser scanners at the supermarket checkout, fiber optic communications, rockets, and satellites circling the earth? Is it people using tools and machines to make their work easier and better? Is it an organization devoted to operating communication, manufacturing, construction, or transportation systems? See Fig. 1-1.

Chances are that you are not sure if it is one, two, or all three of these things. You are not alone in this lack of understanding of technology. Throughout the world, people use technology every day but understand little about it. We go to school for years to learn how to read and write and to study mathematics, science, history, foreign languages, and other subjects. However, few people spend time learning about technology and its impacts on everyday life.

This book has been developed to aid in understanding this vast area of human activity. It will help you become more technologically literate. After this study of technology you will be able to find, select, and use knowledge about tools and materials to solve problems. This leads to an increased understanding of technology as it affects your life as a citizen, consumer, and worker, Fig. 1-2.

PRODUCT

PROCESS

ORGANIZATION

Fig. 1-1. Technology can be described as products, processes (actions), or organizations.

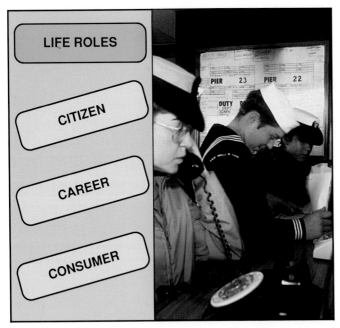

Fig. 1-2. Technology is applied to the roles each individual fulfills during life. (United States Navy)

TECHNOLOGY DEFINED

Almost everyone uses the word "technology," but what does it really mean? To some people it means complicated electronic devices and hard-to-understand equipment. To others, it means the source of the radical changes that are happening in all phases of life. Some people fear it, while others see it as the source of longer and more complete lives. Some people believe it to be a development of the twentieth century. Each of these views is partly correct.

Technology is not necessarily a complex or space-age phenomenon. It can be primitive and crude, or it can be complex and sophisticated. Technology has been here as long as humans have been on earth. *Technology is humans using objects (tools, machines, systems, and materials) to change the natural and human-made (built) environment.* Technology is conscious, purposeful actions by people designed to extend human ability or potential to do work.

Technology increases human capabilities. Through technology we can see more clearly and farther by using microscopes and telescopes. Technology helps us to lift heavy loads by using hoists, pulleys, and cranes. Technology allows us to communicate better and faster by radio, television, and telephone. Technology makes distant places close at hand when traveling by automobiles, trains, and aircraft. Technology makes life more enjoyable with video games, motion pictures, and CD recordings. Technology brings us new products and materials such as microcomputers, acrylic fibers, and artificial human organs.

Fig. 1-3. What is technology?

Technology has four basic features. As shown in Fig. 1-3, these are:

- Technology is human knowledge.
- Technology uses tools, materials, and systems.
- Application of technology results in artifacts (human-made things) or other outputs (pollution and scrap, for example).
- Technology is developed by people to modify or control the environment.

These four characteristics suggest that the products and services available in society are the result of technology. They were designed through technological innovation (development). They were produced by technological means (processes). These means are integrated with people, machines and materials to meet an identified need (management). The products were distributed to customers by technology. They are maintained and serviced through technological actions. When the products fulfilled their function or became obsolete, they were (or should have been) recycled by technological means. As you can see, without technology the world as we know it today would not exist.

Nearly all technology is developed and produced by profit-centered businesses. These businesses were organized by people to make a product or perform a service. Through this action, the owners had a product to sell. They hoped to pay their business expenses (wages, material costs, taxes, etc.) and have money left over–a profit.

To stay competitive, these businesses change and improve their products and services, Fig. 1-4. This competition exists between companies and nations. Each country's economy is directly dependent on its ability to produce products and services that can compete in the world market. The country that produces the best products will have a healthy economy and a highly utilized work force. Countries that fail to develop competitive products will experience high levels of unemployment and growing poverty.

Individual companies must be competitive in the same way. They must produce products and services that meet customer's performance, aesthetic, and price expectations. For example, automobile companies produce new models that they hope will capture a large share of the market. Likewise, cereal companies bring out new products to increase their sales. A competitive edge is gained through employing technology. This includes standardization of products, interchangeable parts, labor-saving practices, and machines.

TECHNOLOGY IS DYNAMIC

All these points suggest that, by its very nature, technology is dynamic. It is changing. It *causes* change. Technology, almost always, seems to improve on existing technology. See Fig. 1-5. The typewriter was an im-

Fig. 1-4. Most technology is developed by profit-centered businesses. (Cincinnati Milacron)

Fig. 1-5. Technology constantly improves upon previous technological devices and systems. (John Deere and Co.)

TECHNOLOGY IS CONSTANTLY CHANGING

provement over writing by hand. The electric typewriter was faster than the manual typewriter. Now the word processor and laser printer outperform the electric typewriter. This will not be the end. New devices will improve on the computer/printer system.

If past technology improved productivity, future technology will increase it even more. There is no turning back. We cannot feed the world's population using the walking plow and mules. Commerce cannot be maintained using covered wagons. People cannot build hand-hewn log cabins fast enough to provide housing for a rapidly expanding population. Technology is necessary for survival and the hope for a better future.

TECHNOLOGY IS
NOT ALWAYS GOOD

Technology, however, also has negative aspects. Poorly designed and used technology pollutes the air we breathe, the water we drink, and can cause soil erosion as shown in Fig. 1-6. Calm sunsets and unspoiled wilderness areas are threatened by inappropriately used technology.

Technology may cause unemployment and radical changes in the ways people live. New technological devices may displace workers from traditional jobs. Families may have to move long distances to find employment. Their new home may be in areas that have vastly different values and lifestyles.

The consequences of technology are now more feared than natural events. Many people are more concerned about nuclear winter, acid rain, and air pollution than they are of earthquakes and tornadoes.

This good news/bad news view of technology requires a new type of citizen. Technology must be developed and used wisely and appropriately. This challenge requires individuals who *understand* and can *direct* new technology. People who have this understanding and ability are called *technologically* literate.

TECHNOLOGY IS
DIFFERENT THAN SCIENCE

Some people think technology is applied science. They are mistaken! Technology and science are closely related but they are different. To see this difference, let's use your imagination. Assume that you plan to see your first glacier in Rocky Mountain National Park. You drive to the parking lot and leave your car. As you climb up the trail the trees are getting shorter. After a while there are no trees. The type of plant life is changing. Lichen is growing on the north side of the rocks. Finally you reach your destination. You notice that the cirque (small lake) in front of the glacier appears to be green. A cold wind is blowing over the continental divide and across the glacier. It is midsummer but snow still covers the ground. If you want to know why these phenomena happen, you must consult **science**, the knowledge of the natural world. See Fig. 1-7. Scientific knowledge is gathered from detached observation. Scientists distance themselves from the phenomena as they try to develop their explanations. They try to explain why something exists or happens in a certain way. Their work can be described as **research**.

Fig. 1-6. Poorly used technology caused this soil erosion.
(U.S. Department of Agriculture)

Fig. 1-7. Science explains the natural world.

Let's return to our make-believe hike. The long hike has made you tired. Now you must the climb down the mountain. As you come around a curve in the trail you see a welcome sight–your car! You get into it, start the engine, and drive along a paved road. You pass a small hydroelectric dam with electric power lines leading in all directions. Along the road there are homes with lights and heating systems. Finally you reach town with its stores, gas stations, and a pizza shop. Food at last! All these things are a result of technology. They are part of the human-made world or the built environment, as shown in Fig. 1-8.

These parts of the world have been developed by technological innovators and technologists. These people work with materials and they use machines and tools to make things happen. Their work can be described as **development**. They design and build products and structures to make our lives better. Technologists are not observers; they become directly involved with the processes they develop and use.

Science and technology are two major types of knowledge. A third type is the **humanities**. This type of knowledge describes the relationships between groups of people. The humanities study how people behave individually (psychology), how they work in groups (sociology), what they value (religion and ethics), and how they express themselves (art, music, and literature). The humanities also look at human behavior over time (history and anthropology).

You can begin to understand the world by using these three types of knowledge: the knowledge of the natural world (science), the human-made world (technology), and human actions within these environments (humanities). Leave out any one of them and you will make false judgments and operational mistakes.

TECHNOLOGY IS EVOLVING

Technology is seen by many people as new and dramatic. It is not. It is as old as humanity. The world is said to be about 5 billion years old. Humans have been on earth for the last 2.5 million years. During this time humans have been distinguished from other species by two major factors: their ability to make tools, and their ability to use those tools.

This indicates that humans are the only species that can develop and use technology. The level of this technology determines the type of life available.

Early humans were said to live in **primitive** conditions. These conditions are determined by nature. Primitive people try to exist with nature. They do not attempt to control nature or improve the natural condition. For example, members of primitive cultures de-

Fig. 1-8. Technology develops and explains the human-made world.

pend on harvesting natural vegetation and hunting game. Drought would severely affect their food supply. However, there is nothing they can do about it. If nature does not provide ample food, the people will starve. Likewise, primitive cultures would use only naturally occurring materials to build shelters and fabricate clothing. There are groups of people on earth today that live in primitive conditions.

CIVILIZATION

A **civilized** condition is much different from primitive existence. Civilized societies introduce the human will to the natural scene. Tools are fabricated, crops are grown, materials are engineered, and transportation systems are developed.

The evolution of civilization is directly related to the tools of the time. The major divisions of the history of civilization are named for the materials used to fabricate tools.

The Stone Age

The earliest period was the **Stone Age**. It started with pebble tools like those found in the Olduvai Gorge of East Africa. These simple stone tools were used to cut and pound vegetables and cut meat from animal carcasses. From these modest beginnings came pointed stone hunting tools. These tools allowed people to more efficiently obtain food. The population became more productive. This meant more people could live in a given area.

Technology: How it Works

SMART HOUSE: A house that allows for computer control of appliances and energy use.

Computer technology has been used to create an entirely new type of house, called a SMART HOUSE. The SMART HOUSE system makes home automation possible. Home automation allows the electrical, telephone, gas, television systems, and appliances in a house to be interactive. A few test homes using SMART HOUSE technologies have been built, and the system is being built into new homes across the U.S., Fig. 1.

The heart of a SMART HOUSE is the System Controller. It serves as the hub for messages between SMART HOUSE products in the house. The system controller constantly checks the system for problems. Electric power, telephone service, and cable television are brought into the house at the service center, where the system controller is installed.

The SMART HOUSE system uses microprocessors, also known as "chips", and a unique wiring system to allow the electrical appliances and products in a house to interact. Participating companies are developing products with SMART HOUSE chips built-in. These products can then interact with the system controller and the rest of the SMART HOUSE system.

SMART HOUSE allows the homeowner to change how their house works. This is done by programming the system controller. Changing what a switch controls is done by re-programming the switch. The heating system can be programmed to vary the home's temperature during the day. There could be cool rooms for sleeping and a warm bathroom when the first person wakes up. The water heater can be set to turn on so that hot water will be available for a morning shower. The electric range can be turned on to cook a meal that is in the oven. The dishwasher could be set to start at night in order to take advantage of lower utility rates.

The wiring system in a SMART HOUSE is unique. SMART HOUSE uses three types of cables, Fig. 2. Hybrid branch cables carry 120-volt AC power and control signals. This cable provides power for standard appliances, such as coffee makers, VCRs, and lamps. Communications cables carry audio, video, computer data, and telephone signals. Applications cables carry 12-volt DC power and control signals. This cable provides power for low-voltage devices such as smoke detectors, security systems, and switches. Large appliances, such as ovens and clothes dryers, use conventional high-amperage wiring.

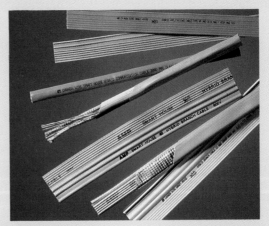

Fig. 2. The wiring used in a SMART HOUSE is special. The photo on the top shows the three types of cables: (from top to bottom) applications, communications, and hybrid branch. The photo on the bottom shows a convenience center. These special outlets can be changed to meet the needs of the homeowner.

Fig. 1. In the 21st century, many homes will use a system called SMART HOUSE to control energy use, security, communications, lighting, and entertainment.

The Bronze Age

As populations grew, new technology was needed to support the demand for food and shelter. This need was coupled with the discovery of copper and the smelting of nonferrous metals. These events allowed humans to enter the second major historical period: the **Bronze Age**. During this period, copper and copper-based metals were the primary materials for tools. Agriculture was developed so that the people did not have to depend on native vegetation and animal life. Better ways for storing food were developed, as was writing, navigation, and a number of other basic technologies.

The Iron Age

The next historical period is called the **Iron Age**. Iron and steel became the primary materials for tools during this period. It was a period of sustained technological advancement. This development continued even during the Dark Ages (about 500 A.D. to 1000 A.D.), when cultural and governmental institutions became less important.

During the last 250 years, technology has dramatically changed the world. This period saw technology widely applied to agriculture. A small percentage of the population could now grow the food and clothing fiber needed to sustain the population. This allowed large numbers of people to leave the farms and migrate to growing towns and cities. In these places, a new phenomenon was developing–the marvel of manufacturing.

Several technological and managerial developments came together here. Special machines had been developed to weave cloth, grind grain, and shape wood and metal. The waterwheel was attached to central drive shafts that could power large numbers of machines. Employees were divided into workers and managers. Each of these groups was given specific tasks to complete. Efficiency of production became an area of serious study.

The Industrial Revolution

This era has been called the **industrial revolution**, Fig. 1-9. It started in England about 1750 and moved to America in the late seventeenth century. Some inventions and developments that spurred this revolution along were Oliver Evans' continuous process flour mill, Edmund Cartwright's power loom, Joseph Jacquard's pattern weaving loom, and Eli Whitney's interchangeable parts. Henry Ford's movable conveyor, and Frederick Taylor's scientific management studies came into being later, early in the twentieth century.

The industrial revolution was further supported by the steam engine, which was followed by the fractional horsepower electric motor. These developments lead to continuous manufacture, Fig. 1-10, which is characterized by:

- Interchangeable parts.
- Dividing the job into parts that are assigned to separate workers.
- Material-handling devices that bring the work to the workers.
- Professional management.

During the industrial revolution, sophisticated transportation and communication systems were

Fig. 1-9. The industrial revolution caused the creation of the factory. (US Craftmaster)

Fig. 1-10. A view of a continuous manufacturing activity. (Maytag Co.)

developed to support the growing industrial activities. Dirt and gravel roads were turned into paved highways. The steam locomotive was replaced by the diesel electric locomotive. The dominance of transportation by the railroad was successfully challenged by the motor truck and the airplane. Pony express letter carriers were replaced by the telegraph, telephone, radio, and television.

Construction practices advanced to provide the factories, stores, homes, and other buildings needed to meet a growing demand for shelter. Log cabins and hand-hewn log homes were replaced by mass-produced dwellings. Metal buildings became the factory of choice. The shopping center and the shopping mall made downtown shopping areas less important.

Efficient production was coupled with rising consumer demand. Extended free time was available for the first time with the 40-hour workweek and annual vacations. Children could stay in school longer, because they were not needed on the farm and in the factories. Universal literacy became a possibility, even though it has never been reached.

TECHNOLOGY AND THE INFORMATION AGE

The industrial revolution moved us from an agrarian era into the industrial era. We moved from a period when most people worked raising food and fiber to a period where people worked in manufacturing. Technology is now moving many nations into a new period. This stage of development is called many different names, including the **information age**.

During the industrial age, the most successful companies processed material better than their competitors. The information age changes this emphasis. It places emphasis on processing information and cooperative working relations between workers and managers. The information age has several characteristics including:

- Wide use of automatic machining and information processing.
- High demand for trained technologists and engineers.
- Declining numbers of low-skill jobs.
- A blurring of the sharp line between workers and managers.
- A constant need for job-related training and retraining of workers.
- High quality, world-class products.
- High energy consumption.
- Technological unemployment (loss of jobs due to technology).
- Technological unemployability (inability of poorly educated workers to find jobs).
- New technologically related jobs, Fig. 1-11.

These factors promise change. However, this is not something new. It takes place with every generation. One generation traveled West in covered wagons during their youth and saw a man circle the globe in their later years. What will you see in your lifetime?

Fig. 1-11. The information age provides a different work environment and job requirements.

SUMMARY

Technology has always been a part of human life. It changes life and is changed as life progresses. It can be described as the use of tools, materials, and systems to extend the human potential for controlling and modifying the environment. It makes life better for many people. However, if it is improperly used, it can cause serious damage to people, society, and the environment. Only technologically literate people can properly develop, select, and responsibly use technology. This is the challenge for today's youth.

WORDS TO KNOW

All of the following words have been used in this chapter. Do you know their meanings?

Bronze Age
Civilized
Development
Humanities
Industrial revolution
Information Age
Iron Age
Primitive
Research
Science
Stone Age
Technology

TEST YOUR KNOWLEDGE

Write your answers on a separate sheet of paper. Please do not write in this book.
1. What is technology?
2. True or false. Technology uses tools, machines, and materials.
3. _____ tries to explain the natural world.
4. True or false. The humanities describes the human-made world.
5. True or false. We are living in the industrial era.
6. A _____ literate person can wisely select and use technology.
7. Scientists use _____ to develop their knowledge while technologists use _____ to develop technology.

APPLYING YOUR KNOWLEDGE

1. Design a simple technological device (a tool to make a job easier) that could be used in your daily activities.
2. Develop a chart like the one shown below. For each problem on the left, list the scientific and technological knowledge that would help solve it.

Problem	Scientific knowledge	Technological knowledge
Depletion of the ozone layer		
Rapid depletion of petroleum		
Rising cost of electricity		
Injuries on the football field		

3. Select three major problems in your community. Develop a chart like the one shown above. For each problem listed on the left, list the scientific and technological knowledge that would help solve it.

Technology creates tools, such as this crane, to solve problems.

2

CHAPTER

TECHNOLOGY AS A SYSTEM

After studying this chapter you will be able to:
- *Discuss technology as a response to human needs.*
- *Describe technology as a human-made system.*
- *Discuss the parts of a technological system.*
- *List and describe the inputs of technological systems.*
- *Describe the two major types of processes used by technology.*
- *Discuss the outputs of technology.*
- *Describe feedback as it is used in technological systems.*

Technology is doing a task by using an object that is not part of the human body. Suppose that you need to crack a walnut. You could put it in your mouth and bite down. This is not technology because the human body is doing the work. Perhaps your jaw is not strong enough to do the job. Maybe you value your teeth and do not want to hurt them. You will need another way to open the nut shell. You could place it on a rock and strike it with another rock. You have now employed technology. The rock extended your potential, or ability, to do a specific task.

The example above suggests that technology is the development and application of knowledge, tools, and human skills to solve problems and extend human potential, Fig. 2-1. The various parts of technological activities work together in a predictable way. This relationship is said to be *systematic.*

All technological systems share some common characteristics:
- They arise out of a human need.
- They are designed and developed by people.

- They integrate resources to produce outputs.
- They have consequences for people, society, and the environment.
- They are evaluated by people.
- In time, they are modified or abandoned.

IMPETUS OF TECHNOLOGY

Technology is a totally human entity. It is created in the human mind. It takes form through human ingenuity and labor. Technology is designed to benefit people. It has positive and negative impacts on the quality of human life. Its future is in the hands of human will.

How much technology can you see in Fig. 2-2? There is none. No technology existed on earth before human life. When humans came on the scene, the need for technology arrived. The human species is poorly fitted for survival in the natural world. We do not have thick hides or coats of fur to keep us warm and cushion blows. Our hands and feet do not make climbing or digging easy. We are not particularly strong and we possess limited stamina. Our senses of sight and hearing

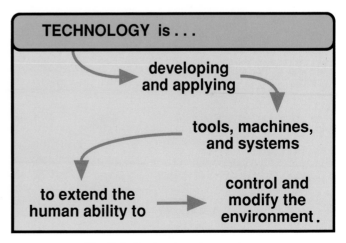

Fig. 2-1. Technology is a system.

Fig. 2-2. Can you see any technology in this photo? Technology does not occur in the natural world.

USE OF FIRE

Most technology can be traced back to a single event– the reasoned use of fire. Every animal has discovered the power of fire. They flee from it as it burns the grasslands and forests. Only humans have developed ways to use fire for their own purpose. No group of humans has been discovered that did not use fire for its own ends. Controlled use of fire allowed our early ancestors to survive the cold of the Ice Ages. They used it to cook and preserve food. Fire provided light to extend the length of the day and support communal living. The use of fire let humans plan and carry out those plans.

With the controlled use of fire, populations grew and so did the need to become more efficient producers. More food was needed; therefore, improved hunting and harvesting tools were developed. However, a predictable supply of natural food was still not available. The availability of game and wild plants was limited and directly affected by natural events such as droughts and disease.

are not sharp. Our teeth are most efficient when eating food that has been changed from its natural state by cooking.

In short, humans have needed help to survive over time. They needed what technology could provide. They needed tools and machines, warm and safe shelter, transportation devices, and communication media. These are the products of need, Fig. 2-3. There is not a single technological artifact (tool, machine, etc.) that came about without a need. The inventor of each item thought that life would be better with the new device. Can you think of some technological device you need that is not on the market? What need would it fulfill?

AGRICULTURE

This situation led to a second critical development: agriculture. About 14,000 years ago, people in what is now the Middle East discovered that they could plant seeds, harvest the crops, and assure themselves an adequate food supply. This discovery led to the development of agricultural tools and irrigation systems which allowed people to grow large quantities of food.

Agriculture allowed humans, for the first time, to produce more food than they needed, Fig. 2-4. Now, selected people could be freed from the daily search for food. They could become writers, poets, philosophers, artists, merchants, and priests.

Fig. 2-3. What human need is the technology shown in this photo designed to meet? (United Airlines)

Fig. 2-4. Modern agricultural technology allows a small number of people to grow vast quantities of food and fiber. (John Deere and Co.)

Ample supplies of food allowed commerce (the trading of goods) to develop. Goods from distant locations were transported and traded. New tools of transport were developed. The wagon and the sailing ship were two of these new inventions.

Expanded commerce led to the development of small factories with crude machines. These new enterprises produced the products needed by people. In more recent years, complex industrial machines have been developed and refined. They allow us to have an abundance of products and services. Today, entire factories turn out products with few humans involved. In the future, computers will probably plan, direct, and monitor most factory operations.

Each of these historical steps in technological development was a response to human needs. There will continue to be new technological advancements designed to meet needs that we cannot envision today. Your great-grandparents did not imagine space travel and humans on the moon, Fig. 2-5. These events were for comics books and science fiction stories. However, you know that they have happened and are no longer questioned. Likewise, thousands of technological advancements will take place in your lifetime. Most of these you cannot even imagine at this time. They will all arise out of someone's definition of a need.

SYSTEM COMPONENTS

Technology is a human-made system. This means that technology has parts and each part has a relationship with all other parts. Typically, as shown in Fig.

Fig. 2-5. Only a few years ago space travel was the work of fiction writers. Technology has made it a reality. (NASA)

2-6, a system has several major components which include:
- Goals.
- Inputs.
- Processes.
- Outputs.
- Feedback.

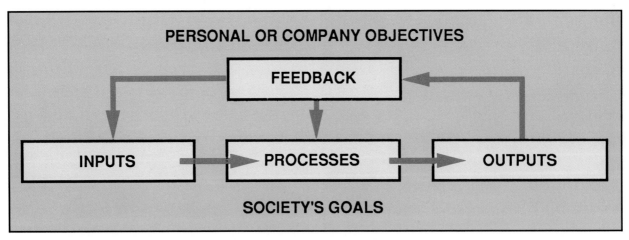

PERSONAL OR COMPANY OBJECTIVES

FEEDBACK

INPUTS → PROCESSES → OUTPUTS

SOCIETY'S GOALS

Fig. 2-6. This is a simple model of a system.

GOALS

We said earlier that all technology was developed to meet an actual or perceived need. This would suggest that all technology is designed to reach a goal. For example, RADAR (*RA*dio *D*etection *A*nd *R*anging) was developed to determine the position and speed of aircraft. RADAR had a major impact on military strategies and actions during World War II. Today this technological innovation contributes to safe, reliable air transportation. However, this is only one goal that RADAR meets. The goals that any technology can meet are generally numerous and complex.

Suppose that a new coal powered automobile is being developed. It is not enough to say the automobile is being created to *transport people*. This is certainly the primary goal. However, it could, also, give the innovators an opportunity to *make money*. Local government

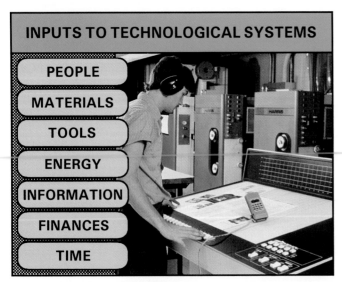

INPUTS TO TECHNOLOGICAL SYSTEMS

PEOPLE

MATERIALS

TOOLS

ENERGY

INFORMATION

FINANCES

TIME

Fig. 2-7. These are the inputs of technological systems. All of these inputs are needed to operate any system, such as the printing press shown in the photo.

leaders could support its development as a means of *economic growth* for their city. The Federal government may provide support to reaffirm *national technological leadership* and improve the *foreign trade deficit*. The general public could support it as an alternative to petroleum powered vehicles and to reduce *dependence on foreign oil*. Workers could support the activity as a way to improve *job security*. Coal mining companies may provide support to *enlarge the market* for their product. Environmental groups would comment on how the car *affects the environment*. Note the number of different goals and concerns that are highlighted by italic type. As you can see, a technological development needs to meet a number of different goals and concerns which are important to different groups.

INPUTS

All natural and human-made systems have inputs. **Inputs** are the elements that flow into the system and are consumed or processed by the system. Technological systems, as shown in Fig. 2-7, have at least seven inputs.

People

People are the most important input to technological systems, Fig. 2-8. Human needs and wants give rise to the systems. Human will and purposes decide the types of systems that will be developed. People, through their local, state, and national governments, make policies that promote or hinder technology. People bring to the systems specific knowledge, attitudes, and skills. They provide the management and technical know-how to design and direct the systems. Their labors make the systems function. Human ethics and values control and direct the systems. Finally, people are the consumers of technological outputs. They use the products and services that the systems provide.

Materials

All technology involves physical artifacts (tools, structures, vehicles, etc.). These artifacts are made from liquid, gaseous, and solid materials. **Materials** are all natural matter that is directly or indirectly used by the system. Some of these materials provide the mass and structure for technological devices. Other materials support the productive actions of the system. They may lubricate machines, contain data, package and protect products, etc.

Tools (Machines)

Technology is characterized by tools and machines. **Tools** are the technical means that must be present before we have technology. These technical means may be simple hand tools that rely on human muscle power for their operation. Tools may also be complex **machines** which amplify the speed, amount, or direction of a force.

The complexity of machines and tools may range from a hammer to a space vehicle or from a can opener to a television transmitter. Tools and machines are devices used to locate and extract resources, make products, build buildings and other structures, communicate information and ideas, convert and transmit energy, and transport people and goods.

Energy

Technology involves doing something, and all technological activities require energy. This energy may be in a number of forms. It can range from human muscle power to nuclear power; from heat energy to sound energy. **Energy** is the ability to do work.

Technological systems require that energy be converted, transmitted, and applied. The energy of falling water may be converted into electrical energy by a hydroelectric generator. The electricity may be carried to a factory over transmission lines. At the factory, motors may convert the electrical energy into mechanical motion, or halogen and fluorescent bulbs may convert it into light energy, or radiant heaters may convert it into heat energy.

Information

The world is full of data, information, and knowledge. Everywhere you turn there are facts and figures which are called *data*. These elements become useful only when they are grouped by type and organized for review. Organized data is called **information**, and is essential for operating all technological systems.

An example will show the difference between data and information. You might measure the size and weight of everything you can find. This would be data because it is random and assorted. However, you could sort the data so the height and weight of all people are grouped together. Now you have information which is organized data. From this information you can see relationships and draw conclusions. Adults are taller than children. Men are generally taller than women. With this final step you have developed **knowledge**. Knowledge is people using information to understand, interpret, or describe a specific situation or series of events. Information lets you say "I know!" while knowledge lets you say "I understand!"

Finances

Technology inputs require people, materials, energy, information, and tools. These resources have value and, therefore, must be purchased. Also, the outputs of technology have value and can be sold. **Finances** are the money and credit necessary for an economic system.

For example, a technological artifact, such as a home, may be needed. Land and materials to build the structure must be purchased. Plans for the dwelling must be obtained. Equipment must be rented or bought and workers must be hired to build the house. All of these actions require money. The finished home may be sold to cover these costs. Generally money will be left after all expenses are paid. This amount of money is called **profit**, which is the goal of most economic activities.

Time

All jobs or activities take **time**. Each person has only 60 minutes in an hour and 24 hours in a day. This time is allotted to the various tasks that need to be done. Likewise, time must be allotted to all technological endeavors. The most important ones will be completed. If time is not available, less critical tasks will be left

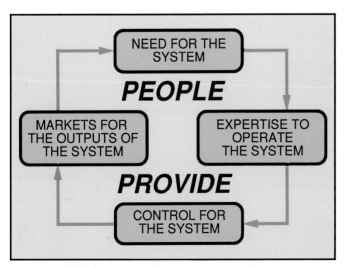

Fig. 2-8. These are the roles humans play in the technological system.

undone or postponed to a later date. Therefore, not all technology that is needed can be developed immediately. Some will have to wait until time and other resources are available.

PROCESSES

All technology is characterized by action. A series of identifiable tasks must be completed. The steps needed to complete these tasks are called **processes**. Technology uses two major types of process: problem solving processes and transformation processes, Fig. 2-9.

Problem Solving Process

In Chapter 1, science was described as the activities that study and interpret the natural world. Scientists use research to develop their descriptions. They carry out their work through a set of procedures called the scientific method. It structures the research so that valid results can be obtained.

Technology, as described earlier, develops and uses tools, machines, and systems to extend the human ability to control and modify the environment. One key word in this definition is *develop*. This part of technology requires creative action. The procedure used to develop technology is called the **problem solving process.** It involves five major steps, Fig. 2-10:

- Identifying the problem. Basic information about the problem and the design limitations are developed
- Developing solutions to the problem. Several possible solutions to the problem are developed and refined through ideation and brainstorming (creative thinking) procedures.
- Isolating and detailing the best solution. The best solution is selected and refined.
- Modeling and evaluating the solution. Physical and/or graphic models of the selected solution are

produced and tested.
- Finalizing the solution. A final solution is selected and prepared for production and use.

Transformation Processes

Once a technological device or system is designed, it must be built and operated. This process may be seen as **transformation**. Resources are used to change or transform a key input into a more usable form. For example, technology may change (transform) petroleum into polyethylene. The polyethylene may be changed through technological actions into packages. Likewise, technology can change trees into lumber which is used to make furniture.

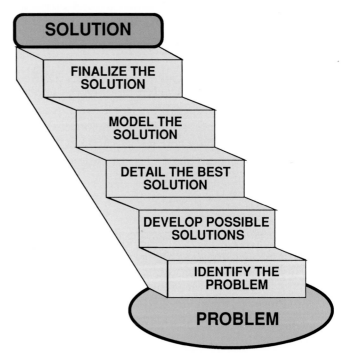

Fig. 2-10. These are the steps in the problem solving process.

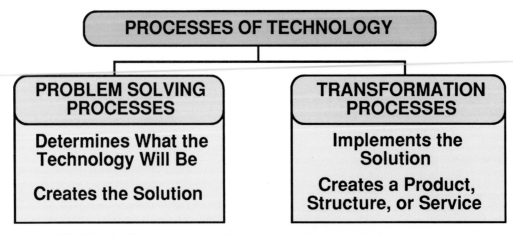

Fig. 2-9. Technology uses problem solving and transformation processes.

TRANSFORMATION PROCESSES

MANAGEMENT PROCESSES

PRODUCTION PROCESSES

Fig. 2-11. The types of transformation processes are management and production. (General Motors Corp.)

Transformation requires two types of technological processes. These, as shown in Fig. 2-11, are production processes and management processes.

Production Processes

Production processes are actions that are completed to perform the function of the technological system. For example, a series of production processes are used to produce an informational booklet, Fig. 2-12. The message or copy to be communicated is written and edited. Photographs to illustrate the document are produced. The photographs and copy are integrated into a page layout. The layout is converted into printing plates from which copies of the booklet are printed.

Production processes are used to change natural resources into industrial materials, to convert materials into products or structures, to transform information into media messages, to convert the form of energy, or to use energy to power transportation vehicles to relocate people or goods. All production processes involve the five major actions shown in Fig. 2-13. These are:

- Analyzing the design for the device (product or structure), service or system. (The design was the output of the problem solving process.)
- Obtaining the resources (inputs) needed to produce the device, service, or system.
- Using the resources to produce the device, service, or system (process).

Fig. 2-12. The production process can convert page layouts (top) into booklets (bottom). (Graphic Arts Technical Foundation)

```
┌─────────────────┐      ┌─────────────────┐      ┌─────────────────┐
│  ANALYZE THE    │ ───▶ │     OBTAIN      │ ───▶ │    PRODUCE      │
│     DESIGN      │      │   RESOURCES     │      │    OUTPUTS      │
└─────────────────┘      └─────────────────┘      └─────────────────┘
```

PRODUCTION PROCESSES

```
┌─────────────────┐      ┌─────────────────┐
│  MAINTAIN AND   │ ◀─── │    DELIVER      │ ◀───
│ RECYCLE OUTPUTS │      │    OUTPUTS      │
└─────────────────┘      └─────────────────┘
```

Fig. 2-13. These are the actions in the production process.

- Delivering the device, service, or system (output).
- Operating, maintaining, servicing and, in time, disposing of or recycling the device or system.

Management Processes

Management processes are all the actions that are used to ensure that the production processes operate efficiently and appropriately. These processes are used to direct the design, development, production and marketing of the technological device, service, or system. Management activities involve four functions, Fig. 2-14:

- **Planning:** Setting goals and courses of action to reach the goals.
- **Organizing:** Dividing the tasks into major segments so that the goals can be met and resources assigned to complete each task.
- **Actuating:** Starting the system to operate by assigning and supervising work.
- **Controlling:** Comparing system output to the goal.

Management processes are used by individuals and groups to organize and direct their activities. You may have a task to complete, such as writing a term paper. First, you must *plan* this activity by selecting a topic, establishing major steps to be completed, and setting deadlines for each task. You then must obtain and *organize* resources. Reference materials, writing or word processing equipment, and time must be obtained. Then, the work must be *actuated* by reading and viewing reference material, taking notes, preparing a draft of the paper, and editing the draft into final form. Finally, the paper will be compared to established standards. The results will be given to you. This step is *control,* which includes evaluation, feedback, and corrective action.

OUTPUTS

All technological systems, are designed to produce specific outputs. These **outputs** may be manufactured products, constructed structures, communicated messages, or transported people or goods. However, it would be inaccurate to suggest that these primary outputs are the only ones that come from technology.

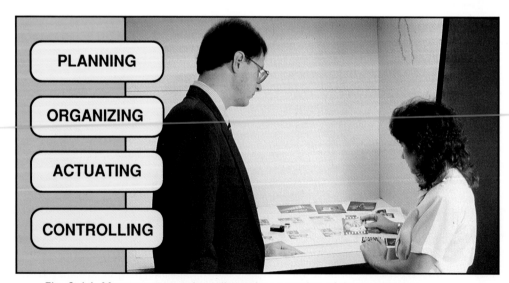

Fig. 2-14. Management actions direct the operation of the production process.

Technology: How it Works

SOLAR FURNACE: A high temperature device that is heated by concentrated solar energy.

The sunlight striking the earth contains a great quantity of energy. The challenge facing advocates of solar energy is capturing this energy for use. It as been estimated that if all solar energy could be collected and used, it would produce as much electricity as 170 million large electric generating plants. This is far more energy than will ever be needed.

Solar collectors can be of two types: solar thermal and solar electric. Solar thermal collectors capture heat from the sun by moving air or a fluid through the collector. The air or fluid absorbs the heat and carries it away for storage or for use. Solar electric collectors use materials that are *photoelectric*. When light strikes a photoelectric material, electrons move and create an electric current.

A large number of collectors can be placed together in a *solar orchard*, Fig. 1. Both solar thermal and solar electric collectors are placed in solar orchards. Solar thermal collectors are used to produce steam. This steam is fed to a generating plant to make electricity. Solar electric collectors can be connected directly to the electric utility grid. Both types of collectors use tracking systems to follow the sun across the sky.

Another method of using solar energy is being tested. This is the solar furnace, which uses the reflected light from a number of mirrors called heliostats. Each of these

Fig. 1. This solar orchard is in Phoenix, Arizona. These collectors use solar electric cells with concentrating lenses to increase the power output.

devices track the sun and reflect the solar energy onto a parabolic mirror. This mirror, in turn, reflects the light from all the heliostats onto a single point. The result is a very intense source of heat. The temperatures can reach 5,400 °F (3,000 °C).

A large solar furnace has been built for research in France, Fig. 2. It uses 63 flat mirrors to collect and reflect the sun's energy.

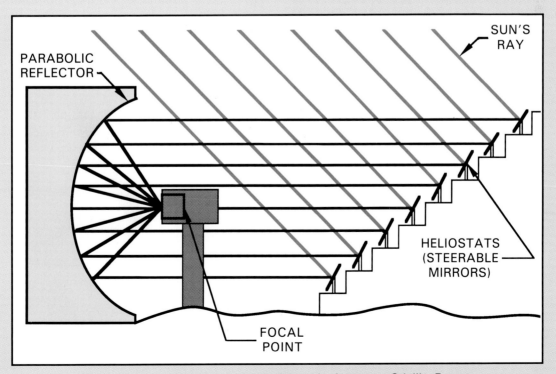

Fig. 2. This is a schematic drawing of the solar furnace at Odeillo, France.

In operating technological systems, other less direct and unwanted outputs are produced. Scrap and waste is generated from manufacturing and construction activities. Chemical by-products, fumes, noise, and other types of pollution are often products of manufacturing operations. Noise, air pollution, and congestion are often produced by cars, airplanes, trains, and trucks. Poorly designed housing developments, industrial parks, and shopping centers can contribute to soil erosion, as can careless farming practices. See Fig. 2-15.

Technological systems also have social and personal impacts. The products of technology shape society and are shaped by society. For example, the automobile was a novelty in the early 1900s. However, the wide acceptance of the automobile has greatly changed where and how we live. We travel more and live farther from our place of work. The small compact town has been replaced by sprawling cities. Shopping is no longer close to the neighborhood. On the other hand, we have shaped the automobile. Rising gasoline prices and government fuel-economy standards have made cars smaller and lighter. Personal buying habits also dictate the types of cars that will be built.

FEEDBACK

All systems are characterized by **feedback**. This process involves using information about the outputs of a process or system to regulate the system. Feedback is used in many common situations. For example, many homes have heating systems that are controlled with a thermostat. The occupant sets the thermostat at a desired temperature. The thermostat measures the room temperature. If it is too low, the thermostat closes a switch that turns on the furnace. As the room is heated, the thermostat monitors the temperature. It is using the output of the furnace to determine the necessary adjustments for the system. When this output (heat) warms the room to the proper temperature, the thermostat changes the system's operation. It opens the switch and stops the furnace. The cycle is repeated when the room cools to a specific temperature.

SUMMARY

Technology is human-made systems that use resources to produce desired outputs. The systems use seven major types of resources: people, materials, tools, energy, information, finances, and time. These are used during the operation of the technological systems. This operation stage is called the process. It involves two major types of processes: problem solving and production. The result is desired and unwanted outputs. Technological systems are used to produce products, construct structures, communicate information, and transport people and goods. These things improve our way of life. These are the desired outputs. However, we also get pollution, scrap and waste, altered life styles, and increased health risks. These are the negative impacts or outputs of technology. The challenge for each person developing or using technology is to maximize the desired outputs and minimize the negative impacts at the same time.

WORDS TO KNOW

All of the following words have been used in this chapter. Do you know their meanings?

Actuating
Controlling
Energy
Feedback
Finances
Information
Inputs
Knowledge
Machines
Management processes
Materials
Organizing
Outputs
People

Fig. 2-15. Negative outputs may be produced by poorly designed or operated technology systems. This soil erosion was caused by the improper use of agricultural technology. (Soil Conservation Service)

Planning
Problem solving
Processes
Production processes
Profit
Time
Tools
Transformation

TEST YOUR KNOWLEDGE

Write your answers on a seperate sheet of paper. Please do not write in this book.

1. What roles do people play in technology?
2. True or false. All technology is developed to meet human needs and wants.
3. Two early steps in technological development were _____ and _____.
4. List and define the seven inputs to all technological systems.
5. True or false. Technology primarily uses the scientific method to develop new technological devices and services.
6. The two transformation processes are _____ and _____.
7. True or false. All technology produce desires and undesired outputs.
8. List three desired outputs of technologies that you use.
9. List three undesired outputs of technology that you have encountered.

APPLYING YOUR KNOWLEDGE

1. Select a technological device that you use and list its inputs, processes, and desired and undesired outputs. Organize your answer in a form similar to the one shown below.

Device: _____
Major inputs:
Production processes used to make the device:

Desired outputs	Undesired outputs

2. Select an early technological advancement (invention) and prepare a short report that includes the inventor's name, nation from which it came, events that led up to the invention, its impact on life at the time, and its later refinements. Include a sketch of the item.

All technological systems work together to make a product or construct a structure, such as this water tower.

TYPES OF TECHNOLOGICAL SYSTEMS

After studying this chapter you will be able to:
- *Describe technology in terms of level of development, economic structure, number of people involved, and type of product or service produced.*
- *Discuss high technology and the development of society.*
- *Describe the three stages of developing and operating technology by a private enterprise.*

Technological systems are everywhere. They support our daily lives, produce the devices that make up the human-made world, and are vital to the economic and political health of every nation in the world.

Technological systems, as shown in Fig. 3-1, can be grouped in at least four ways:
- Level of development.
- Economic structure.
- Number of people involved.
- Type of technology developed and used.

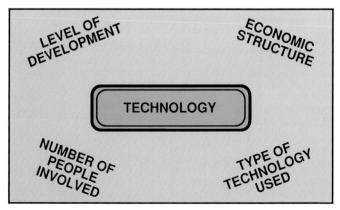

LEVEL OF DEVELOPMENT

ECONOMIC STRUCTURE

TECHNOLOGY

NUMBER OF PEOPLE INVOLVED

TYPE OF TECHNOLOGY USED

Fig. 3-1. This diagram shows the ways to view technology.

LEVEL OF DEVELOPMENT

Technology is a constantly changing phenomena. Simple technologies are constantly being replaced by more sophisticated ones. For example, the sail was used on boats as early as 5000 B.C. in Mesopotamia. This made water transportation more efficient, and took less human effort than rowing the boat. The steam engine was used by Robert Fulton to power a ship in 1803. This advancement soon replaced the sail as the primary power source for commercial shipping. However, the steam engine did not last as long as the sail. It was replaced by the diesel engine in the 1900s. In 1954, nuclear power was first used in the submarine U.S.S. Nautilus and is in use in naval vessels around the world.

OBSOLETE TECHNOLOGY

At each stage of history there are *obsolete, current,* and *emerging* technologies, Fig. 3-2. The obsolete technologies are those that can no longer efficiently meet human needs for products and services. For example, hand spinning and weaving of cloth are obsolete. Some artists and home hobbyists still use this technology to produce works of art and personally designed items. Likewise, artists use the skills of the old-time blacksmith to produce decorative, forged steel items. However, this technology is seldom used to produce products for the mass market.

CURRENT TECHNOLOGY

Current technologies include the range of techniques used to produce most of the products and services today. You can see these technologies everywhere you look. Trucks, trains, and aircraft are the common vehicles we use to transport the majority of goods in this

OBSOLETE TECHNOLOGIES

CURRENT TECHNOLOGIES

EMERGING TECHNOLOGIES

Fig.3-2. These are examples of the three stages of technology. (Left) The gasoline-powered Wright brothers' biplane is obsolete. (Center) Jet-powered passenger airplanes like these are used around the world today. (Right) An efficient high-bypass engine for use on passenger airplanes is now being developed and tested. (United Airlines, NASA)

country. The print and electronic media deliver the majority of the information to the general public. There are also a series of common technologies used to produce products and construct structures.

EMERGING TECHNOLOGY

Emerging technologies are the new technologies that are not widely employed today. They may, however, be commonly used in a later period of time. Today, laser machining, space travel, and electronically delivered magazines and books are emerging technologies. We often call these technologies, **high technology** or **high-tech**. This term is somewhat misleading because many of yesterday's high technologies are now current technologies, Fig. 3-3. Today's high tech processes may be tomorrow's current technology. For example, fewer than 25 percent of all Americans can remember a world without television. However, television was considered high technology in the early 1950s. Fiber-optic communication was unknown to most people until the 1980s. Today, it is used in most telephone systems in the United States. Therefore, it is better to think of high-tech as a moving focus: new and strange today, current and common tomorrow, and obsolete and seldom used at a later date.

Fig. 3-3. This phonograph was a high-tech entertainment device in the early 1900s, but now it is an antique collector's prize.

ECONOMIC STRUCTURE

Technology has been developed by people to serve humankind. It has also become an integral part of the economic system. From this point of view, we can see technology as either public (government) or private (profit-centered).

Most technology is developed and applied by private enterprise. These are the businesses that produce the goods and services that people want. They plan to make a profit by meeting these human needs.

PROFIT-CENTERED COMPANIES

These profit-centered companies generally move technology through three stages as they meet human needs and wants. These stages, as shown in Fig. 3-4 are:

- **Research and development:** Designing, developing, and specifying the characteristics of the product, structure, or service.
- **Production:** Developing and operating systems for producing the product, structure, or service.
- **Marketing:** Promoting, selling, and delivering the product, structure, or service.

These stages are supported by two major activities that provide the resources needed to operate the systems. These are:

- **Industrial relations:** Setting up and managing programs to ensure that the relations between the company and its workers and the public are positive.
- **Financial affairs:** Obtaining the money and physical resources and keeping the financial records needed to manage the system.

GOVERNMENT

Some technology that is important to the general public is too expensive or risky to be developed by private enterprise. In this case the government may address the "general welfare of the citizens" by developing a specific technology. The goal is not to make a profit from the undertaking. The space program with its specific hardware is an example of this type of public-funded technological developments. Out of this type of program come many products and systems that can later be used by individuals and companies. Examples of NASA's (National Aeronautics and Space Administration) space program technological innovations that have become commonplace are:

- Mylar from the early Echo program, which is now used for clothing, packaging tape, and decorative balloons.
- High-temperature composite materials from the Lewis Research Center, which are used in aircraft engines and aircraft structures.
- Satellite communications systems from many programs that now make it possible to communicate around the globe instantly, Fig. 3-5.

Fig. 3-4. These are the stages in developing, producing, and marketing the outputs of technological systems. Magazines and books are shown here.

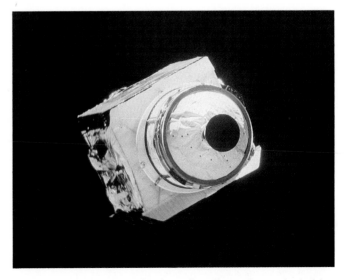

Fig. 3-5. Technology that benefits all people, such as this communications satellite, is sometimes developed by the government. (NASA)

Technology: How it Works

CONTAINER SHIPPING: Using sealed containers to group and contain items for bulk shipment.

Shipping cargo across oceans has become a commonplace activity. In earlier times each crate and box was lifted by itself onto a ship and stored in the hold. This technique was time-consuming and expensive. It tied up the ship in port for a number of days at each end of its trip.

To increase the efficiency of ocean shipping a new type of ship was needed. To meet this need the *container ship* was developed in the 1960s. This type of ship uses steel containers that are 20 or 40 feet (6 or 12 meters) long, 8 feet (2.4 m) high and 8 feet wide (2.4 m). Each container can hold up to 30 tons (27 metric tons) of cargo.

Container shipping works by grouping a number of small shipments going to one place into one large load. The grouping of shipments by their destination is called *consolidation*. Each item is handled only twice: once when it is placed in the container, and another time when it is removed from the container.

Container shipping has several steps, Fig. 1. First, the cargo is picked up from the shipper. A number of different shipments destined for the same location are put together. These shipments are loaded into a container. The container is then transported to the dock. Containers can be placed on a set of wheels to form a semi-trailer that is pulled by a truck. The container can be pulled by the truck to the dock. Containers are sometimes hauled by truck to a rail yard. Here the container is lifted off the set of wheels and set onto a railcar. It makes the rest of its journey by rail.

Often, entire trains are made up of rail cars loaded with containers. The train moves the containers to the port where they are stored in large stacks. When the ship comes to port, the containers are moved to the dock. They are lifted and placed on the ship by large cranes. Using this process dock workers can load an entire ship in less than a day. Using the older loading techniques could take up to five days to unload a ship and another five days to load it.

In addition to being faster, container shipments do not require warehousing at dock sites. Each container is a small, weather-tight "warehouse."

Container shipping is now being adapted for air transportation, Fig. 2. The development of the wide-body airplanes allows the use of large containers. Containerized shipping is popular with overnight parcel companies.

LOADING THE CONTAINER

MOVING THE CONTAINER TO THE PORT

CONTAINERS ARRIVE AT THE PORT

LOADING THE CONTAINER ONTO A SHIP

Fig. 1. These are the steps used in container shipping by sea. Once the ship reaches its destination port, these steps are reversed. The containers are unloaded from the ship, and then moved to their individual destination.

Fig. 2. Wide-body aircraft, such as this 747, allow for containers to be shipped by air.

- Riblet film with its v-groove design that reduces water and wind resistance. It was developed for increasing fuel efficiency of jet aircraft and was used on the America's Cup winning yacht, *Stars and Stripes*.
- A corrosion-resistant coating that was developed for launch pads, was applied to the refurbished Statue of Liberty.

NUMBER OF PEOPLE INVOLVED

Technology can be described by the number of people involved and how they work with the technological system. In isolated cases, technology is developed by people for their own use. A cattle rancher may develop an automatic feeding system for his or her livestock. A home gardener may develop a unique hoe that meets his or her personal needs, as shown in Fig. 3-6. A home craftsperson may design and build a desk for his or her own use. This type of personal technology is not common. It accounts for a small portion of all the technology in use today.

CORPORATE PARTICIPATION

Most technology is developed and applied by groups or teams of people. The key to this activity is cooperative attitudes. Often it is said that "We live in a competitive society." This belief is partly true. We compete to get on a sports team or to get a job. Companies compete to sell products and gain market share. Nations compete for prestige. However, within each of these groups

Fig. 3-6. This home-built garden hoe was developed by a gardener for his own use.

is cooperation. The winning sports team is a study in cooperative attitudes. The most profitable companies exhibit high levels of management and worker cooperation, Fig. 3-7. Nations cooperate in organizations such as the United Nations. Once a person gets a job, cooperation will assure success.

This spirit of cooperation in developing and operating a technological system can be described as corporate participation. This does not mean that the corporate form of ownership must be present. It means that the people are united and combined into one body. They share a common vision, work toward similar goals, and share in the success of the enterprise.

Fig. 3-7. Most technology requires a high level of cooperation between people for its development and efficient operation. (Northern Telecom)

TYPE OF TECHNOLOGY DEVELOPED AND USED

The three ways of looking at technology discussed so far are useful in a general way. However, they give only part of the picture. A more meaningful approach to understanding technology categorizes the area by the type of technology developed and used. It looks at technology as human actions and groups them accordingly. This grouping, as shown in Fig. 3-8, includes four types of technology:

• Communication technology.
• Construction technology.
• Manufacturing technology.
• Transportation technology.

These are the technologies that humans have used to control and modify their environment throughout history. The study of past cultures, a view of present life, and predictions of future civilizations suggest people use technology to:

• Communicate ideas and information.
• Construct structures for housing, business, transportation, and energy transmission.
• Manufacture products from materials.
• Transport people and goods from one location to another.

TRANSFORMATION

These technologies each use the seven basic inputs (people, materials, tools, energy, information, finances, and time) to transform one or more specific resources. For example, **manufacturing** transforms *materials* into products. **Construction** transforms *materials and manufactured goods* into structures. **Communication** technology transforms *information* into signals and media messages. **Transportation** uses *energy* to power vehicles that move people and goods.

These systems can be looked at individually. However, this gives an inaccurate view. They are all closely related and are part of a single effort: to help humans live better. These systems work together to support one another. In one case manufacturing may be the focus with the other systems supporting. In another case manufacturing may be the supporting technology. To see this relationship, look at Fig. 3-9. Let us follow one

TYPES OF TECHNOLOGY

COMMUNICATION

CONSTRUCTION

MANUFACTURING

TRANSPORTATION

Fig. 3-8. These are the types of technology.

OBTAINING RESOURCES	PRODUCING INDUSTRIAL MATERIALS	PRODUCING PARTS AND PRODUCTS	THE FINAL PRODUCT

Fig. 3-9. These photos show the steps involved in the production of a manufactured product. Each of the steps, from raw material to finished product, depends on communication, construction, manufacturing, and transportation. Each works toward the goal of producing a product.

material from its natural state to a finished product that is designed to make life better for us. We will follow iron ore on its journey to become a steel crowbar.

Our story starts with *manufacturing technology*. This technology involves three major activities:

- Locating and extracting the raw materials (iron ore, limestone, and coal) to make steel.
- Producing bars of steel from the raw materials.
- Making (forging) the crowbar from the steel bars.

These manufacturing activities could not exist without the other technologies. *Communication technology* played a role at every point in the product's development. The need for the tool was communicated through sales orders. The specifications for the steel and the crowbar were communicated through engineering drawings and specification sheets. The availability of the tool was communicated to potential customers through advertising. Sales reports communicated product success to the company's management.

Likewise, *construction technology* was essential. The roads to the iron ore and coal mines and the limestone quarries were constructed. Workers traveled to the steel mill and the tool factory on constructed roads, and worked in constructed buildings. Constructed power lines brought electricity to the various manufacturing sites. Water was available because of constructed dams and pipelines.

Transportation systems were also widely used. They moved the raw materials to the steel mill and the steel rods from the mill to the tool manufacturer. Finished products were delivered to the hardware store by transportation systems. Customers used private cars or public transportation systems to visit a store to purchase the product.

The focus of the rest of this book will be on the four major technological systems. It will focus on the system components for each of these systems and explore the productive processes in greater depth.

SUMMARY

Technology has been part of human existence as far back as history goes. As civilization became more advanced, technology has evolved, and it continues to evolve. Older technologies are replaced with newer, more efficient technologies.

Most present-day technology is developed and applied by industrial and business enterprises. These technologies are developed by research and development personnel, applied through production activities, and the outputs are sold through marketing efforts. The system is driven by a desire to profit from the development and application of technologies. Technologies are commonly developed to help people construct structures, communicate information, manufacture products, and transport people and cargo.

WORDS TO KNOW

All of the following words have been used in this chapter. Do you know their meanings?

Communication
Construction
Financial affairs

High technology
Industrial relations
Manufacturing
Marketing
Production
Research and development
Transportation

TEST YOUR KNOWLEDGE

Write your answers on a seperate piece of paper. Please do not write in this book.

1. List and describe the three stages of technology present at any point in history.
2. What is high-technology?
3. Designing, developing, and specifying the characteristics of the product, structure, or service is called _____.
4. True or false. Developing systems for producing the product, structure, or service is called marketing.
5. _____ promotes, sells, and delivers the product, structure, or service to a customer.
6. Which of the four technologies completes the following tasks?

    _____ Erects structures for housing, business, transportation, and energy transmission.
    _____ Makes products from materials.
    _____ Carries people and goods from one location to another.
    _____ Moves ideas and information from senders to receivers.

    A. Communication.
    B. Construction.
    C. Manufacturing.
    D. Transportation.

APPLYING YOUR KNOWLEDGE

1. Develop a chart similar to the following one, and list five technological devices from each type of technology that you use daily.

Technology	Technological device
Communication	
Construction	
Manufacturing	
Transportation	

2. Identify a new technological device that you could use, and sketch what it would look like and how it would work.
3. Ask an older person (a grandparent or neighbor) to tell you about five technological devices that they used in their lifetime that are no longer around. List them (obsolete technologies) and the devices that have replaced them (current technologies) and ones that you think will replace the current ones (emerging technologies) on a chart like the one below.

Obsolete technology	Current technology	Emerging technology
(example) 78 rpm records 45 rpm records 33 1/3 rpm records	Cassettes tapes Compact discs	Digital audio tapes
1.		
2.		
3.		
4.		
5.		

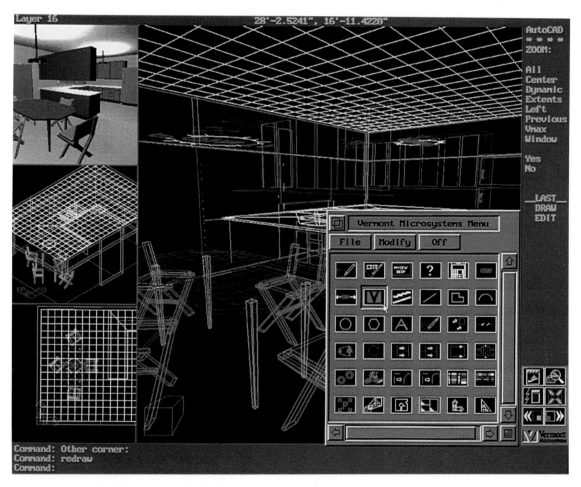

Computers are useful tools for designing systems, products, and structures. (Vermont Microsystems)

Section One - Activities

ACTIVITY 1A - DESIGN PROBLEM

Background:

Technology is the application of knowledge to create machines, materials, or systems to help us make work easier, to make life more comfortable, or to help control the natural or human-made environment.

Situation:

The Easy-Play Game Company has developed a new board game for two players. The game uses five red marbles and five green marbles as the playing pieces. They find that counting out the marbles one by one is too costly.

Challenge:

Design a technological device (machine) that will count five marbles at a time from a box containing a large number of marbles.

ACTIVITY 1B - FABRICATION PROBLEM

Background: All technology has been developed to meet a *need* or *opportunity*. Most early technology answered a functional need. Some of the earliest technology dealt with producing clay containers for transporting liquids and storing food.

Challenge:

Divide your class into two groups. Each group will use different clay forming techniques to produce a clay pot and lid. *Group One* will form the two parts without the aid of any external devices. *Group Two* will use a form to aid in the production process. The form group two will use is a frozen orange juice concentrate can. See Fig. 1-A.

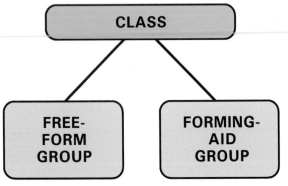

Fig. 1-A.

The product should be the diameter of the orange juice can and 1 1/2″ tall. The lid should be 1/2″ tall.

Materials and Equipment:
   Potter's clay
   Frozen orange juice concentrate cans
   Sponge
   Table knife

Procedure:

Look at Fig. 1-B. Produce a clay pot and lid using the procedure for your group.

FREE-FORM GROUP

1. Obtain the supplies.
2. Separate a portion of clay.
3. Roll several 3/8″ diameter strips of clay.

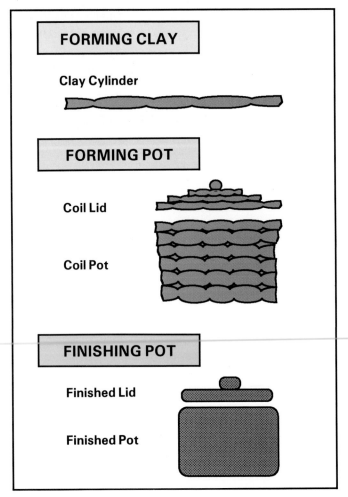

Fig. 1-B.

4. Coil a strip of clay flat on the table top until a disk the diameter of the juice can is produced. This is the bottom of the pot.
5. Build the wall of the pot by layering strips of clay until the 1 1/2″ height is reached.
6. Use a damp sponge and your fingers to smooth the clay and form the finished pot.
7. Coil a strip of clay flat on the table top until a disk the diameter of the juice can is produced. This is the lid for the pot.
8. Gently shape the disk into a concave shape that is 1/2″ tall.
9. Form a handle (1/2″ ball) and attach it to the center of the lid.
10. Use a damp sponge and your fingers to smooth the clay and form the finished lid.
11. Allow the pot to air dry.
12. Fire the pot in a kiln. (Optional)

FORMING-AID GROUP

1. Obtain the supplies.
2. Separate a portion of clay.
3. Roll several 3/8″ diameter strips of clay.
4. Coil a strip of clay flat on the table top until a disk just a bit larger that the diameter of the juice can is produced. This will become the bottom of the pot.
5. Use the can like a cookie cutter to cut a perfectly round disk.
6. Place the can on the table top with the open end down.

7. Build the wall of the pot by loosely coiling strips of clay around the can until the 1 1/2″ height is reached. Do not push the clay against the can because it will stick.
8. Carefully slide a table knife between the can and the pot to separate them.
9. Remove the can.
10. Place the pot wall on top of the bottom.
11. Use a damp sponge and your fingers to smooth the clay and form the finished pot.
12. Coil a strip of clay flat on the table top until a disk the diameter of the juice can is produced. This will become the lid for the pot.
13. Gently shape the disk into a concave shape that is 1/2″ tall.
14. Form a handle (1/2″ ball) and attach it to the center of the lid.
15. Use a damp sponge and your fingers to smooth the clay to form the finished lid
16. Allow the pot to air dry
17. Fire the pot in a kiln. (Optional)

Analysis:

Meet as a class and analyze the two clay-forming techniques. Answer the following questions:

1. Do you think the free-forming method is a common technique for producing modern ceramic products? Why or why not?
2. Which method was the easiest to use?
3. Which method produced a higher quality pot?

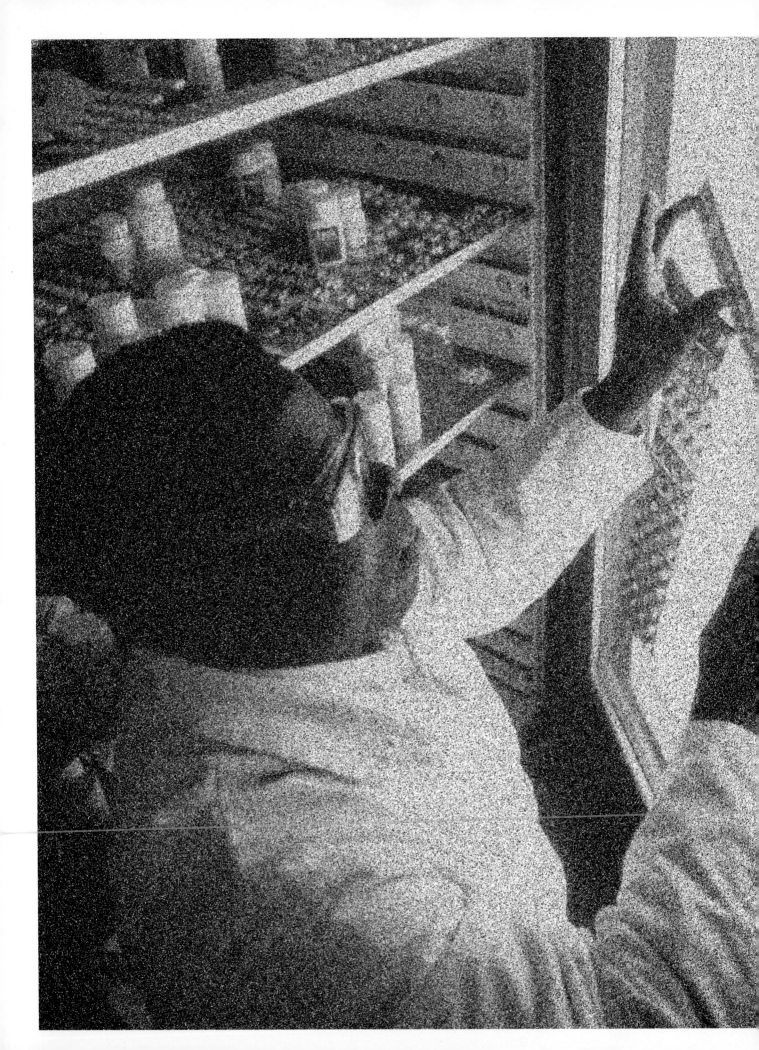

Technological System Components

4. Inputs to Technological Systems

5. Technological Processes

6. Outputs and Control of Technology Systems

Section 2 Activities

People are the most important input to a technological system.

4
CHAPTER

INPUTS TO TECHNOLOGICAL SYSTEMS

After studying this chapter you will be able to:
- Describe the seven inputs to technological systems.
- Explain the types of skills and knowledge people bring to technological systems.
- Describe the role and types of energy as inputs to technological systems.
- Describe data, information, and knowledge as inputs to technological systems.
- List and describe the types of machines and tools used as inputs to technological systems.
- Explain how time is an input to technological systems
- Describe the sources of finances that are used as inputs to technological systems.
- Describe the types and sources of materials that are inputs to technological systems.
- Describe the necessity to intelligently use and conserve the resources that are the inputs to technological systems.

Humans have lived on the earth about 70,000 years. This may seem like a long time. However, it is relatively short considering the earth is about 5 billion years old. In the short span of human history people have developed a special ability. They have learned how to build and how to use tools. You have learned to call this process technology. This ability to develop technology has led to many kinds of technological systems. You learned in Chapter 2 that all systems include four parts. These parts are: inputs, processes, outputs, and feedback. This chapter will explore the *inputs* that are common to all technological systems.

WHAT ARE INPUTS?

All of these inputs may be grouped into seven major categories, as shown in Fig. 4-1. These are:
- People.
- Machines and tools.
- Materials.
- Information.
- Energy.
- Finances.
- Time.

These inputs are the resources that are used to make the system operate. They are the elements that are changed by technological processes or are used by technology to change other inputs. Let's look at each one separately.

PEOPLE

Humans should be considered as the most important input to technological systems. These systems are a

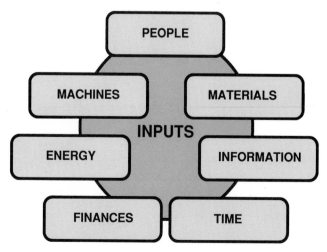

Fig. 4-1. These are the seven types of inputs used in technological systems and products.

product of people. Their minds create and design the systems and their outputs. They use their skill to fabricate and operate the systems. Their management abilities make systems operate efficiently. Their needs and wants are satisfied by the systems. Therefore, it is easy to see why people are fundamental to technology.

Creative People

There are a number of specific skills and abilities that people bring to technological systems. These include **creative abilities** to design systems and products, Fig. 4-2. Designers develop ideas for the product and

Fig. 4-2. This person used her creative abilities to design a greeting card. (American Greetings)

decide what size, shape, and color the product will be. Their decisions can add beauty to the world through well designed products and structures. Their creativity also add zest to the printed word and electronic media presentations–radio and television shows.

Scientists

Other people involved are the scientists, engineers, and technologists who design, specify, and test products, structures, and components of technological systems. **Scientists** generally develop the basic knowledge needed to develop products and processes, Fig. 4-3. Often this is physics, materials science, geological, and chemistry knowledge.

The work of the scientists is applied by two types of technology experts: engineers and technologists. Both use their knowledge of efficient and appropriate action to create and operate technological systems.

Engineers

Engineers apply scientific and technological knowledge in designing products, structures, and systems, Fig. 4-4. They determine appropriate materials and processes needed to produce products or perform services. For example, civil engineers will determine the correct structure for a bridge to carry vehicles across a river. Mechanical engineers will determine the correct material for automotive parts. They will specify its proper size and shape. Electrical engineers design circuits for computers. Manufacturing engineers design manufacturing processes and quality control equipment.

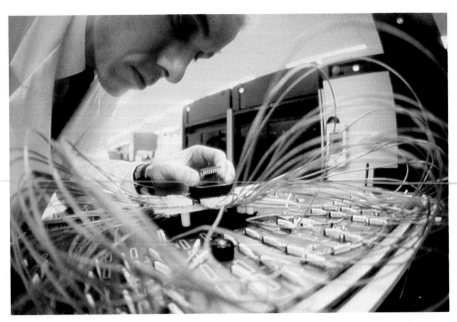

Fig. 4-3. Scientists generate knowledge to develop technological systems. (Northern Telecom)

Fig. 4-4. These engineers are developing product applications for a new composite material. (Reynolds Metal Co.)

Technologists

Industrial and **engineering technologists** are a new type of worker. They have entered the work force in the last 20 to 30 years. They bridge the gap between engineering and operations personnel. They help manufacturing and construction personnel apply engineering designs and solve production problems.

Still other people build products and structures, produce communication messages, and transport people and cargo. They also maintain the systems that result in these outputs. We often call these people workers, technicians, mechanics, or craftspeople. They use technical skill and knowledge to physically produce the output of the technological system.

Workers

Production workers in manufacturing and construction are often described as unskilled, semi-skilled, and skilled workers. **Unskilled workers** perform tasks that require only a minimum of training. Fast food cooks, custodians, and product packagers are examples of unskilled workers. **Semi-skilled workers** perform tasks that require a limited amount of training. They may run machines, assemble products, and service equipment. These workers are generally trained on the job. **Skilled workers** are highly skilled individuals that have had extensive training and work experience. They commonly operate computer systems, set up machines, custom-build products, maintain complex products

and systems, and prepare drawings of products and structures. They may receive their training in special schools or through apprenticeship (receiving on-the-job and classroom instruction from experienced workers).

Skilled workers in laboratories and product-testing facilities are usually called **technicians**. Skilled workers in service operations are often called mechanics. Examples of people with this title are auto mechanics and aircraft mechanics.

Management

Another category of people establish and operate businesses. They are called managers and entrepreneurs. **Managers** organize and direct the work of others. They set goals, structure tasks to be completed, assign work, and monitor results. **Entrepreneurs** are the people that create business. They have vision of what can be done and are willing to take risks to see it happen. Often an entrepreneur will form a company and guide it in its early life. As the company grows, trained managers are hired to direct segments of the company's operations. Entrepreneurial skills and managerial abilities are not the same. In many cases, entrepreneurs are not good managers of day-to-day operations. They enjoy the challenge of starting an enterprise but either lack the skill or patience to attend to day-to-day details.

Still another group of people bring business knowledge to the company. They keep the financial records, maintain sales documents, develop personnel systems, keep inventory records, and maintain the business plans. They are not managers because they do not direct the work of others. They provide the support needed to see that products and structures are built, communication media are produced, and people and cargo are transported. They have titles like accountant, salesperson, and employment office clerk. We often call these people *white-collar* workers.

Consumers

The final group of people involved in technological systems are **consumers**, Fig. 4-5. They are the reason for the system in the first place. Their need for products, structures, information, and mobility gives rise to the system. Their feelings about styling, price, and service directly impacts the systems' outputs. Their money is spent on the products or services. Therefore, they financially support the system.

MACHINES

Humans are the only species on earth that can develop and use technology. As we said earlier, this is based on fact humans are tool builders and tool users. Fundamental to all technology are the technical means which we call **tools**, **machines** or equipment. They include the machines in factories, the equipment on construction sites, the vehicles used to transport people and cargo, and the devices used to produce and deliver communication messages. We sometimes call these technical resources capital equipment or simply capital.

Fig. 4-6. Humans have long used tools. The museum diorama shows a person using a tool to start a fire.

Tools

The "tool box" of humans has grown over history. Early people had a few crude tools, Fig. 4-6. There were rocks lashed to sticks to make crude hammers. Chipped stones attached to shafts became arrows and spears. Sharpened stones made crude scrapers and knives. Today there is a tool for almost every task. There are curved and straight claw hammers, ball and cross peen hammers, rivet hammers, soft faced hammers, wood and rubber mallets, sledge hammers, etc.

Fig. 4-5. Fundamental to all technological systems are consumers. They provide the reason for technology, and buy the products of technological systems. (VICA)

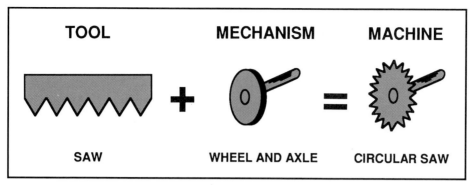

TOOL

SAW

+

MECHANISM

WHEEL AND AXLE

=

MACHINE

CIRCULAR SAW

Fig. 4-7. People use tools, mechanisms, and machines to extend their ability to do work. Tools are combined with mechanisms to make machines.

Technical means include three major things: tools, mechanisms, and machines. People use tools, mechanisms, and machines to increase their ability. These devices help to make us stronger, hear better, see farther, and sleep warmer. The difference between the three types of technical means are shown in Fig. 4-7.

Tools are simple devices that are used to perform a basic task. Each major activity has its own tools. Microscopes, analytical balances, and telescopes are scientific tools. Footballs, pitching machines, and tennis rackets are recreational tools. Paint brushes, sculpture chisels, and silkscreens are artist's tools. Postal scales, volume measurements, and billing machines are tools of commerce.

Likewise, there are tools of technology. Manufacturing, construction, communication, and transportation have their own tools. The band saw, automobile, radio transmitter, and the cement mixer are examples of technology tools.

Common Hand Tools. Almost every technology uses a common set of hand tools to produce, maintain, and service products and equipment. These tools may be grouped into six major categories as shown in Fig. 4-8. These are:

- Measuring tools: used to determine the size and shape of materials and parts.
- Cutting tools: used to separate materials into two or more pieces.
- Drilling tools: used to produce holes in materials.
- Gripping tools: used to grasp and, in many cases, turn parts and fasteners.
- Pounding tools: used to strike materials, parts, and fasteners.
- Polishing tools: used to abrade and smooth surfaces.

Mechanisms

Mechanisms are basic devices that are used to power or adjust equipment and machines. In physics the mechanisms are called simple machines. These mechanisms

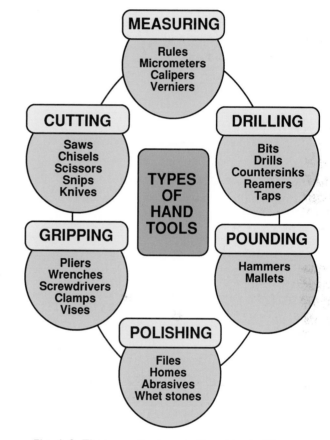

MEASURING
Rules
Micrometers
Calipers
Verniers

CUTTING
Saws
Chisels
Scissors
Snips
Knives

DRILLING
Bits
Drills
Countersinks
Reamers
Taps

TYPES OF HAND TOOLS

GRIPPING
Pliers
Wrenches
Screwdrivers
Clamps
Vises

POUNDING
Hammers
Mallets

POLISHING
Files
Homes
Abrasives
Whet stones

Fig. 4-8. These are the types of hand tools. They are grouped by the action they perform.

work on two basic principles and can be grouped under six categories. They are shown in Fig. 4-9 and include:

- Lever.
- Wheel and axle.
- Pulley.
- Inclined plane.
- Wedge.
- Screw.

The first three, the lever, the wheel and axle, and the pulley operate on the basic principle of the lever. The second three, the inclined plane, wedge, and the screw, operate on the principle of the inclined plane.

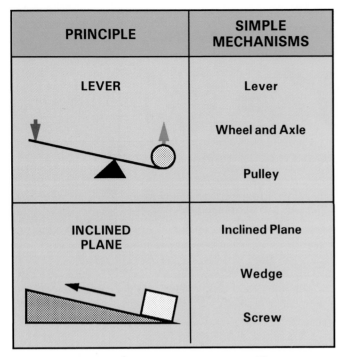

PRINCIPLE	SIMPLE MECHANISMS
LEVER	Lever
	Wheel and Axle
	Pulley
INCLINED PLANE	Inclined Plane
	Wedge
	Screw

Fig. 4-9. There are six types of mechanisms. There are two groups based on the principle behind them.

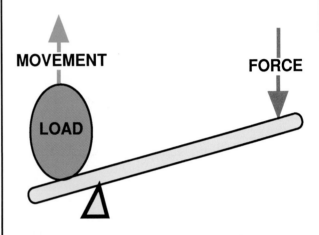

Fig. 4-10. Levers use a *lever arm* and a *fulcrum* to which a force is applied and a load is moved. The crowbar shown at left uses this principle.

Lever Mechanisms. Almost everyone has used a lever. If you have used a crow bar to pry open a crate or pulled a nail with a claw hammer, you have used a lever. A **lever** includes two major parts: a *lever arm* and a *fulcrum*. In addition, it involves a load to be moved and a force to be applied. These elements are shown in Fig. 4-10.

Levers are grouped as *first class*, *second class*, and *third class*, Fig. 4-11. Each class applies force differently to move the load. An example of a first class lever is a pry bar. A wheelbarrow uses the principle of a second class lever. A person moving dirt with a shovel applies the principle of a third class lever.

Like all other simple machines, the lever can be either a force or a distance multiplier. A *force multiplier* lever places the fulcrum close to the load. The force is applied at the other end of the lever arm. This application will move a heavy load with a light force. However, the force must move a greater distance that the load.

A *distance multiplier* lever is just the opposite. The fulcrum is close to the force. The load is at the other end of the lever arm. The load will move a greater distance

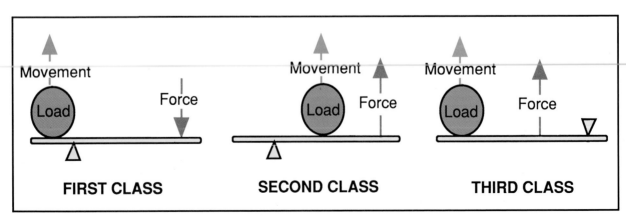

Fig. 4-11. There are three classes of levers.

than the force. However, a large force is required to move a light load. These two applications of levers are shown in Fig. 4-12.

A **wheel and axle** is a shaft attached to a disk. It acts as a second class lever. The shaft or axle acts as the fulcrum. The circumference of the disk acts as the lever arm. If the load is applied to the disk, the wheel and axle becomes a force multiplier, Fig. 4-13. Automotive steering wheels use this principle. A 15″ wheel attached to a 1/4″ shaft will multiply the force by 60 times.

If the load is applied to the shaft, the wheel and axle become a distance multiplier. Automobile transaxles use this type of wheel and axle mechanism. One revolution of the axle will cause the wheel to revolve one time. However, the circumference of the wheel is many times that of an axle. Therefore, the vehicle moves a considerable distance down the road for each revolution. For example, a 20″ wheel attached to a 1/2″ shaft will multiply the distance for each revolution of the shaft 40 times.

Pulleys are grooved wheels attached to an axle. They also act as second class levers. Pulleys can be used for

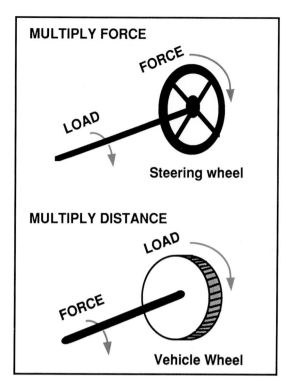

Fig. 4-13. A wheel and axle can multiply force or distance.

three major purposes as shown in Fig. 4-14. A single pulley can be used to change the direction of a force. Two or more pulleys can be used as force and distance multipliers. The number and diameters of the pulleys used will determine the mechanical advantage (force multiplication) of a pulley system.

Inclined Plane Mechanisms. Inclined plane mechanisms are sloped surfaces which are used to make a job easier to do. The principle of the three mechanisms (inclined plane, wedge, and screw) is that it is easier to move up a slope than up a vertical surface, Fig. 4-15. The simplest application of this principle is the **inclined plane**. It is used to roll or drag a load from one elevation to another. A common use of inclined planes are roadways in mountains and ramps to load trucks.

A second application of this principle is the **wedge**. The device is used to split and separate materials and to grip parts. A wood chisel, a firewood splitting wedge, and a doorstop are example of devices using this principle.

The screw is the third mechanism using an inclined plane. A **screw** is actually an inclined plane wrapped around a shaft. The screw is a force multiplier. Each revolution of the screw moves it into the work only a short distance. For example, a 1/2 x 12 machine screw is 1/2″ in diameter and has 12 threads per inch. With one revolution of the screw, the circumference moves about 1 1/2 inch but into the work only 1/12″ of an inch. In this example, the force is multiplied about 18 times.

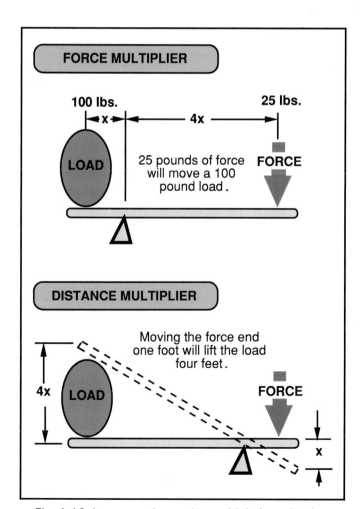

Fig. 4-12. Levers can be used to multiply force (top) or distance (bottom).

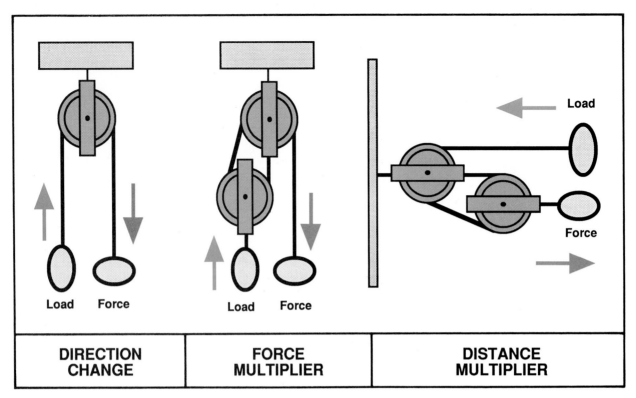

| DIRECTION CHANGE | FORCE MULTIPLIER | DISTANCE MULTIPLIER |

Fig. 4-14. Pulleys can be used to change the direction of a force, multiply force, or multiply distance.

Mechanisms Combine to Make Machines. These mechanisms provide the building blocks to construct a wide variety of technological devices which we call machines. Machines are combinations of tools and mechanisms to form complex devices. The major technological machines can be classified as:

- Material processing machines.
- Information processing machines.
- Energy processing machines.

These groups of machines will be the focus of Chapter 8: Productive Tools.

MATERIALS

Our world is a material world. Look around you. Everywhere you look there are materials. They come in all sizes, shapes, and types. They possess a number of specific properties. All this material is made up of one or more of the 96 elements that occur naturally on earth. These elements combine to produce literally thousands of compounds. To understand materials a person must know about the types of materials and the properties that materials exhibit, Fig. 4-16.

Types of Materials

One way to classify the materials around us is by their origin, Fig. 4-17. Materials are composed of either living matter or nonliving matter. Materials that come from living organisms are called **organic materials**. Wood, cotton, and flax are products of plant fibers. Wool and leather are products of animals. Petroleum, coal, and natural gas are the products of decayed and fossilized organic materials. All materials that do not come from living organisms are called **inorganic materials**. Metals and ceramic materials are inorganic.

LOW SLOPE - EASY LIFT

Small force → Load

Lift

STEEP SLOPE - HARD LIFT

Large force → Load

Lift

DIRECT - DIFFICULT LIFT

Very large force

Load

Lift

Fig. 4-15. An inclined plane is a sloped surface that makes lifting a load easier.

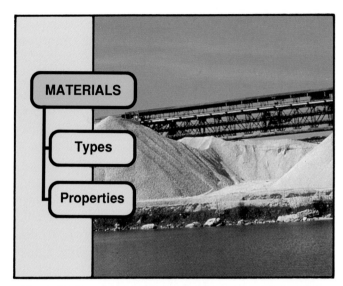

Fig. 4-16. People must understand the types of materials available and their properties in selecting a material for a job. (United States Steel)

Natural and Synthetic

Another way to view materials are from their natural state. Many materials occur naturally on earth. These materials are often called natural resources. Iron, carbon, petroleum, and silica are examples of **natural materials**. They can be refined and combined to make products.

Other materials are human-made or **synthetic materials**. The most common synthetic materials are plastics. They are developed and produced from cellulose (vegetable fibers), natural gas, and petroleum.

Exaustible. Materials may also be classified by their ability to be regenerated. Some materials naturally occur on earth in a specific amount. These materials cannot be replaced by human action or nature. There is a finite (limited) quantity of these materials on earth. Once they are used up there will be no more. They are also called **exhaustible materials**. Metal ores, coal, petroleum, and natural gas fall in this category of materials.

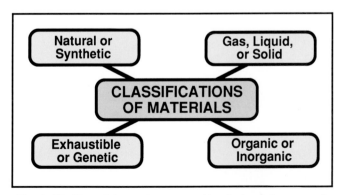

Fig. 4-17. Materials can be grouped by their origin, natural state, ability to be regenerated, and physical state.

Genetic. Other materials have a life cycle. They can be produced by nature or by human action. These materials are called **genetic materials** because they are produced by living things. These materials are the result of farming, forestry, and fishing activities. The technology associated with producing these materials is called bio-related technology. Wood, meat, wool, cotton, and leather are all genetic materials.

Still other materials are the product of natural reactions. Carbon dioxide and oxygen are the by-product of natural processes. Water is purified through natural processes in lakes, wetlands, and rivers. Salt is produced around salt water by natural evaporation.

Physical State

An important way of grouping materials is by their physical state. Materials are either gases, liquids, or solids. **Gases** are materials that easily disperse and expand to fill any space. They have no physical shape but do occupy space and have volume. Gases can be compressed and put into containers. Gases meet many human needs. They form the air we breathe, fuel for rockets, carbonation for beverages, and compressed air to inflate tires.

Liquids are visible, fluid materials that will not normally hold their size and shape. They cannot be easily compressed. Common liquids include the water we drink and bathe with, fuels for transportation vehicles, the sea water that supports boats and ships, and coolants for industrial processes.

Solids are materials that hold their size and shape. They have an internal structure that causes them to be rigid. They can support loads without losing their shape. Solid materials are often called engineering materials because they are used in engineering products and structures.

Engineering materials may be divided into four categories, as shown in Fig. 4-18:
• Metallic materials (metals): Inorganic substances that have a crystalline structure. They are the most widely used of all engineering materials. They are generally used as an alloy: a mixture of a base metal and other metals or nonmetallic materials. For example, steel is primarily an iron-carbon alloy and brass is a copper-zinc alloy.
• Polymeric materials (plastics): These synthetic materials contain complex chains of hydrogen-carbon (hydrocarbon) molecules. These materials are either thermoplastic (soften when heated) or thermosetting (made rigid by heat).
• Ceramics: Mostly inorganic crystalline materials, such as clay, cement, plaster, glass, abrasives, or refractory material.

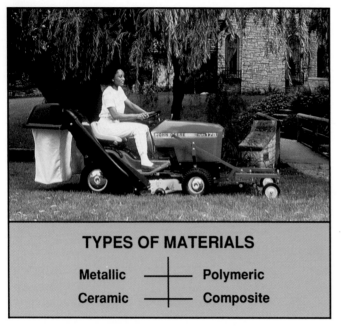

TYPES OF MATERIALS

Metallic ——————— Polymeric

Ceramic ——————— Composite

Fig. 4-18. Many complex products, like this lawn tractor, contain metallic, polymeric, ceramic, and composite materials. (John Deere and Co.)

- Physical: The characteristics due to the structure of the material: including size, shape, density, moisture content, and porosity.
- Mechanical: The reaction of the material to a force or load. This property affects the material's strength (ability to withstand stress), plasticity (ability to flow under pressure), elasticity (ability to stretch and return to the original shape), ductility (ability to be bent), and hardness (ability to withstand scratching and denting).
- Chemical properties: The reaction of the material to one or more chemicals in the outside environment. This property is often described in terms of chemical activity (degree that the material will enter into a chemical action) and corrosion resistance (ability to resist attack from other chemicals).
- Thermal properties: The reaction of the material to heating and cooling. This property is expressed in thermal conductivity (ability to conduct heat), thermal shock resistance (ability to withstand fracture from rapid changes in temperature), and thermal expansion (change in size due to temperature change).
- Electrical and magnetic properties: The reaction of the material to electrical and magnetic forces. This property is described in terms of electrical conductivity (ability to conduct electrical current) and magnetic permeability (ability to retain magnetic forces).
- Acoustical properties: The reaction of the material to sound waves. Acoustical transmission (ability to conduct sound) and acoustical reflectivity (ability to reflect sound) are measures of acoustical properties.
- Optical properties: The reaction to visible light waves. Optical properties include color (waves that are reflected), optical transmission (ability to pass light waves), and optical reflectivity (ability to reflect light waves).

- Composites: A combination of two or more kinds of materials. One material forms the matrix or structure. The other material fill the structure. Fiberglass is a composite with a glass fiber structure filled with a plastic resin. Wood is a natural composite with a cellulose fiber structure filled with lignin, which bonds the structure together.

Properties of Materials

All materials exhibit a specific set of properties. For example, the properties of iron are different from those of oak. These properties are considered as materials are selected for specific uses. The common properties can be grouped under seven categories, as shown in Fig. 4-19.

PROPERTIES OF MATERIALS

Physical

Mechanical

Chemical

Thermal

Electrical and Magnetic

Acoustical

Optical

Fig. 4-19. These are the seven properties of materials.

Technology: How it Works

FLIGHT CONTROLS: A series of movable control surfaces that can be adjusted to cause an airplane to climb, dive, or turn.

An airplane is the only transportation vehicle in common use that has three degrees of freedom, Fig. 1. Consider how many directions a car can move compared to an airplane. A car can move forward and backward (one degree of freedom), or turn from side to side (2nd degree). An airplane must move forward in order to fly, but it can turn to go back in the opposite direction. Remembering this, we can say that airplanes can move forward

(1st degree), turn from side to side (2nd degree). However, an airplane also can climb and dive, providing the third degree of freedom.

Airplanes change their direction as they fly by using a system of flight controls. These controls are operated by the pilot. Flight controls change the *lift* that acts on the wings and other control surfaces.

Lift is what makes an airplane fly. The engine pushes or pulls the airplane through the air. Air flows over the wings, creating lift. If not enough air is flowing over the wings, no lift is produced, and the plane loses altitude. The airplane must maintain forward motion in order to fly.

Turns are made by using two types of control surfaces, the rudder and the ailerons. The rudder is a movable flap on the vertical portion of the tail assembly. The rudder is controlled by two pedals. The ailerons are flaps on the rear (or trailing edge) of each wing, Fig. 2. To make a turn, the rudder is moved to one side. The aileron on that side is lifted; and the one on the other wing is lowered. This causes the plane to bank (turn) in the direction the rudder is pointed.

Climbing and diving are controlled by surfaces on the horizontal stabilizer. These control surfaces are called elevators. If the elevators are tilted downward, the tail is forced upward. The nose is pointed downward, so the plane will lose altitude or dive. If the elevators are tilted upward, the tail will be forced down, and the nose points up. The plane will gain altitude or climb, Fig. 3.

Fig. 1. Airplanes use control surfaces to govern their direction in flight. Control surfaces are located on the wings, tail, and horizontal stabilizer.

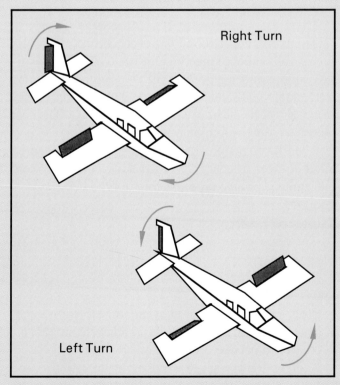

Right Turn

Left Turn

Fig. 2. This diagram shows how the rudder and ailerons are set for left and right turns.

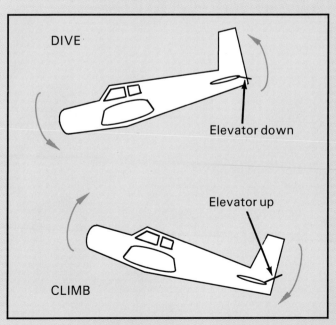

DIVE

Elevator down

Elevator up

CLIMB

Fig. 3. Climbing and diving are controlled by the elevators.

INFORMATION

Humans are greatly different from other living species. People have the ability to think, reason, and enter into articulate speech. This means that people can observe what is happening around them, make judgments about those observations, and explain them to other people. This unique human ability requires knowledge, which is derived from data and information, Fig. 4-20.

Data are all the raw facts and figures that are collected by people and machines. (Remember: the word data is plural and, therefore, we write or say "The data are...") Data are the raw materials for information.

Information is data that has been sorted and categorized for human use. Data processing is the computer based actions of collecting, categorizing, and presenting data so that humans can interpret it.

We can classify information in three groups:

- Scientific information: Organized data about the laws and natural phenomena in the universe. Scientific information describes the natural world.

- Technological information: Organized data about the design, production, operation, maintenance, and service of human-made products and structures. Technological information describes the human-made world.

- Humanities information: Organized data about the values and actions of individuals and society. Humanities knowledge describes how people interact with society and the values held by individuals and groups of people.

Information that is learned and applied by people is called **knowledge**. It is the result of reasoned human action. Knowledge guides people as they determine which course of action to take. It can be described as scientific knowledge, technological knowledge, and humanities knowledge. However, in reality all these types of knowledge must be brought to bear on any problem needing a solution. Scientific knowledge may provide a theoretical base for the solution. Technological knowledge is used to implement the solution. Humanities knowledge will tell us if the solution is acceptable to society.

ENERGY

Human existence is based on converting one type of energy to another. Energy powers our factories, heats and lights our homes, propels our vehicles, drives our communication systems, and supports our construction activities.

There are a number of types of energy. Everyone uses human energy to complete tasks. However, human energy falls short of meeting our needs. There is a limited supply and we don't want to use it all on work. Also, some tasks cannot be done with human energy alone. Try to heat a house with human energy. The heat radiated from the human body is not enough.

Therefore, people throughout history have used other sources of energy. We can look at energy from two vantage points: type and source.

Types of Energy

There are seven major types of energy, Fig. 4-21. These are:

- Chemical: energy stored in a substance and released by chemical reactions.
- Electrical: energy created by moving electrons.
- Thermal (heat): energy that comes from the increased molecular action caused by heat.
- Radiant (light): energy produced by the sun, fire, and other matter, which includes light, radio waves, X-rays, and ultraviolet and infrared waves.

Data
Laws, facts, and figures

Information
Organized (sorted and categorized) data

Knowledge
Learned and applied information

Fig. 4-20. Data are processed into information so that humans can gain knowledge.
(Amoco, AC-Rochester, Goodyear Tire and Rubber Co.)

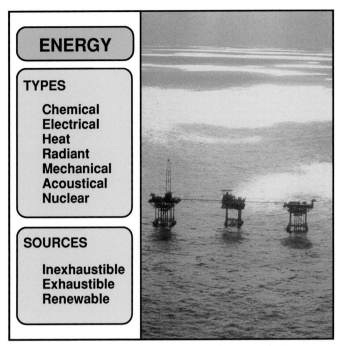

ENERGY

TYPES

Chemical
Electrical
Heat
Radiant
Mechanical
Acoustical
Nuclear

SOURCES

Inexhaustible
Exhaustible
Renewable

Fig. 4-21. The two main factors in understanding energy are the type of energy and the source it comes from. (American Petroleum Institute)

- Mechanical: energy produced by moving water, animals, people, and machines.
- Acoustical (sound): energy associated with audible sound.
- Nuclear: energy produced by splitting atoms or uniting atomic matter.

Sources of Energy

Energy is available from three major sources. These sources are grouped in terms of the supply. The first source of energy is the sun. The sun is the fundamental source of energy for our solar system. Its actions warms the earth, causes the wind to blow, creates weather and generates lightning, and indirectly with our moon creates the ocean tides. This form of energy has always been with us and will continue to be so. It is said to be **inexhaustible**.

A second source of energy comes from living matter. Human and animal muscle power can provide valuable sources of energy. Wood and other plant matter can be burned as a fuel. These sources are called **renewable**. They are used up but can be replaced with the normal life cycle of the energy source. An example of this replacement is that new trees can be grown, but this takes many years.

The third source of energy is **exhaustible**. These sources are limited by the quantity found on earth. The sun or human action will not create additional supplies. Coal, petroleum, and natural gas are exhaustible energy sources.

The goal of wise energy utilization is to maximize the use of inexhaustible sources, recycle the renewable sources, and use the minimum of exhaustible energy.

FINANCES

Technological systems require people, machines, materials, and energy. These are generally purchased. People are paid wages or salaries for their labor and knowledge. Materials are purchased. Machines are bought or leased. Energy is purchased. All this requires money, Fig. 4-22, which provides the financial foundation of the technological activity.

Money to develop and operate technological systems can be obtained in two ways. The technological system or the company can be sold to people. They can become part owners of the operation. If one person owns it, the operation is called a *sole proprietorship*. A few owners can form a partnership in which each person owns a portion of the company. A legal entity, called a corporation, can be formed to own the operation. Shares, which are certificates of ownership in the corporation, can be sold. In all three cases, the money was raised by selling equity–a portion of the company. Therefore, raising money in this way is called **equity financing**.

Money can also be borrowed from other sources. Banks, insurance companies, or investment groups can loan money to support the activities of a company. This loan constitutes a debt that must be repaid. This type of financing is called **debt financing**.

TIME

Humans have constantly been aware of time. There was only so much daylight in which to hunt and gather food. The growing season determined what crops could be grown. People lived to a certain age.

This has not changed. However, technology has accelerated the use of time. We allocate machine time,

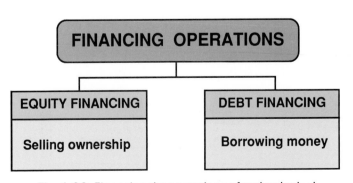

Fig. 4-22. Financing the operations of technological systems and companies is done through equity and debt financing.

computer time, sales response time, etc. At one time we measured time in years, months, and days. Then hours and seconds were observed. Now engineers worry about nanoseconds (billionths of a second) in computer processing. Time is becoming an even more valuable resource for technological systems.

SUMMARY

All technological systems involve inputs which are processed into outputs. The quality and quantity of these inputs will often determine the type of technology a society can use. Many of these inputs are in limited supply. Therefore, everyone designing, building, or using technology should use appropriate inputs and recycle as many of these as possible.

WORDS TO KNOW

All of the following words have been used in this chapter. Do you know their meanings?

Consumer
Creative abilities
Data
Debt financing
Engineer
Entrepreneur
Equity financing
Exhaustible energy source
Exhaustible material
Gas
Genetic material
Inclined plane
Inexhaustible energy source
Information
Inorganic material
Knowledge
Lever
Liquid
Machine
Manager
Mechanism
Natural material
Organic material
Pulley
Renewable energy source
Scientist
Screw
Semi-skilled worker
Skilled worker
Solid
Synthetic material
Technician
Technologist
Tool
Unskilled worker
Wedge
Wheel and axle

TEST YOUR KNOWLEDGE

Write your answers on a seperate sheet of paper. Please do not write in this book.

1. A person who starts a new business is called an _____.
2. List the six common mechanisms.
3. True or false. Liquids are often called engineering materials.
4. Which of the following are genetic materials: (a) wool, (b) copper, (c) lumber, (d) wheat, (e) natural gas, and (f) asphalt?
5. True or false. We should quit eating hamburger because beef is an exhaustible material.
6. Define (a) data, (b) information, and (c) knowledge.
7. List the mechanisms that a wheelbarrow uses in its operation.

APPLYING YOUR KNOWLEDGE

1. Look around the room you are in. List the human abilities that were used to design, construct, and decorate it and the products within it. Under each category list three ways each ability was used. Use a chart similar to the one below.

ABILITY	TASK WHICH USED THE ABILITY

2. Design and sketch a simple device that uses at least three of the six simple machines (mechanisms) to do a job.
3. Select a product that is made from exhaustible materials or a task that uses an exhaustible energy source. Describe how that product could be made or the task could be completed using renewable materials or renewable energy sources.

TECHNOLOGICAL PROCESSES

After studying this chapter you will be able to:
- Define and describe the major types of technological processes.
- List and describe the steps in the problem solving/design process.
- Explain the major production steps used in the communication, construction, manufacturing, and transportation processes.
- Describe management as a technological process.
- Diagram the relationship between problem solving, production, and management processes.
- Apply technological processes to everyday activities.

You have learned that a technological system has inputs, processes, and outputs. Inputs are the resources that go into the system. Processes are what happen within a system. They take inputs and change them into outputs. Processes are used to *design* products, structures, communication messages, and transportation systems. They are also used to:
- *produce* products and structures.
- *communicate* information and ideas.
- *transport* people and cargo.

Additionally, managerial processes see that the technological system runs efficiently, produces quality products and few unwanted outputs.

From this discussion, you can see that there are three major kinds of processes. These, as shown in Fig. 5-1, are *problem solving* or *design* processes, *production* processes, and *management* processes. The first two can be grouped as transformation processes. They

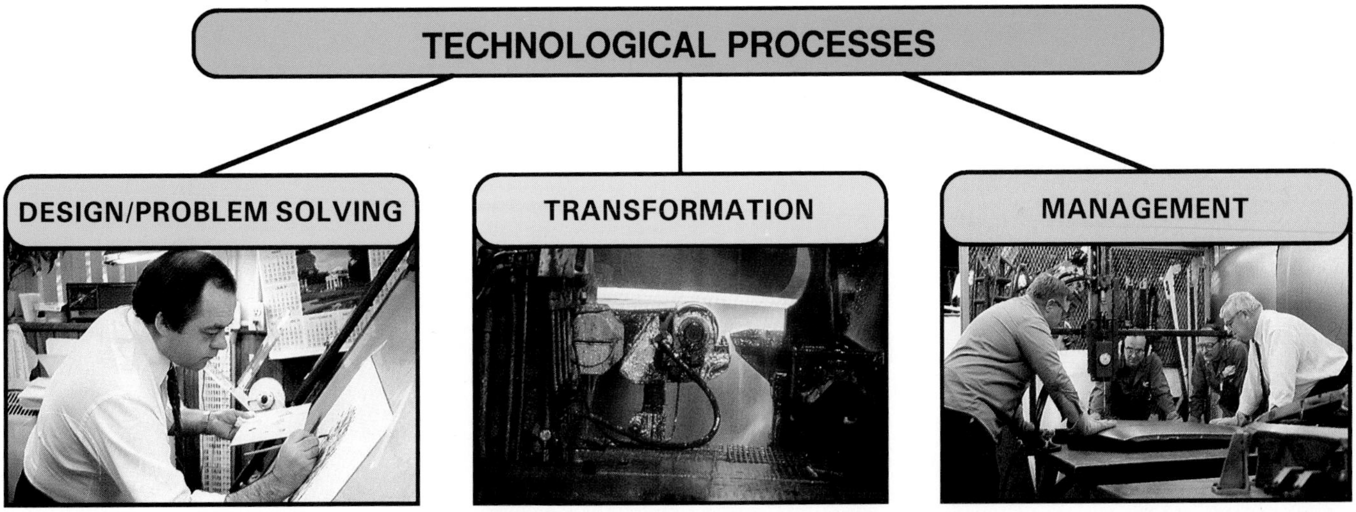

TECHNOLOGICAL PROCESSES

DESIGN/PROBLEM SOLVING

TRANSFORMATION

MANAGEMENT

Fig. 5-1. The design/problem solving, production, and management processes are used to develop and operate technological systems. (Inland Steel Co.)

directly transform an idea into a tangible output. This chapter will explore each process separately and show how they relate to one another.

PROBLEM SOLVING OR DESIGN PROCESSES

All technology has been developed to meet human needs and wants. Each device or system is designed to solve a problem or to meet an opportunity, Fig. 5-2. Early technology was almost always problem oriented. Tools were designed to make work easier. New housing technology made living more comfortable. New transportation devices helped humans move loads from place to place. New communication methods made information exchange easier.

Many modern technological devices are designed for the same reason: to help solve a problem. But other technologies are developed to meet an opportunity. Spacecraft are not designed to solve a pressing transportation problem. However, they give us the opportunity to study the earth from orbit. Likewise, many products are designed to make money. How many people really need an electric toothbrush? Do standard toothbrushes fail to meet our needs? The electric model was most likely designed to give the inventors and manufacturers an opportunity to *profit* from their idea. There is nothing wrong with this motive. In fact, the drive to make money has brought us many things that we now take for granted. It simply suggests that our civilization has progressed a great deal. We now have time to apply technology to luxuries. We are beyond the survival mode of our early ancestors.

STEPS IN PROBLEM SOLVING AND DESIGN

Designers and problem solvers follow a common procedure in designing and developing technology. The procedure is the same for solving problems or meeting opportunities. The five most important steps in the problem solving/design process are shown in Fig. 5-3. They are:

- Identifying the problem or opportunity.
- Developing multiple solutions to the problem or opportunity.
- Isolating and detailing the best solution.
- Modeling and evaluating the selected solution.
- Communicating the final solution.

Each of these steps leads to the next one. However, remember that all technological processes have feedback. The results from one step may cause the designer to retrace her/his steps. For example, a prototype built

Fig. 5-2. Technology is designed to solve a problem or meet an opportunity. (Zenith Corp.)

in the modeling step may show major problems with a design. The solution may have to be modified by changing it and selecting a new "best" solution. Now, let's look at each of these steps.

Identifying the Problem or Opportunity

As we said, technology starts with a problem or an opportunity. However, these problems and opportunities are seldom clearly seen or felt. The first step in the design/problem solving process is to describe the task that is being undertaken. The problem or opportunity must be identified.

This process includes defining a problem or opportunity. Along with the definition will be a list of limita-

Fig. 5-3. The design process defines, develops, and specifies solutions to problems.

tions. These can include appearance, operational, manufacturing, marketing, and financial criteria. These criteria communicate the expectations for the solution.

Developing Several Solutions

Once the problem or opportunity is defined, a designer seeks solutions. Designers create many possible answers. Their sketches show ways to meet the challenge. They first make rough sketches to capture thoughts. Then refined sketches mold thoughts into more specific solutions. The process of letting the mind create solutions is called **ideation**, Fig. 5-4.

Isolating and Detailing the Best Solution

Refined sketches of possible solutions are studied. Ideas are analyzed to select the most promising. This *best* solution may not meet all the criteria perfectly.

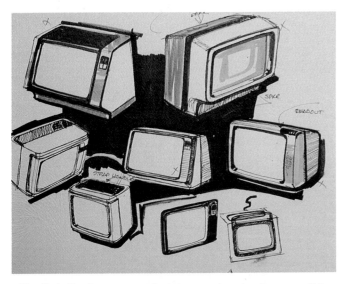

Fig. 5-4. Designers use sketches as they explore possible solutions for design problems or opportunities.

However, it must be a functional (work properly) product, structure, communication message, or transportation system that can be produced and marketed within cost limits.

Once the solution is chosen it must be detailed. Specifications and general characteristics of the final solution must be established.

Modeling and Evaluating

Quite often a model of the expected solution is produced, Fig. 5-5. A model allows designers and managers to review the solution's performance. Models may be physical, graphic, or mathematical models. Often, graphic models based on mathematics are first developed on a computer. These models are refined to optimize the solution. Then, physical (working or appearance) models may be constructed.

Communicating the Design

The final solution must be carefully specified for production. Drawings showing its size, shape, and component arrangement must be developed. Specifications for the materials to be used must be produced. A material list or bill of materials must be formulated. These documents communicate the characteristics of the product, structure, media, or system.

The design/problem-solving process ends with a specified solution. Section 4 of this book contains a more detailed discussion of this process. The solution chosen by the design or problem-solving process then moves to the next step: the production process.

PRODUCTION PROCESSES

Production processes are actions that create the physical solution to the problem or opportunity. They

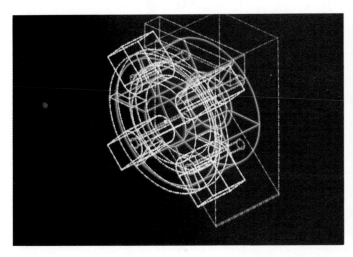
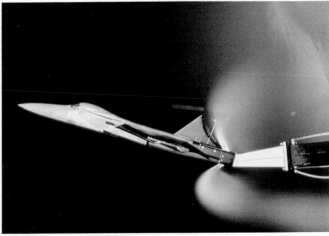

Fig. 5-5. Models are used to test product and structure designs. Models can be made on computers (left), or physical as is this airplane model in a wind tunnel (right). (Daimler-Benz)

produce products, construct structures, generate communication messages, or transport people or cargo. The design/problem-solving process is used across system lines. The process generates the description of the solution. The same processes or sequence of events can be used to design production systems, structures, communication systems, and transportation systems.

However, each of the technological systems has its own unique production processes. Manufacturing processes are different from construction processes. Communication processes are different from transportation processes. These processes move the solutions described by the design process into tangible solutions. This relationship between the generic design processes and the unique production processes are shown in Fig. 5-6. You will note that management processes, which will be discussed later, are also used across system lines.

COMMUNICATION PROCESSES

Humans have always exchanged information and ideas. This process is called communication. The simplest communication processes involve spoken language. This is not technology since no technical means are used. However, as civilization grew, additional communications techniques were developed. These involved technology. Technological means are used to help us communicate better. We now use technology to produce printed, graphic, and photographic media. Also, telecommunication technology allow us to communicate using electromagnetic waves.

Graphic Communication

The communication process described above includes two major headings. The first are the **graphic communication** processes. Its messages are visual and two-dimensional. Included are technical graphic messages such as technical illustrations and engineering drawings. Also under this heading are the printed messages commonly found in books, magazines, owners' and service manuals, and promotional flyers. The final graphic communication media is photographic communication. This group includes the film and print media coming from photographic processes.

Telecommunications

The second type of communication technology is telecommunications. These media depend on electromagnetic waves to carry their messages. **Telecommunication** techniques include broadcast (television and radio), hard-wired (telephone and telegraph), and surveillance (radar and sonar) systems.

All these technologies, which will be discussed in depth in Section 6 of this book, involve five major steps, Fig. 5-7:
- Encoding.
- Transmitting.
- Receiving.
- Storing/Retrieving.
- Decoding.

Communication technology organizes information so that people can receive it, Fig. 5-8. The communication process starts with an information source. Most often this source is the human brain. In some cases it can be machines. The first step in the communication process is to **encode** the information. Encoding involves placing the information in a format or pattern that the communication technology can use. This process may involve producing images on film, pulses on a light wave, electrical charges on a tape, or graphics on paper.

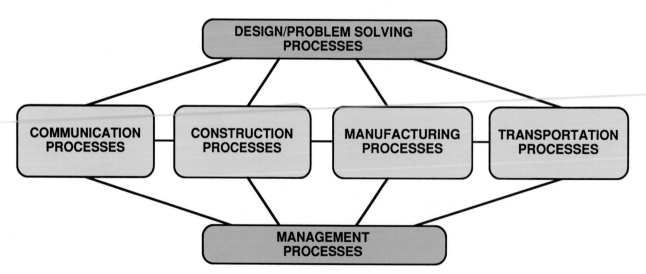

Fig. 5-6. The same design/problem solving processes and management processes are used for all four technological areas.

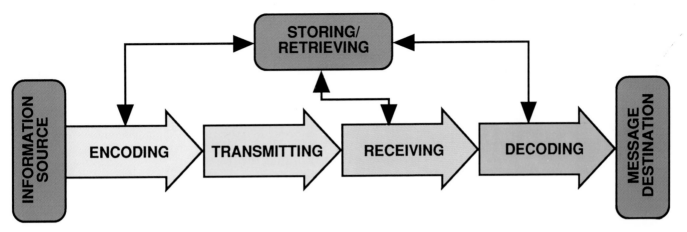

Fig. 5-7. These are the steps in the communication process.

The coded message must then be **transmitted** from the sender to the receiver. Transmission may include moving printed materials, broadcasting radio and television programs, transmitting telephone messages along wires or fiber optic strands, and using pulses of light to send messages between ships at sea.

Transmitting a message is not enough. Someone or something must receive it. **Receiving** the message requires recognizing and accepting the information. A radio must recognize the coded radio waves. The telephone receiver must recognize the pulses of electricity. The human eye must see the semaphore flags (ship-to-ship communication).

The received message must then be **decoded**. This means that the coded information must be changed back into a recognizable form. The decoded informa-tion is presented in audio or visual format that humans can understand. Radio waves are changed into sound waves by the radio receiver. The human brain places meaning to the position of the semaphore flags. The telephone changes electrical impulses to sound.

Throughout the communication process information may be stored and retrieved. Storage processes allow the information to be retained for later use. Books may be shelved in libraries, recorded music may be stored on tapes or compact discs, television programs may be stored on tapes, data may be stored on computer disks or tapes, and pictures may be stored in file folders. Later this information may be retrieved (brought back). It can be selected and delivered back into the communication process. Storage and retrieval can happen at any time in the communication system. It is commonly stored at the source or at the destination.

CONSTRUCTION PROCESSES

Humans first used construction technology to produce shelter. This allowed early humans to move out of caves into crude homes. This humble beginning took the form of huts and tents. We now construct many types of buildings. But construction technology does not stop with buildings. We also construct many types of structures to support other activities. These include:

- Roads, canals, and runways for transportation systems.
- Factories and warehouses for manufacturing activities.
- Studios, transmitter towers, and telephone lines for communication.
- Dams and power lines as a part of our electricity generation and distribution systems.

Constructed works are everywhere.

Constructed works can be grouped into two major categories: buildings and civil structures, Fig 5-9.

Fig. 5-8. Communication technology allows us to send, receive, and store data and information. (Harris Corp.)

Fig. 5-9. Construction technology is used to build buildings (top) and civil structures (bottom).

roads, dams, communication towers, railroad tracks, pipelines, airport runways, irrigation systems, canals, aqueducts, and electricity transmission lines.

Most construction projects include several steps which will be presented in more depth in Chapters 17 and 18. These, as shown in Fig. 5-10, include:

- Preparing the site.
- Setting the foundation.
- Building the framework.
- Installing the utilities.
- Enclosing the framework.
- Finishing the project.
- Completing the site.
- Servicing the structure.

Site Preparation

Not all construction projects use all the steps listed above. However, most projects start with **site preparation**. This task includes removing existing buildings, structures, brush, and trees that will interfere with locating the new structure. The site is then rough graded. The desired slope of the site is established. Then surveyors locate the exact spot for the new structure.

Foundation

The base or **foundation** for the structure is then constructed. This generally involves digging pits or trenches. Then concrete or rock is placed in the holes. If the hole extends to solid rock, additional foundations may not be needed. Foundations provide a stable surface onto which the building can be built.

Buildings are all the structures that are erected to protect people and machines from the outside environment. Buildings can be used for three major purposes. There are **residential structures** that people live in. These structures can be homes, town houses, condominiums, and apartment buildings. Buildings are also used as **commercial structures**. They are the stores and offices that are used to conduct business. Governmental buildings such as schools, city halls, and state capitol buildings can be placed in this category. Finally, buildings can be **industrial structures**. They are the power plants, factories, transportation terminals, and communication studios used by major companies.

The second major type of construction produces **civil or heavy engineering structures**. These are the structures that are primarily designed with the knowledge of the civil engineer. Common civil structures are

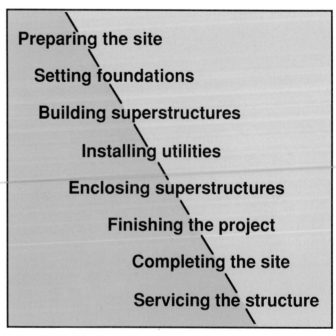

Preparing the site

Setting foundations

Building superstructures

Installing utilities

Enclosing superstructures

Finishing the project

Completing the site

Servicing the structure

Fig. 5-10. Most construction projects move through the eight steps shown here.

Superstructure

The **superstructure** of the project is constructed on the foundation. This may include the framework of the building or tower. Superstructures are also the pipes for pipelines, surfaces for roads and airport runways, and the tracks for railroads.

Utilities

In many cases the constructed structure will include **utilities**. This will require installing water, gas, and waste pipes; electrical and communication wire; and heating and cooling ducts. Also, many runways, railroads, and highways require lighting and communication systems to be installed.

After the utilities are installed, the superstructure must be enclosed. Walls will need interior and exterior skins. Roofs, ceilings and floors must be covered. Doors and windows must be installed to close openings.

Completing the Structure and Site

Finally, the structure must be completed. Walls, ceilings, doors, windows, and building trim must be painted. Likewise, communication towers must be coated to protect them from the natural elements. Runways and roadways must have traffic stripes painted on them and signs installed.

Also, the site must be completed. It must be graded to reduce erosion and increase its beauty. Shrubs, grass, and trees are added to landscape the site and keep the soil in place.

Servicing

The result of this sequence of steps is a new constructed structure. During its life it will require **servicing**. This means that it will be maintained, repaired, and reconditioned. Surfaces will be periodically painted, roofs will be replaced, and utility systems components will be replaced. Servicing will attempt to keep the structure in good working order.

MANUFACTURING PROCESSES

You use manufactured products for everything you do. You ride to school in a manufactured car or bus. You look out of the building through a manufactured window. You wear manufactured clothing. You are reading a manufactured book, sitting on a manufactured chair in a room lighted with manufactured fixtures. It is hard to imagine a world without manufactured products. Each product has moved through three stages, as shown in Fig. 5-11.

Obtaining Resources

Manufacturing activities start by obtaining material resources. This involves searching for or growing materials that can be harvested or extracted from the earth. Farming and forestry grows the trees, plants, and animals needed to support the manufacturing processes. Exploration companies will locate petroleum, coal, and metal ore reserves.

Once the resources are found, they must be gathered or obtained. This activity will use one of three major processes:

- **Harvesting**: Gathering genetic materials (living materials) from the earth or bodies of water at the proper stage of their life cycle.
- **Mining**: Obtaining materials from the earth through shafts or pits
- **Drilling**: Obtaining materials from the earth by pumping them through holes drilled into the earth.

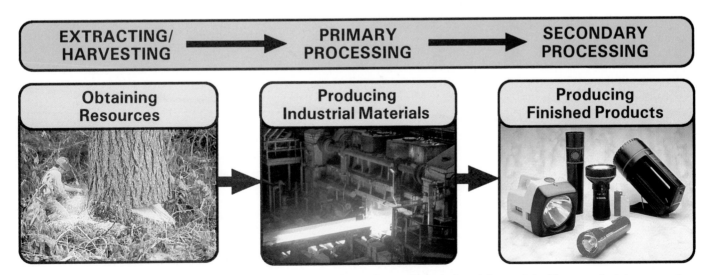

Fig. 5-11. Manufacturing obtains material resources, which are changed into industrial materials. These materials are used to make products. (Weyerhauser Co., American Iron and Steel Institute, Eveready Battery Co.)

Technology: How it Works

FLEXIBLE MANUFACTURING: Using a set of automatic, programmable machines to produce products in low-volume production runs.

Computer control has greatly changed how manufacturing systems work. The industrial revolution was based on producing large numbers of uniform products using a *continuous manufacturing* system. Any change in the design of a product makes it necessary to change the machines that make the product. These changes are very expensive to make.

The advent of computers has changed how products are designed. Products can be designed, tested, and refined on a computer. The design for a product is expressed in a *computer language*. This language then can be used to control machines, and inspect product quality, Fig. 1. Computers are also used to control how products move through a sequence of operations. This level of computer control has led to computer-integrated manufacturing (CIM), or flexible manufacturing systems.

These systems use a number of technologies, including computer-aided design, to develop and specify the product's size and shape. Robots are used to load and unload machines and move products from machine to machine, Fig. 2. Computer numerical control is used to guide machine operation. Electronic measurement and computer data processing help to improve inspection and quality control, Fig. 3.

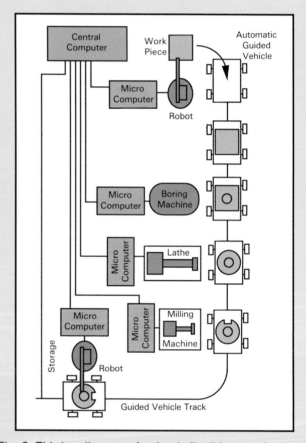

Fig. 1. Flexible manufacturing is based on computer control.

Fig. 2. This is a diagram of a simple flexible manufacturing system. Each part of the system is under computer control.

MACHINING AND ASSEMBLING

MATERIAL HANDLING

INSPECTING

Fig. 3. Computers are used to direct the operation of machine tools and assembly machines (left), material handling systems (center), and inspection systems (right).

Primary Processing

Material resources are changed to industrial materials in **primary processing** plants. Steel mills will change iron ore, limestone, coke and other materials into steel sheets and bars. Forest products plants will change trees into paper, lumber, plywood, particleboard, and hardboard. Smelters will change aluminum and copper ore into usable metal sheets, bars, and rods.

Typically these materials are refined using heat (smelting, melting, etc.), mechanical action (crushing, screening, sawing, slicing, etc.), and chemical processes (oxidation, reduction, polymerization, etc.). The result is an industrial material or standard stock. Examples are a sheet of plywood, glass or steel, a plastic pellet, or metal rods and bars.

Secondary Manufacturing Processes

Most industrial materials have little value to the consumer. What would you do with a sheet of plywood, a bar of copper, or a bag of plastic pellets? These materials become valuable when they are made into products. They are the material inputs to the **secondary processes** of manufacturing. These processes change industrial materials into industrial equipment and consumer products, Fig. 5-12.

There are six basic ways materials are changed during secondary processing. They are given size and shape by casting and molding processes, forming processes, and separating processes.

- **Casting and molding** introduce a liquid material into a mold cavity. There the material solidifies into the proper size and shape.
- **Forming** uses force applied from a die or roll for reshaping materials.
- **Separating** uses tools to shear or machine away unwanted material. This shapes and sizes parts and products.
- **Conditioning** uses heat, chemicals, or mechanical forces to change the internal structure of the material. The result is a material with new, desirable properties.
- **Finishing** coats or modifies the surface of parts or products to protect them, or to make them more appealing to the consumer.
- **Assembling** brings materials and parts together to make a finished product. They are bonded or fastened together to make a functional device.

The result of obtaining resources, producing industrial materials, and manufacturing products are the devices that we use daily. More information about these manufacturing activities is in Chapters 14, 15, and 16.

Fig. 5-12. Secondary manufacturing produces industrial equipment (top) and consumer products (bottom). (Minster Machine Co., Northern Telecom)

TRANSPORTATION PROCESSES

Humans have always wanted to move around. They have wanted to move themselves and their possessions from one place to another. This movement is called transportation. Transportation takes place in three basic modes with a fourth one on the horizon, Fig. 5-13.

Transportation Modes

The earliest movement was on **land**. People first walked the land. Then they tamed animals to help them move for one place to another. Finally they developed a number of land vehicles. These range from the horse-drawn wagon of old to the magnetic levitation trains now in development.

Fig. 5-13. Transportation can be on land, over or under water, and through the air. Later, we will probably use space transportation systems more. (Norfolk Southern Corp, OMI Corp., and NASA)

The second transportation medium is **water**. Again, water transportation has a long history. It ranges from the dug-out log canoes to nuclear submarines.

The last major transportation medium is **air**. This medium is of recent vintage. It can be traced back to early balloon travel and now extends to supersonic airplanes.

Now we are starting to develop a fourth transportation medium, **space**. Today we only explore the near reaches of space, Fig 5-14. However, the future holds the promise of space transportation systems.

Transportation Sub-systems

Most transportation systems contain two major subsystems: vehicular systems and support systems. The vehicular systems are the on-board technical systems that make a vehicle work. Generally there are five of these systems, as shown in Fig. 5-15:

- **Structure**: The framework of the vehicle. This includes passenger, cargo, and power system compartments.
- **Propulsion**: Generates motion through energy conversion and transmission.

Fig. 5-14. Space travel presently is used for experimental work and study. In this rendering, astronauts test a construction system. (NASA)

- **Guidance**: Gathers and displays information so that the vehicle can be kept on course.
- **Control**: Makes changes in speed and direction of the vehicle possible.
- **Suspension**: Keeps the vehicle held in or onto the medium being used (land, water, air or space).

Additionally, transportation systems require support systems. These include pathways and terminals, Fig. 5-16.

- **Pathways** are the structures along which the vehicles travel. They can include roads, railways, waterways, and flight paths.
- **Terminals** are the structures that house passenger and cargo storage and loading facilities. A complete discussion of transportation technology is included in Section 7 of this book.

Fig. 5-16. Transportation support systems include pathways like the road in the top picture, and terminals like the railway terminal shown below. (Amtrak)

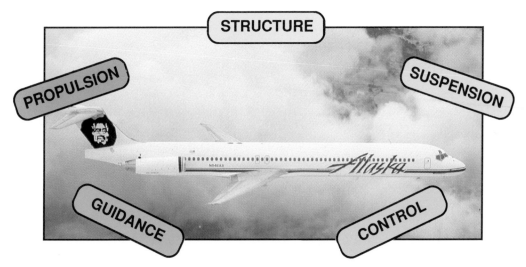

Fig. 5-15. Most vehicles are made up of five systems. (Alaska Airlines)

MANAGEMENT PROCESSES

The third major type of process used in technology is management. These are the processes designed to guide and direct the other processes. Management provides the vision for the activity. Specifically, management includes four steps, Fig. 5-17. The first step is **planning**, which involves developing goals and objectives. These may be overall goals or specific goals for activities such as production, finance, or marketing.

Once plans are developed, the activity must be structured. Procedures to reach the goal must be established. Lines and levels of authority within the group or enterprise must also be drawn. These actions are called **organizing.**

After the activity is organized, actual work must be **actuated**, or started. Workers must be assigned to tasks. They must be motivated to complete the tasks accurately and efficiently.

Finally the entire operation must be **controlled**. The outputs must be checked against the plan. Control is the feedback loop that causes management activities to be adjusted.

Sometimes we think of management only in terms of companies. Every human activity is managed. Goals are set. A procedure for finishing the task is established. Work is started and completed. Results are evaluated. Try to think of a productive activity that does not contain these elements. A coach plans, directs, and evaluates the plays a football team makes. The Girl Scout troop leader manages cookie sales. You develop a plan to mow a lawn.

It is easy to see that we manage personal and group activities. Some of these group activities are with social, religious, and educational groups. Others are with technological enterprises. The management of technological activities and companies will be explored in Section 9 of this book.

SUMMARY

All technological systems have processes. They include processes to design and engineer the outputs and the systems themselves. Often these processes use problem solving methods. Problem solving techniques are to technology what the scientific method is to science. It guides the creative activities of the field.

Once the system and outputs are designed, production processes are used. These transform inputs into outputs. Materials, human abilities, machines, information, energy, finances, and time are used to produce products, structures, media messages, and transported people and cargo.

Finally, all technological systems are managed. They are planned, organized, actuated, and controlled. Through these processes the human-made world around us was created and is being sustained.

WORDS TO KNOW

All of the following words have been used in this chapter. Do you know their meanings?

Actuating
Air transportation
Assembling
Building
Casting and molding
Civil engineering structures
Commercial structure
Conditioning
Control
Controlling
Decoding
Drilling
Encode
Finishing
Forming
Foundation
Graphic communication
Guidance
Harvesting
Heavy engineering structures
Ideation
Industrial structure
Land transportation
Mining
Organizing

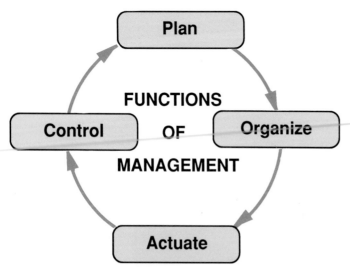

Fig. 5-17. Management involves setting goals, establishing a structure to meet them, assigning and supervising work, and monitoring the results.

Pathways
Planning
Primary processing
Propulsion
Receiving
Residential structure
Retrieve
Secondary processes
Separating
Servicing
Site preparation
Space transportation
Storage
Superstructure
Suspension
Telecommunication
Terminals
Transmit
Utilities
Vehicle structure
Water transportation

TEST YOUR KNOWLEDGE

Write your answers on a seperate sheet of paper. Please do not write in this book.

1. List the four modes in which transportation vehicles operate.
2. Buildings people live in are called _____ structures.
3. True or false. Setting goals for a system is called organizing.
4. List the four steps used to move information from a source to a destination.
5. Which of the following is NOT a way of obtaining materials resources? Harvesting, Casting, Drilling, or Mining.
6. The buildings used to receive passengers and cargo in transportation systems are called _____.
7. True or false. Secondary manufacturing processes change industrial materials into products and equipment.
8. The part of a constructed work that supports the superstructure is called a _____.
9. List the three major types of processes used in technology.

APPLYING YOUR KNOWLEDGE

1. Select a product or structure that you know about. Develop a chart similar to the one below and list five considerations that were used in each of the technological processes used to produce the item.

Product or structure name:	
Process	**Factors considered**
Design/ Problem solving	
Production	
Management	

2. Select a problem that exists in the world that can be partially solved through technology. List the role each technology would play in solving the challenge.
3. List and describe how you have used the production processes in one of the technological areas (communication, construction, manufacturing, or transportation). Use a chart similar to the one below.

Product or structure name:	
Technology	**Contributions to solving the problem**
Communication	
Construction	
Manufacturing	
Transportation	

The proper design and control of an industrial process is very important if a quality product is to be produced.

OUTPUTS, FEEDBACK, AND CONTROL

After studying this chapter you will be able to:
- *Describe the major types of outputs of technology systems.*
- *Differentiate between desired and undesired outputs of technological systems.*
- *Define and describe feedback.*
- *List the major types of control that are exercised on technology systems.*
- *Describe the parts of a control system.*
- *Explain how people and society control technology.*
- *Discuss the positive and negative aspects of control on technological systems.*
- *Discuss the relationships between outputs, feedback, and control.*

All technological systems are purposeful. They are designed to meet specific needs and wants of people. Therefore, there is a direct relationship between the need (the reason for the system) and the output (the satisfying of the need). These outputs can be categorized in three ways, as shown in Fig. 6-1:
- Desirable–undesirable.
- Intended–unintended.
- Immediate–delayed.

DESIRABLE AND UNDESIRABLE OUTPUTS

The needs for and outputs of technology are countless. We have already classified technological systems as communication, construction, manufacturing, and transportation. Each of these technologies has a general type of output, Fig. 6-2. Communication technology satisfies the need for information. Its outputs are media messages. Construction technology satisfies the need for shelter and support structures for other activities. Its outputs are buildings and civil structures. Manufacturing technology satisfies the need for tangible goods. Its outputs are consumer and industrial products. Transportation technology satisfies a desire

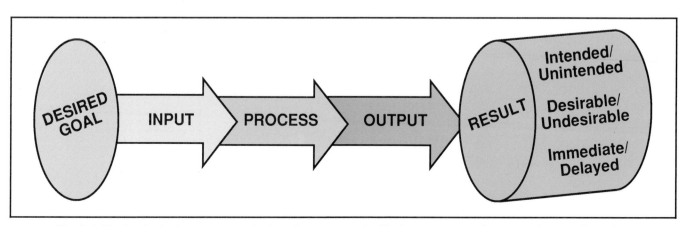

Fig. 6-1. Technological systems are designed to meet goals. Their outputs can give several types of results.

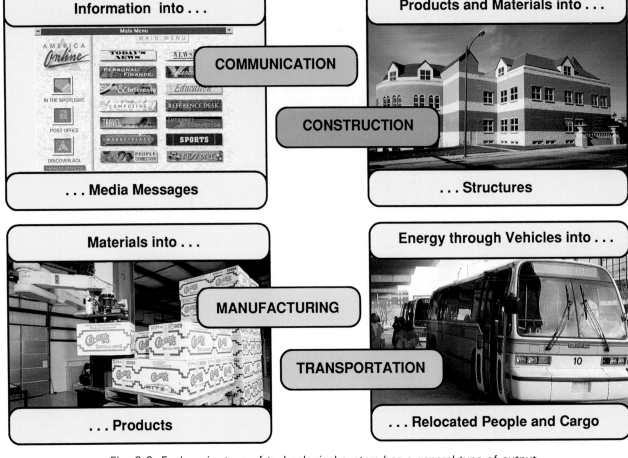

Information into...	Products and Materials into...
COMMUNICATION	CONSTRUCTION
...Media Messages	...Structures

Materials into...	Energy through Vehicles into...
MANUFACTURING	TRANSPORTATION
...Products	...Relocated People and Cargo

Fig. 6-2. Each major type of technological system has a general type of output. (America Online, Kolbe and Kolbe Millwork Co., Inc., FANUC Robotics North America, Inc.)

to move humans and things. Its output is the movement of people and goods (cargo). These are the **desirable outputs** of the systems, Fig 6-3.

However, there are also **undesirable outputs.** For example, some manufacturing activities produce fumes and toxic chemical by-products. If improperly treated these outputs poison the air and water around us. Poorly planned construction projects can cause soil erosion. The result may be the loss of valuable top soil and increased flooding. Communication technology can use billboards which some people find unsightly. Also, people differ on the value of some communication messages. Every communication message, to some people, is unsightly or is considered noise. Finally, transportation systems create outputs we do not want. Exhaust fumes from cars and trucks pollute the air. Airport and highway noise impinge on residential areas. New rail lines and roads can separate neighborhoods or divide farms. All of these are examples of the undesirable outputs of technology.

INTENDED AND UNINTENDED OUTPUTS

As we said, technological systems are designed to produce specific outputs. These may be refrigerators, weekly magazines, high-rise apartments, municipal bus

DESIRABLE OUTPUT

UNDESIRABLE OUTPUT

Fig. 6-3. Outputs of technological systems can be seen as desirable or undesirable. (United Airlines)

transportation, or any of thousands of other products and services. These are the **intended outputs.**

Depending on your point of view, the intended outputs may be desirable or undesirable. For example, suppose you planned to make 10″ diameter plywood disks. You would have to cut them from rectangular sheets of plywood. Cutting circles from rectangular material will produce waste material. You can plan for the waste, consider it in pricing your product, and reduce it to a minimum. However, you cannot totally eliminate it. Both the disks and the waste were intended. They were in the plan for the system. The disks are the intended *desirable* output and the waste is the intended *undesirable* output.

Sometimes, technological systems produce outputs that were not considered when the system was designed. For example, some heating systems produced in the 1940s through the 1960s used asbestos to insulate pipes. The material was considered an excellent insulator. Later, people who worked with asbestos developed lung cancer. This was an undesirable, **unintended output** of the technological system.

IMMEDIATE AND DELAYED OUTPUTS

Most technological systems are designed to produce a product or service for **immediate** use. We don't produce steel for use in year 2051. We produce it for use this year. This practice reduces inventory costs and loss due to theft and spoilage. This is an example of an immediate, intended, and desirable output.

However, some outputs have **delayed** effects. The chemicals used as propellants in aerosol (spray) cans and as refrigerants are now affecting the ozone layer around the earth. The accumulation of this matter has taken decades to reach a dangerous level. The same is true of the sulfur dioxide produced by coal-fired power plants. The resulting acid rain is killing forests in the United States and Canada. These are examples of delayed, unintended, undesirable outputs of technology.

The relationship of these three views of outputs is shown in Fig. 6-4. You will note that it constitutes a series of blocks. The lower left block in the front row includes all the desirable, immediate, intended outputs. The upper right block at the front contains the unintended, immediate, undesirable outputs. Can you think of examples of outputs that would fit into each of the eight blocks shown?

FEEDBACK AND CONTROL

You know by now that technology systems are designed to benefit humans. Therefore, the people

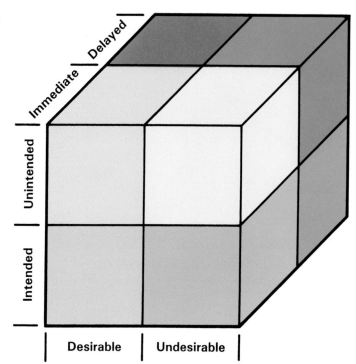

Fig. 6-4. The eight types of outputs are shown as boxes inside a cube. Can you think of an example for each box?

designing and using technology should strive to maximize the desirable results. The undesirable outputs can never be totally eliminated. They need to be held to a minimum. To do this, most technology has control systems. These systems are used to compare the results with the goals, Fig. 6-5.

Controlling systems is often viewed at the operating level. This view explores how processes are controlled so that outputs meet specifications. This is an important type of control. However, a broader view is also required to understand technological system control. Political actions, personal value systems, international competition and other factors outside the system also exercise control. We will explore both areas of control in this chapter.

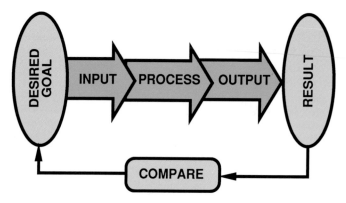

Fig. 6-5. Feedback compares the results with the goal and identifies reasons for any differences.

Fig. 6-6. Technology uses open-loop and closed-loop control systems. The food processing equipment on the left runs at a setspeed (open loop), while the power generators shown on the right automatically adjust to increased demand for electricity (closed loop). (FMC Corp., American Electric Power)

TYPES OF CONTROL SYSTEMS

Two major types of control systems are used in technology, Fig. 6-6. The simplest control system is the open-loop. This system uses no feedback to compare the results with the goal. The system is set and then operates without the benefit of output information. Suppose you decide to drive a car to an adjoining town. You could start it and hold the gas pedal at a specific point. The car will accelerate to a certain speed. It will stay at that speed until you reach a hill. It will slow down as it takes more power to move the vehicle. As you crest the hill, the car will speed up as it coasts down the other side. As long as you do not look at the speedometer and adjust the gas pedal the car will change speed as it reaches each new road condition. This type of control is **open-loop control**. Output information is not used to adjust the process in this type of control system.

However, most technological systems use feedback in their control systems. This means that output information is used to adjust the processing actions. Feedback can be used to control various stages or factors within a system or the entire system. Systems using feedback are called **closed-loop control** systems.

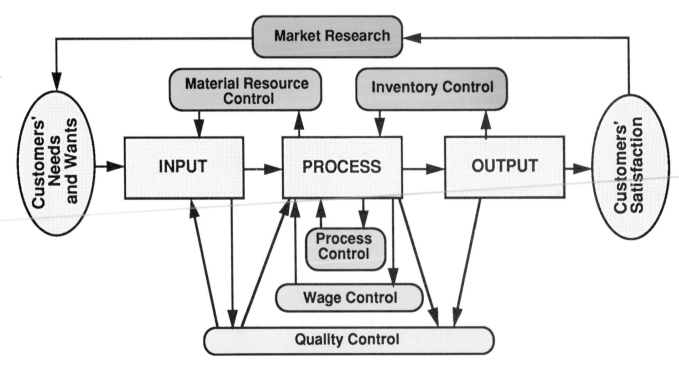

Fig. 6-7. These are some selected types of control used in manufacturing. Note the feedback. Information enters the control system, is processed, and is then fed back into the system.

Let's look at closed-loop control in only one technology: manufacturing, Fig. 6-7. Customer reactions are used to evaluate present and future product designs. This information is gathered through *market research* and the analysis of customer complaints. Parts, assemblies and finished products may be compared with engineering standards. The processes is called *quality control*, Fig. 6-8. It makes sure the product performs with in an acceptable range. Finished goods in warehouses and products in process are compared with sales projections. This is called *inventory control*, Fig. 6-9. The goal is to closely match product production with product demand. The operation of machines and equipment is closely monitored. This is called *process control*. The goal is to see that each manufacturing process produces the proper outputs. The need for materials and the quantity on order is monitored. This is called *material resource control* (or planning). The goal of this activity is to reduce the raw material inventory. The hours worked by employees producing products is monitored. This is called *wage control*. The goal of this activity is to keep the labor content of a product within planned limits.

Control is not limited to manufacturing. Automobiles have emission control systems. Engine have governors to control their operating speeds. Some magazines are sent to people who need only that kind of information. Control systems monitor the ink being applied to paper during the printing process. Thermostats control

Fig. 6-9. This worker is checking the inventory of tread plates used in making truck bodies. (Reynolds Metals Co.)

the temperatures in buildings, ovens, and kilns. Fuses and circuits breakers control the maximum amount of electric current in a circuit. Control is everywhere.

COMPONENTS OF CONTROL SYSTEMS

We want transportation services, products, information, and structures to meet our needs. To ensure that this happens we design control systems with three major parts, Fig. 6-10. These are:

- Monitoring devices.
- Data comparing devices.
- Adjusting devices.

Monitoring Devices

The first part of a control system gathers information about the action being controlled. It **monitors** inputs, processes, outputs, or reactions.

The information gathered can be of many types. These include material size and shape, temperatures of enclosures, speed of vehicles, and impacts of advertising. It can be very specific mathematical data, or more general opinions of customers. It may be gathered and recorded in short intervals, or over long periods of time.

The type of information gathered during the monitoring step depends on how the data will be used. Process performance data will be very narrow and focused. Data about the impacts of technological systems will be much more general. For example, data needed to determine the emission levels of a motorcycle engine are more specific than the data needed to determine the impact of the air transportation system on business.

Fig. 6-8. Data about product characteristics is gathered and compared to standards by quality control personnel. (Goodyear Tire and Rubber Co.)

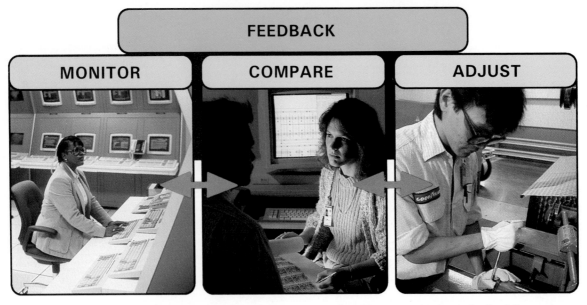

Fig. 6-10. Control systems monitor performance, compare it against standards, and adjust the system to ensure the output meets the goal. (General Motors, AT&T, Goodyear Tire and Rubber Co.)

Process operating information can be gathered using several types of sensors. These include:

- **Mechanical sensors**, which can be used to determine position of components, force applied, or movement of parts.
- **Thermal sensors**, which can be used to determine changes in temperature.
- **Optical sensors**, which can be used to determine the level of light or changes in the intensity of light, Fig. 6-11.
- **Electrical** or **electronic sensors**, which can be used to determine the frequency of or changes in electric current or electromagnetic waves, Fig. 6-12.
- **Magnetic (electromagnetic) sensors**, which can be used to determine if there are changes in amount of current flowing in a circuit.

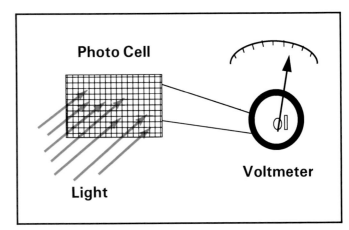

Fig. 6-11. A photographic light meter is a common optical sensor. Light striking the photo cell generates electricity which is measured by a voltmeter. Strong light produces more voltage than weak light.

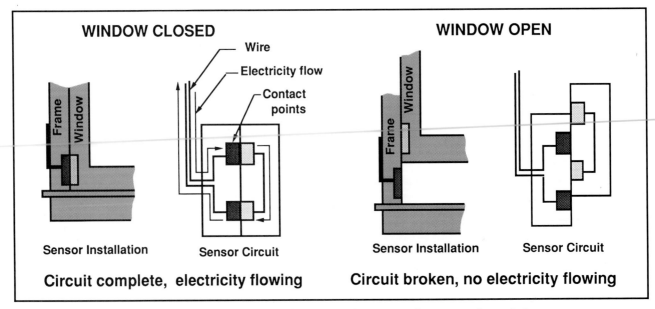

Fig. 6-12. This electrical sensor is part of a home security system. It senses when windows are open.

Technology: How it Works

INTEGRATED CIRCUIT (microchip): A piece of semiconducting material in which a large number of electronic components are formed.

We are living in the information age. We receive information in all forms instantly from all over the world. The information revolution began with the invention of the transistor. A transistor is an electronic device that can be used in switching and amplifying electrical signals quickly and cheaply. The device allowed engineers to eliminate bulky electronic devices such as the vacuum tube. Technology has made even greater advancements in the field of electronics, expanding transistor technology to create the integrated circuit (IC), or microchip.

An IC is a thin piece of *pure* semiconductor material, usually silicon. The material is made in a variety of small rectangular sizes. Impurities are introduced into the semiconductor material to affect its electronic properties. Tens of thousands of electronic devices and their interconnections can be produced in one IC, Fig. 1.

Integrated circuits must be manufactured in a very clean environment. Often, the individual electronic devices on the chip are as small as three microns (one micron equals 40 millionths of an inch). The parts must align within one micron, Fig. 2.

Integrated circuits are very complex and require careful design. This requires a high level of skill and can be very time-consuming. Consequently, computer-aided design programs have been produced to allow people to design complex circuits quickly and accurately

There are several methods used to manufacture ICs. Most begin with a slice of pure silicon. A common procedure used for turning the silicon into an integrated circuit includes:

1. Oxidizing the surface (or coating the surface with a protective oxide). This prevents impurities from entering the silicon.

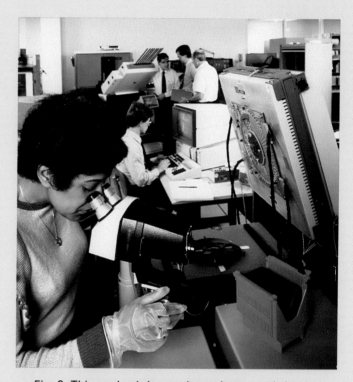

Fig. 2. This worker is inspecting an integrated circuit.

2. Coating the surface with a photoresist. A photoresist is a light-sensitive chemical. When developed, it prevents the oxidized layer of the IC from being etched away.
3. Placing a pattern over the photoresist and exposing it to light. The unexposed areas are then washed away.
4. Etching the IC with acid. This creates windows of unprotected areas on the silicon. Impurities can be introduced through these windows.
5. Introducing impurities into the silicon in a process called *doping*. These impurities create the tiny transistors, diodes, capacitors, and resistors in the IC, giving the chip its electronic properties.
6. Connecting the components with small aluminum leads.
7. Placing the IC in a protective casing. The IC's internal components are connected to the external leads of the casing.

The integrated circuit is responsible for the enormous growth of the computer industry in the latter half of the twentieth century. Currently, ICs can be found in everything from a simple pocket calculator to the most complex space vehicles.

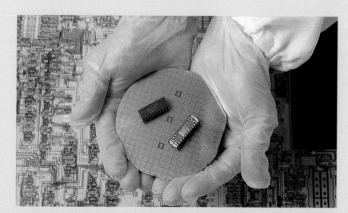

Fig. 1. Microchips are contained inside these cases. The electrical connections extend out of the case.

Comparing Devices

The information gathered by sensing devices is matched against expectations. The results are *compared* to the intent of the system. This is much like a balance scale. The results should balance with the goal.

Comparisons can be analytical or judgmental. **Analytical methods** mathematically or scientifically make the comparison. Statistics may be used to see if the output meets the goal. The temperature of the oven can be compared to the setting. The size of a part can be checked with lasers and compared with data in a computer file. The position of the ship can be compared the location of radio beams. This type of comparison is commonly used in closed loop control systems.

Analytical methods also can be used for managerial control systems. The number of products produced may be directly compared with the production sched-ule. The number of passengers buying bus tickets may be compared with sales projections. The number of people listening to a radio station may be compared with the total listeners for all stations. Analytical methods try to remove human opinion from the evaluation process. The information is reduced to percentages, averages, or deviation from expectations.

Judgmental systems use human opinions and values to enter into the control process. Open-loop control systems require human judgment. The speed you drive a car or the movement you use to cut on a line is based on your judgment. You do not apply mathematical formulas to decide when to accelerate and to brake as you drive on a mountain road.

Likewise human opinion is used in determining value of system outputs. Judgments are used in the design phase of technological systems. It is almost impossible

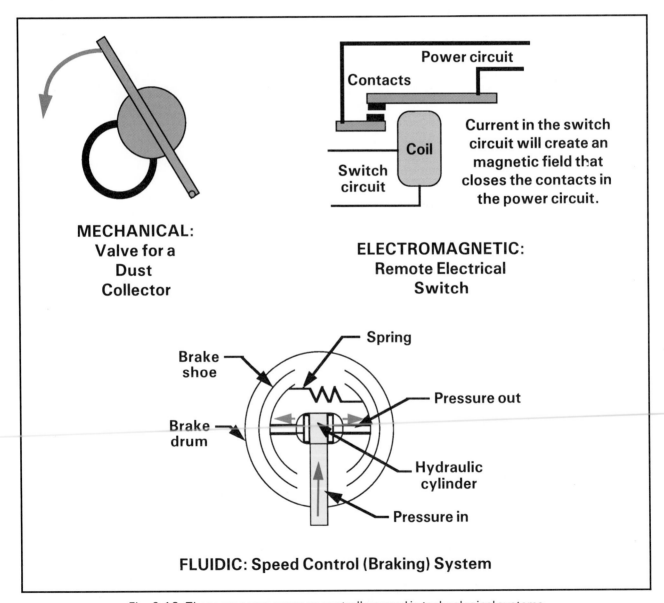

Fig. 6-13. These are some common controllers used in technological systems.

to select a design by analytical means. People decide the styling, color, and shape of most products and buildings. They select the content of magazines and telecommunications programs. After the output is available, analytical means may be used to compare the results. After a television program is produced, the number of viewers watching the program can be measured. The success of the program can be subjected to mathematical analysis. Likewise, a furniture manufacturer can mathematically compare the sales of Early American tables to the sales of French Provincial tables. However, the decision to produce the show or manufacture the particular style of table was a judgment.

Adjusting Devices

Comparing outputs to expectations may indicate that the system is not operating correctly. The output may not meet specifications. The system will need adjusting. This adjusting is done by controllers. They may speed up motors, close valves, increase the volume of burning gases, or a host of other actions. The goal is to cause the system to change and, therefore, produce better outputs.

System controllers can be several types. These, as shown in Fig. 6-13, include:
- **Mechanical Controllers:** using cams, levers, and other types of linkages to adjust machines or other devices.
- **Electromechanical Controllers:** using electromagnetic coils and forces to move control linkages and operate switches to adjust machines or other devices.
- **Electrical and electronic controllers**: using electrical (switches, relays, motors, etc.) and electronic (diodes, transistors, integrated circuits, etc.) devices to adjust machines or other devices.
- **Fluidic controllers:** Using fluids, including oil (hydraulic) and air (pneumatic), to adjust machines or other devices.

OPERATING CONTROL SYSTEMS

Technological control systems can be classified as manual and automatic. **Manual control systems** require humans to adjust the processes. This manual action is intended to correct problems identified by the control system, Fig. 6-14.

Manual Control

A good example of a manual control system is the automobile. The driver uses various speed and direction controls to guide the vehicle to the destination (goal). As the vehicle enters a city the driver eases up on the accelerator to reduce the speed to the new speed limit. Brakes may be applied to meet the new condi-

Fig. 6-14. This pilot is using a manual control system to land an airplane. His eyes monitors the approach to the runway. His mind compares the actual flight path with the desired one. His hands and feet change control surfaces and speed to land the plane safely.
(United Airlines)

tions. The steering wheel is turned as the road curves. In this example, the human eye gathers information. It sees a new speed limit sign. The eye also scans the speedometer to determine the speed of the vehicle. The brain compares the posted speed limit with the speedometer reading. The brain then commands the foot to change its position to allow the speed to be adjusted until it matches the speed limit. In this example, the system is *monitored* by the human eye. Its actual and desired performance are *compared* by the brain. The system is physically *adjusted* by the foot. Manual control systems are everywhere. We use them in riding a bicycle, adjusting the temperature of a gas stove when cooking, setting the focus on a camera, or checking a fence post for plumb.

Automatic Control

Many technological systems have **automatic control**. Technological devices are used to monitor, compare, and adjust the system without human interference. A simple example of automatic control is the thermostat in a heating system, Fig. 6-15. The device measures the room temperature. It compares the temperature of the room to the desired temperature. As long as the temperatures are within a preset range, no action is taken. When the room temperature drops below the set temperature range, a switch is activated. It turns on the heating unit. The unit will operate until the upper limit of the temperature range is reached. At this point the thermostat turns off the unit.

Automatic doors in stores, electronic fuel injection systems in vehicles, and programmable VCRs are other

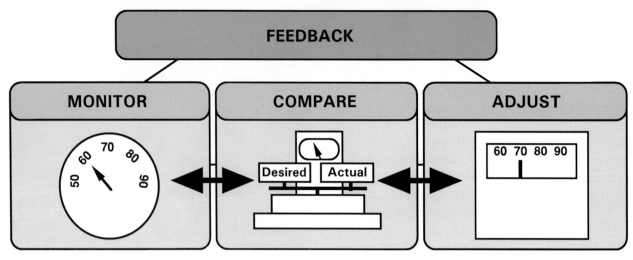

Fig. 6-15. A thermostat is a common automatic control system found in many homes and buildings.

examples automatic control systems you see every day. More sophisticated automatic control systems help pilots land the new generation of jet airliners, control the output of electric generating plants, and help contractors lay flat concrete roads.

OUTSIDE INFLUENCES AS CONTROL

We have looked at control primarily as a internal system component. As we said earlier, control also comes from outside the system, Fig. 6-16. For example, establishing the 40 hour workweek was done through labor union influence and government action. Environmental responsibility is a societal goal that technological systems must address. Systems are designed to reduce air, water, and noise pollution. Likewise, working conditions in factories and offices, on

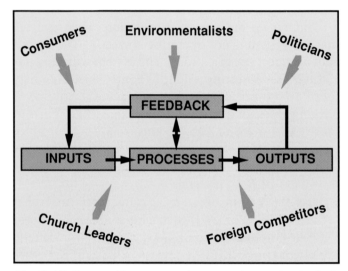

Fig. 6-16. External forces exercise control on technological systems.

Fig. 6-17. These workers are trying to influence governmental action.
(United Steelworkers of America)

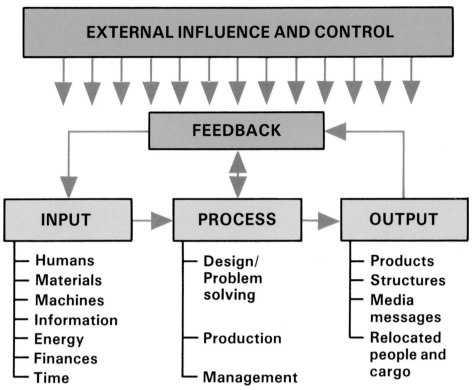

Fig. 6-18. This is a summary model of technological systems.

construction sites, and at transportation terminals and warehouses are controlled through political and labor union actions, as shown in Fig. 6-17.

Public opinion is also a control factor. The opinions of civic groups, churches, labor unions, and various associations affect public policies. Our news media constantly report on issues that will control technological systems. The debate between environmentalists and forest products companies over timber cutting is one example. Another is the dialogue between the pro-nuclear and anti-nuclear power groups. The controversy over the types of fishing nets that should be used is still another example of external control.

This discussion is not designed to list all the important public issues that exercise control on technology. It suggests, however, that looking only at the physical controls built into technological systems gives an incomplete view.

SUMMARY

Technology is a series of complex human-made systems that change resources into useful items and services, Fig. 6-18. The outputs of the systems can be desirable or undesirable, intended or unintended, and immediate or delayed. Control systems are designed to increase the probability that the outputs will be desirable. These control systems may be a part of the system or outside forces that influence the system.

WORDS TO KNOW

All of the following words have been used in this chapter. Do you know their meanings?

Adjusting devices
Analytical systems
Automatic control
Closed-loop control
Comparing devices
Delayed outputs
Desirable outputs
Electrical sensors
Electrical and electronic controllers
Electromagnetic controllers
Electromechanical controllers
Fluidic controllers
Immediate outputs
Intended outputs
Judgmental systems
Manual control
Mechanical controllers
Mechanical sensors
Monitor
Open-loop control
Optical sensors
Thermal sensors
Undesirable outputs
Unintended outputs

TEST YOUR KNOWLEDGE

Write your answers on a seperate sheet of paper. Please do not write in this book.

1. True or false. Technology produces only desirable outputs.
2. Match the outputs with the proper technology systems.

SYSTEM	OUTPUT
_____ Communication.	A. Relocate people.
_____ Construction.	B. Buildings.
_____ Manufacturing.	C. Media messages.
_____ Transportation.	D. Products.

3. Driving a car is an example of a _____ control system.
4. Which kind of control system (mechanical, electrical/electronic, electromechanical, or fluidic) are each of the following?

   Electric light switch _____.

   Solenoid operated starter switch _____.

   Hand brakes on a bicycle _____.

   Automobile brakes _____.

   Timer on a VCR _____.

5. The occurrence of black lung disease among coal miners is what kind of an output?

APPLYING YOUR KNOWLEDGE

1. Select a technology that you use every day. Complete the following chart for it.

OUTPUTS	
DESIRED	UNDESIRED
INTENDED	UNINTENDED
IMMEDIATE	DELAYED

2. For the same technology, diagram the control system used and list the ways information is gathered, comparisons that are made, and adjustment techniques that are used.
3. For the same technology, list the outside influences that helped determine its design and that help control its operation.

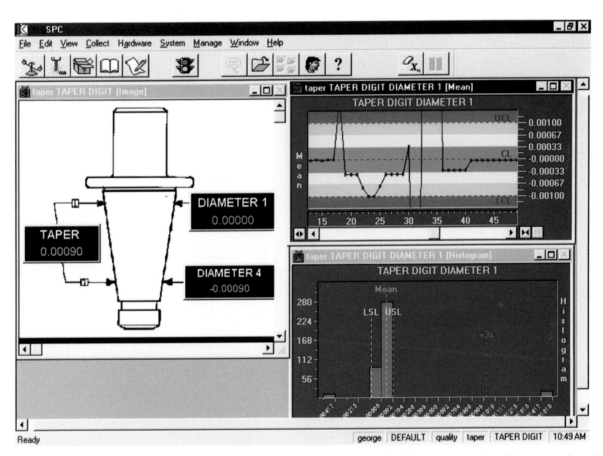

This computer software package is used in the manufacturing plants to automatically measure parts as they are made and determine whether they are within specification. The software also functions as a closed loop control system, using feedback from measuring instruments to adjust the manufacturing process. (Federal Products Co.)

Section Two - Activities

ACTIVITY 2A - DESIGN PROBLEM

Background:

Technology uses productive tools. These tools often employ one or more of the six basic mechanisms. These mechanisms are the lever, the wheel and axle, the pulley, the inclined plane, the wedge, and the screw.

Situation:

The *Science Experiments Company* markets instructional kits for use in elementary and middle school science programs. They have hired you as a designer for their physics kits.

Challenge:

Design a kit that contains common materials, such as tongue depressors, mouse traps, rubber bands, thread spools, small wooden blocks, or string. Develop a technological device (machine) using these materials that contains a power source and employs at least three of the basic mechanisms to lift a tennis ball at least six inches off a table.

ACTIVITY 2B - FABRICATION PROBLEM

Background:

Technological devices are often built to increase human ability. Some devices multiply our ability to lift loads, while others help us to see better. Still other devices improve our ability to hear, move people and products from place to place, or communicate over long distances.

Challenge:

Construct the device that is shown in the drawings below. Test the device to determine how well it helps you to lift a load. Then analyze how well the device worked.

Equipment:

Test stand
1/2″ diameter plastic syringe
3/4″ or 1″ diameter plastic syringe
12″ - clear tubing
Baseball or other load

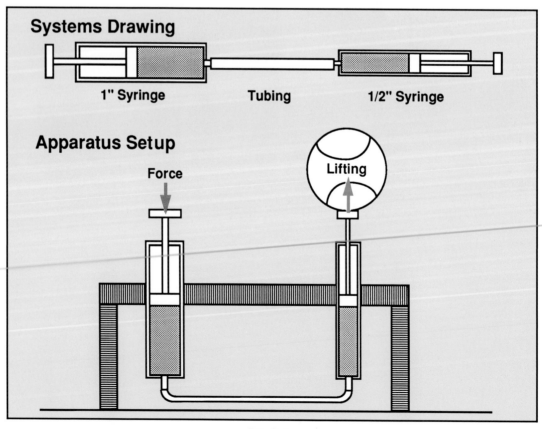

Fig. 2-A.

Procedure:
1. Obtain the supplies.
2. Assemble the hydraulic system as shown in Fig. 2-A.
3. Place the load on the small syringe.
4. Press the large syringe plunger to cause the load to rise.
5. Reverse the load, placing it on the large diameter plunger.
6. Press the small syringe plunger to cause the load to rise.
7. Complete the analysis of the activity that follows.

Analysis:

Analyze the two circuits by answering the following questions:
1. Which arrangement of cylinders lifted the load with the least effort?

_____ Force on large cylinder/load on small cylinder.
_____ Force on small cylinder/load on large cylinder.

2. Which arrangement of cylinders moved the load the greatest distance when the force cylinder was depressed?

_____ Force on large cylinder/load on small cylinder.
_____ Force on small cylinder/load on large cylinder.

3. Which arrangement is a force multiplier?
4. Which arrangement is a distance multiplier?
5. Which type of multiplier would you use as a automobile lift in a service station?
6. Which type of multiplier would you use for a hydraulic elevator in a four-story building?

Tools of Technology

7. **Production Tools in Technology**

8. **Measurement Tools**

Section 3 Activities

MITSUBISHI

Many different tools are required to build a complex product such as this airliner. (McDonnell Douglas)

PRODUCTION TOOLS IN TECHNOLOGY

After studying this chapter you will be able to:
- List the categories of production tools used in technology.
- Contrast tools and machines.
- Explain the interrelationship of natural resources, human energy, and tools in producing material wealth.
- List and describe the major types of material processing tools.
- Explain cutting and feed motions of machine tools.
- Describe the major types of cutting tools.
- List and describe the major types of energy processing tools.
- List and describe the major types of information processing tools.
- Describe the importance of using tools properly and safely.

different tools and machines in use today. It would be impossible to understand them all. Instead, this chapter will introduce you to selected tools used in three major types of processing:
- Material Processing.
- Energy Processing.
- Information Processing.

Fig. 7-1. Tools give us the material world around us. Name as many things as you can that you have built with tools this year. (Faro)

The human ability to design and use tools provides the foundation for technology. It is through tools that we have products and buildings, structures and vehicles, communication media and energy conversion machines, Fig. 7-1.

Tools can make our lives better or threaten our existence. We travel in comfort in modern automobiles, but they pollute the atmosphere. High-rise buildings allow us to live and work in cities, yet social problems and crime in many cities frighten people. Television allows us to be instantaneously touched by events around the globe, but threatens family interaction.

The challenge is to design and use the tools of technology wisely. To do this we must understand the tools and machines around us. There are thousands of

MATERIAL PROCESSING MACHINES

The world around us is full of artifacts (objects made by human activity). Each was made using **material processing** tools and machines. They were the result of actions that changed the form on materials. Tools were used to cast, form, and machine materials into specific shapes. They also helped assemble products and apply protective coatings.

Fundamental to all material processing is a group of tools called **machine tools**. These are the machines used to make other machines. Machine tools have some common characteristics. As shown in Fig. 7-2, these include:

- A method of cutting materials to produce the desired size and shape. This is called a cutting tool.
- A series of motions between the material and the tool. This movement causes the tool to cut the material.
- Support of the tool and/or the workpiece (the material being machined).

Machines can be studied by looking at any or all of these three elements. Let's look at each element separately; then look at the common types of machines that use them.

CUTTING TOOLS

Most cutting actions require a **cutting tool**. Tools come in many sizes and shapes. However, they must meet certain requirements. First, they must be harder that the material they are cutting. A diamond will cut steel, steel will cut wood, and wood will cut butter. Second, the tool must have the proper shape. This means that it needs a sharpened edge and relief angles, as shown in Fig. 7-3. The sharpened edge allows the tool to

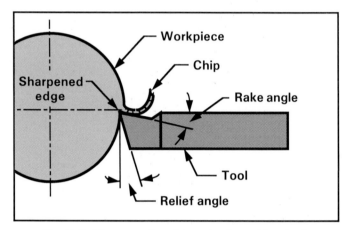

Fig. 7-3. These are the elements of a cutting tool.

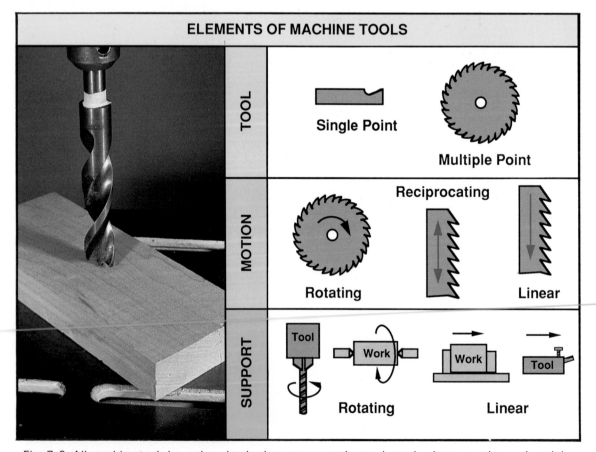

Fig. 7-2. All machine tools have three basic elements: a cutting tool, motion between the work and the cutting tool, and a method of supporting the tool and/or the workpiece.

cut into the material. The relief angles keep the sides of the tool from rubbing against the material as it is cut. The rake angle also helps to create a chip as the material is cut. This action allows waste material to be carried away efficiently.

Two basic types of cutting tools are used in all hand tools and machines. These are single-point and multiple-point tools.

The **single-point tool** is the simplest cutting device available. It has a cutting edge on the end or along the edge of a rod, bar, or strip. Common hand tools that use single points are knives, chisels, and wood planes, Fig. 7-4. Most lathe-turning processes use single point tools.

Multiple-point tools are a series of single-point tools arranged on a cutting device, Fig. 7-5. Most often, these single points are arranged in a set pattern. For example, the teeth of a circular saw are evenly spaced around the circumference of the blade. Likewise, the teeth on band saw and scroll saw blades are spaced along the edge of a metal strip. In some cases there is no pattern for the arrangement of cutting points. The cutting edges or points have a random arrangement. Abrasive papers and grinding wheels are good examples of this type of cutting tool.

MOTION

There must be movement between the tool and the workpiece before cutting can take place. In all cutting operations there are two basic motions: cutting motion and feed motion.

Cutting motion is the action that causes material to be removed from the work. It causes the excess material to be cut away. **Feed motion** is that action that brings

Fig. 7-5. The multiple-point tools shown above include circular, band, and scroll saw blades, grinding wheels and milling cutters for machines. Also shown are hand files, abrasive paper, and drill bits.

new material into the cutter. It allows the cutting action to be continuous.

To understand these movements, consider a band saw, Fig. 7-6. In your mind, imagine a piece of wood placed against the blade. When the machine is turned on, the first tooth will cut the material. This is the cutting motion. The next tooth follows the path of the first tooth. It will not produce a chip. The material in the kerf (the slot cut by a blade) has already been removed by the first tooth. The cutting action will continue only if the wood is pushed into the blade. This makes new material available for the next tooth to cut. Moving the wood is the feed motion.

Cutting and feed motions can be rotating, linear, or reciprocating. **Rotating motion** uses round cutters or spins the work around an axis. **Linear motion** moves a

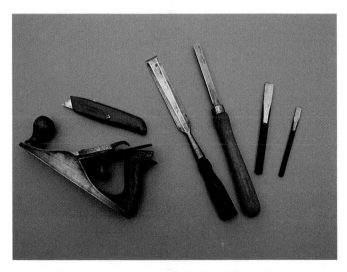

Fig. 7-4. These are some common single-point hand cutting tools.

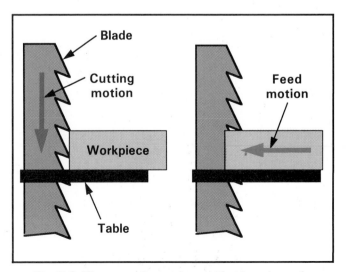

Fig. 7-6. These are the cutting and feed motions of a band saw.

cutter or work in one direction along a straight line. **Reciprocating motion** moves the tool or the work back and forth. Look back to Fig. 7-2 for a diagram of these three types of motion.

SUPPORT

The final element present in all machine tools is support for the tool and the workpiece. The types of cutting and feed motions used determine the support system needed. Rotating motion requires a holder that will revolve around an axis. Chucks are used to hold and rotate drills and router bits. Arbors are used to hold table saw blades and milling cutters. Parts are placed between centers on lathes and cylindrical grinders.

Linear motion is produced in several ways. A tool may be clamped in a holder or held on a rest. It can then be moved in a straight line. This is a common practice for wood and metal lathes and hand wood planes. Band saw blades travel around two or three wheels to produce a linear cutting motion. Material may be pushed through a saw blade while it is supported by the machine table.

Reciprocating motion is common with scroll saws and hacksaws. The blade is clamped at both ends into the machine. Then the blade is caused to move back and forth. Likewise, a workpiece can be clamped to a table that reciprocates under a cutter. This action is used in a surface grinder. People move a workpiece back and forth across the face of a grinding wheel. This action is used to sharpen hand tools, lawn mower blades, and machine tool cutters and knives.

TYPES OF MACHINE TOOLS

There are hundreds of different machines. Look at any machine tool catalog and you can find many of

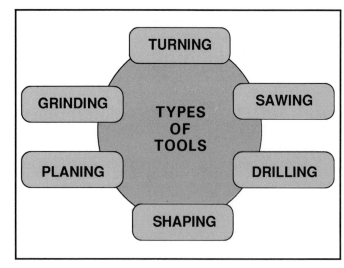

Fig. 7-7. There are six types of machine tools.

them. However, all machine tools can be grouped into six categories, as shown in Fig. 7-7. These are:
- Turning machines.
- Sawing machines.
- Drilling machines.
- Shaping machines.
- Planing machines.
- Grinding machines.

TURNING MACHINES

Turning is a process in which a workpiece is held and rotated on an axis. This action is produced on machines called **lathes**, Fig. 7-8. All lathes produce their *cutting motion* by rotating the workpiece. The *feed motion* is generated by linear movement of the tool.

Lathes are primarily used to machine wood and metal. Plastics can be machined on lathes as well. Wood and metal lathes contain four main parts or a system. A headstock, which contains the machine's

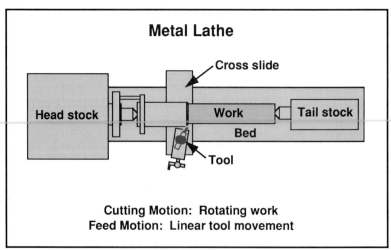

Fig. 7-8. Lathes are used for turning operations.

power unit, is the heart of the lathe. This unit has systems to rotate the workpiece and to adjust the speed of rotation. At the opposite end of the machine is a *tailstock*. This unit supports the opposite end of a part that is gripped at the headstock. The headstock and tailstock are attached to the *bed* of the lathe. Finally, a *tool rest* or *holder* is provided to support the tool. Tool rests on metal lathes clamp the tool in position and feed it into or along the work. Wood lathes commonly have a flat tool rest along which the operator moves the tool by hand.

The work can be held in a lathe in two basic ways. It can be placed between centers. One center is in the headstock and the other in the tailstock. These centers support the workpiece and can be of two types. *Live centers* rotate with the workpiece, while *dead centers* are fixed as the work rotates around them.

Both wood lathes and metal lathes have centers, but they differ in the way the turning force is applied to the workpiece. Wood lathes apply the force through the headstock using a spur center. Metal lathes use a device called a *dog* to rotate the workpiece.

A small workpiece may be rotated using only the headstock in a device called a *chuck*. Three-jaw, four-jaw, or collet chucks are commonly used to grip and rotate the part. Chucks are found on metal lathes.

Lathe Operations

Lathes are used to do a large number of operations, Fig. 7-9. The most common are:

- Turning: Cutting along the length of a workpiece. This operation will produce a cylinder of uniform diameter.
- Tapering: Cutting along the length of a cylinder at a slight angle to produce a cylindrical shape with a uniformly decreasing diameter.
- Facing: Cutting across the end of a rotating workpiece. This operation will produce a true (or square) end on the workpiece.
- Grooving: Cutting into a workpiece to produce a channel with a diameter less than the main diameter of the workpiece.
- Chamfering: Cutting an angled surface between two diameters on the workpiece.
- Parting: Cutting off a part from the main workpiece.
- Threading: Cutting threads along the outside diameter or inside a hole in the workpiece.
- Knurling: Producing a diamond pattern of grooves on the outside diameter of a portion of the workpiece. This produces a surface that is easier to grip and turn.

In addition, drills and reamers can be placed in the tailstock to produce and finish holes in the workpiece.

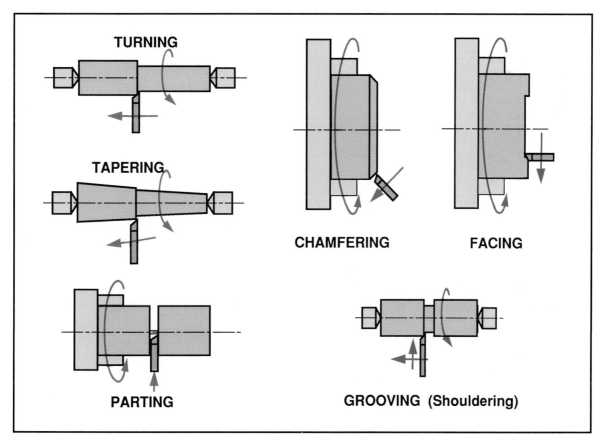

Fig. 7-9. These are common turning operations performed on lathes.

Technology: How it Works

VIRTUAL REALITY: Computer interface that allows a user to interact with three-dimensional, computer-generated images.

Computer technology has taken one giant step after another during the past 30 years. As the computer has become more complex, it has branched off into many different areas. Virtual reality is one of the more spectacular areas of computer technology, Fig. 1.

Virtual reality was born in the 1960s, though in a very crude form. In 1981, the first practical application of a virtual reality system was exhibited. The Visually Coupled Airborne Systems Simulator trained pilots to fly complex, high-speed aircraft. However, the term "virtual reality" did not appear until the mid-1980s. The term was coined by Jaron Lanier, the founder of VPL Research.

Virtual reality allows a user to interact with an environment created by a computer. Therefore, complex equipment is needed in addition to the computer. Computer software, a headset, and some tool that relays the users movements are required. This tool is often a glove that transmits its relative position and many finger movements to the computer.

Uses for virtual reality are many, with new uses being developed everywhere. The military has used virtual reality to train fighter pilots for some time. Companies use virtual reality to train new employees. Architects allow clients to tour new homes during the design stage, Fig. 2. Clients can make changes based on what they see *before construction begins.* And, of course, virtual reality has recreational uses as well. Complex systems

Fig. 1. Employees manipulate a virtual front-end loader. (Caterpillar Inc.)

can be found in expensive arcades. Simpler systems are available for home video games.

There are impressive prospects for the future of this technology. Soon surgeons may have virtual reality training facilities. They could train on virtual patients with virtual scalpels. Scientists, as well, could do research with virtual reality. Areas such as molecular modeling and engineering could see significant benefits.

Fig. 2. Virtual objects can be moved about in this virtual office space.

SAWING MACHINES

Sawing machines use teeth on a blade to cut material to a desired size and shape. These machines are designed to perform a number of different cutting actions. These actions, as shown in Fig 7-10, include:

- Crosscutting (or cutoff): Reducing the length of a material.
- Ripping (or edging): Reducing the width of a material.
- Resawing: Reducing the thickness of a material.
- Grooving, dado, notching: Cutting rectangular slots in or across a part.
- Chamfering and beveling: Cutting an angled surface between two primary surfaces of a material.

Sawing machines can be grouped according the type of blade they use and the methods used to produce the cutting action. This grouping identifies three basic types of saws, Fig. 7-11:

- Circular saws.
- Band saws.
- Scroll saws.

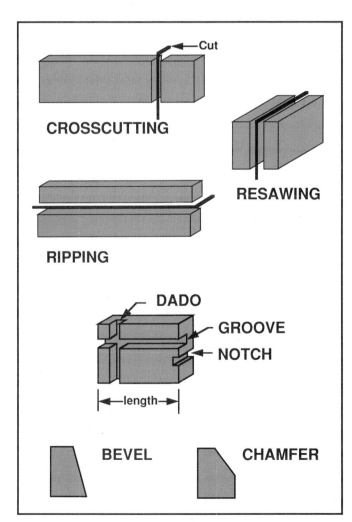

Fig. 7-10. These are examples of typical sawing operations.

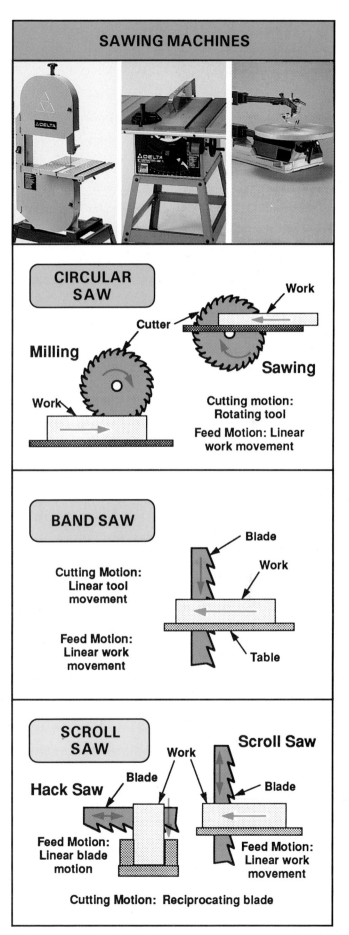

Fig. 7-11. These are three types of sawing machines. (Delta)

Circular Saws

Circular saws use a blade in the shape of a disk with teeth arranged around the edge. These teeth vary in shape and arrangement depending of the operation to be performed. Common blades are available for crosscutting, ripping, and combination cutting (crosscutting or ripping).

There are three basic types of circular saws. These are the table saw, radial saw, and chop saw, Fig. 7-12. All three of these machines generate the cutting motion by rotating the blade. However, their feed motions are different. The **table saw** uses a linear feed of the material. The workpiece is pushed into the rotating blade to generate the cut. The machine operator will manually feed the material in low-volume production settings. Automatic feeding devices are used to increase the speed, accuracy, and safety of many high-volume sawing operations.

Many other machines use cutting and feed motions that are like the circular saw. These include metal milling machines, and wood shapers, routers, jointers, and planers. Each of these machines feeds the workpiece into a rotating cutter.

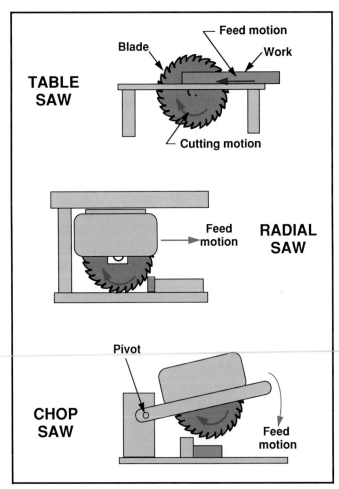

Fig. 7-12. There are three types of circular saws.

The **radial saw** moves the rotating blade across the workpiece. The workpiece is positioned on a stationary table. The rotating blade produces the cutting motion. The rotating blade is manually drawn across the material for the feed motion. The direction of the blade travel is determined by the position of the arm of the machine. Generally the machine can cut a square angle and any angle up to 45° on each side of square.

The **chop saw** is used to cut narrow strips of material to length. The blade is attached to a pivot arm. The material is placed on the table of the machine. The blade assembly is pivoted down to generate the feed motion. Many cutoff saws have saw blade units that can be rotated so that the material can be cut off at a specified angle. This type of saw, called a power miter box, is widely used to cut trim moldings for homes and the parts of picture frames.

Band Saws

A **band saw** uses a blade made of a continuous strip or band of metal. Most of these bands have teeth on one edge. However, large band saw blades used in lumber mills have teeth on both edges.

The band generally travels around two wheels which gives a continuous linear cutting action. The band may be vertical or horizontal. Horizontal band saws are usually used as cutoff saws for metal rods and bars. They have replaced hacksaws as the primary method of cutting metal stock to length. Horizontal machines hold the material stationary in vises or clamps. They produce the feed motion by allowing the blade unit to pivot into the work.

Vertical band saws are widely used to cut irregular shapes from wood or metal sheets. The material is placed on the machine table. Then, it is fed into the blade to produce the cut. Most often, the operator manually feeds the material to produce the desired cut.

Scroll Saws

Scroll saws use a straight blade that is a strip of metal with teeth on one edge. The blade is clamped into the machine. The machine then moves the blade up and down to produce a reciprocating cutting motion. The material is placed on the table and manually fed into the blade.

Portable scroll saws grip only one end of the blade. Then the saw is fed into the work to produce the feed motion.

DRILLING MACHINES

Holes in parts and products are a very common feature. They can be produced in a number of ways. Holes can be punched into sheet metal using punches and dies. Holes can be burned into plate steel using an oxyacetylene cutting torch. Holes can be produced in the part directly as it is cast from molten material. Holes can be produced with powerful beams of light (laser machining) or electrical sparks (electro-discharge machining)

Many cylindrical holes, however, are produced through drilling operations. **Drilling** produces or enlarges holes using a rotating cutter. Generally, the *cutting motion* is produced as the drill or bit rotates. The drill is moved into the work to produce the *feed motion*. The drill press, Fig. 7-13, is the most common machine using these cutting and feed motions.

Drilling may also be done by rotating the work to produce the cutting motion. Then the stationary drill is fed into the work. This practice is common with drilling on a metal lathe or a computer-controlled machining center.

Drilling Operations

A large number of different operations can be completed on drilling machines. The most common are shown in Fig. 7-14. These operations are:

- Drilling: Producing straight, cylindrical holes in a material. These holes can be used to accommodate bolts, screws, shafts, and pins for assembly. They can also be a functional feature of a product. For example, holes are essential for the functioning of furnace burners, automobile carburetors, and compact disc recordings.
- Counterboring: Producing two holes around the same center point. The outer hole has a larger diameter than the inner hole. Counterbores are used to position shafts, recess heads of fasteners, or for a number of other purposes.
- Countersinking: Producing a beveled outer portion of a hole. Most often, countersinking is used with flathead wood and metal screws. Countersinking holes allows screw heads to be flush with the surface of the part.
- Reaming: Enlarging the diameter of a hole. This action is generally done to produce an accurate diameter for a bolt hole.

Drilling operations use a number of different drilling tools. These include twist drills, spade bits, and forstner bits. **Twist drills** are shafts of steel with points on the end to produce a chip. These chips are carried from the work on helical flutes that circle the shaft. **Spade bits** are flat cutters on the end of a shaft. The bottoms of the cutters are shaped to produce the cut. **Forstner bits** are two-lipped wood cutters that produce a flat-bottomed hole. Hole saws, which use sawing machine action, and fly cutters, which use lathe-type tools can also be used to produce holes. Several of these bits and drills are shown in Fig. 7-15.

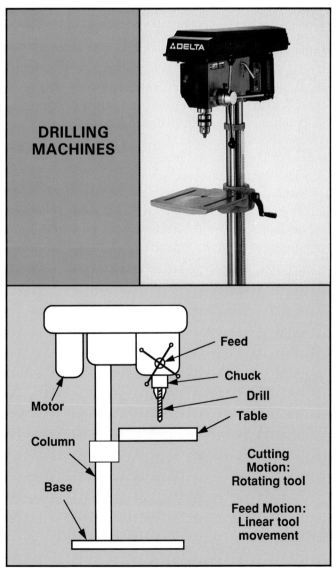

Fig. 7-13. A drill press is the most common drilling machine. (Delta)

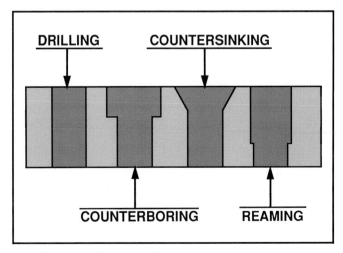

Fig. 7-14. These are the most common operations performed on drilling machines.

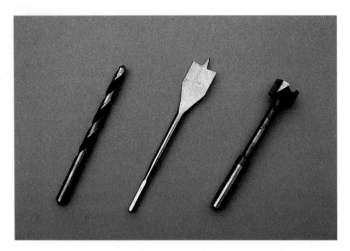

Fig. 7-15. Common drilling tools are (shown from right to left): twist drill, spade bit, forstner bit.

SHAPING AND PLANING MACHINES

Two metal working machine tools that produce flat surfaces are the shaper and the planer, Fig. 7-16. These machines should not be confused with the woodworking shaper and planer which operate on the same principles as sawing machines.

Both the metal shaper and planer use single-point tools and reciprocating motion to produce the cut. Their difference lies in the movements of the tool and the workpiece. The planer reciprocates the workpiece under the tool to generate the cutting motion. The tool is moved one step across the work for each cutting stroke.

The metal shaper moves the tool back and forth over the workpiece to produce the cutting motion. The work is stepped over after each forward cutting stroke to produce the feed motion.

Both shapers and planers can cut on the face or side of the part. They can also be used to machine grooves into the surface. Both machines have limited use in material processing.

A machine that is closely related to these machines is a broach. This machine uses a tool with many teeth. Each tooth sticks out slightly more that the previous tooth. As the broach tool is passed over a surface, each tooth cuts a small chip. However, when all the teeth pass over the work, a fairly deep cut is possible. Broaches are often used to machine a keyway (rectangular notch) in a hole. Keyways are widely used to assemble wheels and pulleys to axles.

GRINDING MACHINES

The last type of machine tool is grinders. These are machines that use bonded abrasives (grinding wheels) to cut the material. These wheels have random cutting surfaces that remove the material in the form of very small chips or particles.

Grinders are basically adaptations of other machine tools. The two most common are cylindrical grinders and surface grinders, Fig. 7-17. **Cylindrical grinders** use the lathe principle to machine the material. The workpiece is held in a chuck or between centers and rotated. A grinding wheel is rotated in the opposite direction. The opposing rotating forces produce the *cutting motion*. The grinding wheel is fed into the work to produce the *feed motion*.

The **surface grinder** works on the metal planer principle. A rotating grinding wheel is suspended above the workpiece. The work is moved back and forth under the wheel to produce the *cutting motion*. The work is

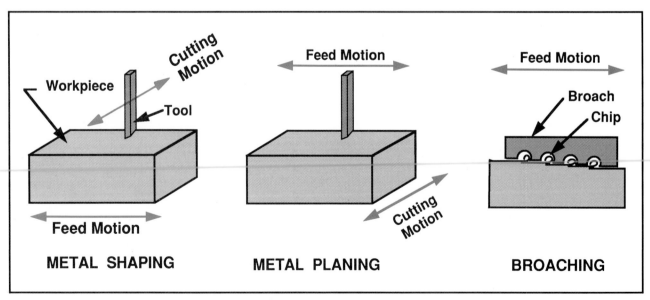

Fig. 7-16. This is a diagram of the feed and cutting motions for shaping and planing machines.

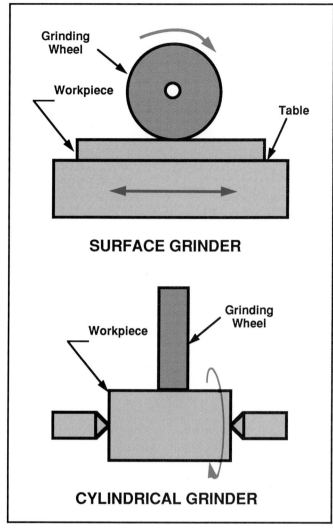

Fig. 7-17. Two common grinding machines are the cylindrical grinder and the surface grinder.

moved slightly or indexed after each grinding pass to produce the *feed motion*.

A third type of grinder that is widely used is the pedestal grinder. The rotating grinding wheel produces the *cutting motion*. The operator manually moves the workpiece across the face of the wheel to produce the *feed motion*.

ENERGY PROCESSING MACHINES

At the heart of all technological systems is energy. It powers most devices that process materials and information. The laws of physics tell us that energy can neither be made nor destroyed. It can, however, be converted and applied to do work. This is the task of energy conversion or **energy-processing machines**. There are literally hundreds of these machines. For example, energy-processing machines can convert:

- Mechanical energy into electrical energy–an electric generator.
- Electrical energy into mechanical energy–an electric motor.
- Radiant energy into thermal energy–a solar heater.
- Heat energy into mechanical energy–an internal combustion engine.

We have named just a few types of energy-processing machines.

The study of human history quickly identifies three major energy processing machines that have helped to shape life as we know it. These are the steam engine, the electric motor, and the internal combustion engine. Of course, these are not the only important energy converters. They will, however, be the focus of this chapter on Production Tools in Technology. Additional energy converters will be discussed in Chapter 27.

STEAM ENGINE

The steam engine was the foundation of the industrial revolution of the 1800s. Earlier converters simply translated one form of mechanical power to another. Animal, water, and wind power reduced human labor. These power sources were still mechanical power devices. The steam engine, however, was the first effective device to convert energy from one form to another. It converted heat energy to mechanical energy.

The steam engine changed our early dependence on moving water to power factories. This allowed people to locate manufacturing plants away from fast-moving rivers. Steam power was also applied to river, ocean, and rail transportation, Fig. 7-18. This application allowed us to move more cargo over longer distances.

The first practical steam engine was developed by Thomas Newcomen in the early 1700s. It applied available knowledge about pistons, steam condensation,

Fig. 7-18. Steam engines were widely used to power locomotives until after World War II. (Norfolk Southern Corp.)

atmospheric pressure, vacuums, and mechanical linkages. The result was a large engine that was widely used to pump water from mines in England and Europe.

The engine was based on the fact that water takes up less space than does steam (hot water vapor). The engine, as shown in Fig. 7-19, operated through the following steps:

1. All valves were closed.
2. A valve was opened to allow steam to move from the boiler into the cylinder. This caused the piston to move upward which, in turn, moved the pump end of the rocker arm downward.
3. The steam valve was closed and a catch caught the piston in the upward position.
4. The cold water valve was opened. This allowed a jet of cold water to enter the cylinder. The steam quickly condensed as the cylinder cooled. This caused a vacuum inside the cylinder.
5. The catch holding the piston was released, allowing the piston to be drawn downward by the vacuum. With this movement, the rocker arm was rotated, causing the pump end to rise.

This five-step cycle was then repeated, causing a continuous pumping action. However, the Newcomen atmospheric steam engine was very inefficient. It took a large amount of fuel to pump a small amount of water. A more efficient design was developed by James Watt.

His engine used condensing steam to drive the piston in both directions. This resulted in a reciprocating steam engine that was widely used to power industrial machinery and transportation vehicles.

The Watt Steam Engine

Look at the drawings of the Watt double-acting steam engine in Fig. 7-20. In the top drawing the steam enters the engine from the boiler. It flows to the left causing the piston to move to the right. This action moves the piston rod which turns the flywheel. As the fly wheel turns it moves the slide valve to the left. Its new position is shown in the lower drawing. Now the steam flows to the right, forcing the piston to the left. Again the piston rod moves the flywheel. The slide valve is moved back to the position shown in the top drawing. The cycle starts over again. The reciprocating motion of the piston continues to turn the flywheel. This wheel can be connected to any device which needs a rotating power source.

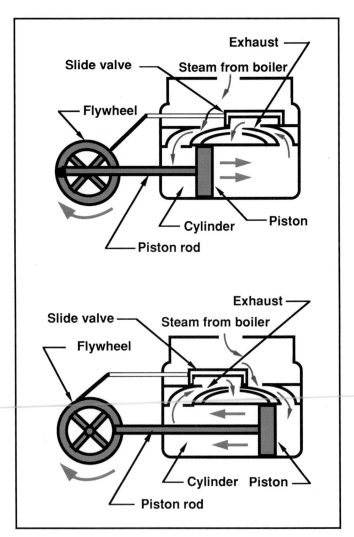

Fig. 7-20. These diagrams show the forward and backward motion of the piston in a Watt double-action steam engine.

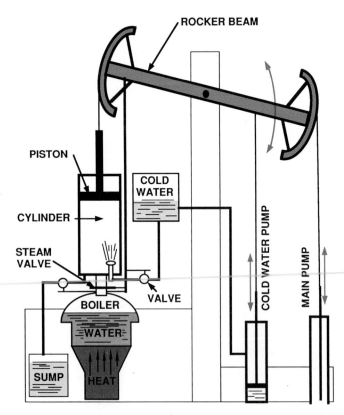

Fig. 7-19. This is a diagram of a Newcomen steam engine.

INTERNAL COMBUSTION ENGINE

The steam engine has been almost totally replaced by other energy converters. The internal combustion engine, which was invented in 1876, now powers most land transportation vehicles.

The common internal combustion engines are gasoline and diesel engines. Both of these engines change heat energy into mechanical energy. They drive a reciprocating piston by igniting a fuel. The piston's reciprocating motion is changed into rotary motion by a crankshaft, Fig. 7-21.

Gasoline engines can be either two-stroke cycle engines or four-stroke cycle engines. The four-stroke cycle engine is the engine most people come into contact with, and will be the focus of this discussion.

A **stroke** is the movement of a piston from one end of a cylinder to another. A **cycle** is a complete set of motions needed to produce a surge of power. A two-stroke cycle engine moves the piston up and back once (two strokes: one up and one down) to produce a power stroke. A four stroke engine moves the piston up and back twice (four strokes) to produce a power stroke.

Look at Fig. 7-22 to see these four strokes: intake, compression, power, and exhaust. During these strokes the following actions take place:

- Intake stroke: The piston moves downward to create a partial vacuum. The intake valve opens. Atmo-

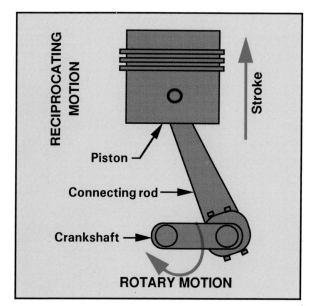

Fig. 7-21. The reciprocating piston in an internal combustion engines turns a crankshaft.

spheric pressure forces a fuel and air mixture into the cylinder.

- Compression stroke: The intake valve closes and the piston moves upward. As the piston moves up, the fuel-air mixture is compressed in the small cavity at the top of the cylinder. The area forms a combustion chamber for the power stroke. The compression ratio of the engine will tell you how much the fuel-air

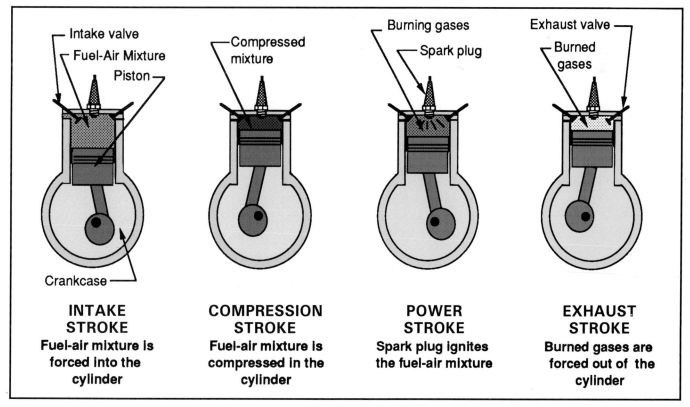

Fig. 7-22. The four strokes of a four-stroke cycle internal combustion engine.

mixture has been compressed. The standard compression ratio is about 15:1 (fifteen to one). This ratio tells you that the cylinder volume at the beginning of the compression stroke is 15 times larger than at the end of the compression stroke. Therefore, the fuel-air mixture has been reduced to about 6.6 percent of its original volume.

- Power stroke: An electrical spark is produced between the two points of the spark plug. This action ignites the compressed fuel-air mixture. The burning gases produce temperatures as high as 4000°F (about 2200°C). This temperature expands the gas and generates pressure up to 1000 psi. or 6895 kilopascals (KPa). The pressure forces the piston downward in a powerful movement.
- Exhaust stroke: When the piston reaches the bottom of the power stroke, the exhaust valve opens. The piston moves upward to force exhaust gases and water vapor from the cylinder. These are the products of the combustion during the power stroke. At the end of this stroke the cylinder is ready to repeat the four strokes.

Single cylinder engines are common for low horsepower applications such as lawn mowers, cement mixers, and portable conveyors. For more demanding applications, several cylinders are combined into one engine. The cycle of each cylinder is started at a differ-

ent point of time so that the engine has a series of closely spaced power strokes. Four-cylinder, six-cylinder, and eight-cylinder engines are common.

ELECTRIC MOTOR

The last energy processing device that will be discussed in this section is the electric motor. This device is probably the most universally used source of power. The average home has over 40 motors in it. They are in clocks, refrigerators and freezers, video recorders, tape and compact disc players, furnaces and air conditioners, clothes washers and dryers, shavers, hair dryers, and many other appliances. They are also on construction sites and in factories, Fig. 7-23. They play a vital part in every transportation and communication system.

The electric motor is based on the laws of magnetism and electromagnetism. These laws, as shown in Fig. 7-24, suggest that:

- Like poles of a magnet repel and unlike poles attract one another.
- Current flowing in a wire creates an electromagnetic field around the conductor.

Look at Fig. 7-25 to see how these laws are applied. You will note that there are two magnets. The outer magnet is stationary and is called the field magnet. The

Fig. 7-23. All the equipment in this metal stamping line is powered by electric motors (Minster Machine Co.)

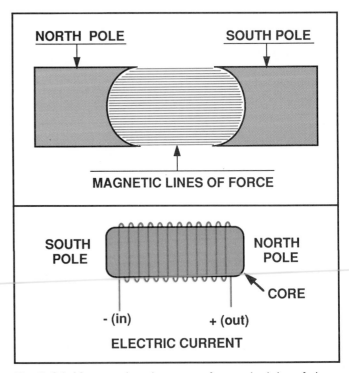

Fig. 7-24. Motors take advantage of two principles of physics. 1. Magnetic lines of force travel between opposite poles of a magnet (top). 2. Current flowing through wires wrapped around an iron core will produce magnetic poles at the ends of the core. (bottom).

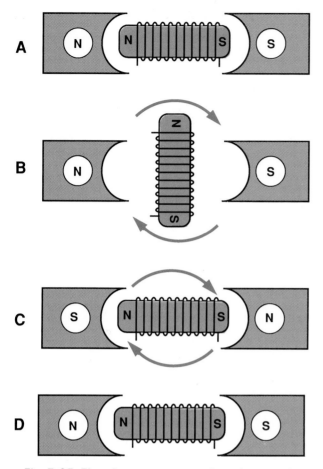

Fig. 7-25. Electric motors operate by using a stationary magnet and an armature that switches its magnetic poles.

inner magnet is an electromagnet that can rotate and is called the armature.

In View A, electrical current is allowed to pass through the armature. The current induces magnetism in the armature core. A north pole will be on one end and a south pole on the other. The direction the wire is wrapped around the core and the direction of the current will determine which end is the north pole.

View A shows the north field pole and the north armature pole next to each other. These are like poles and will repel each other. Therefore, the armature spins one-forth turn to the position shown in View B. Now the unlike poles will attract each other and the armature will continue rotating another one-fourth turn. When the armature is in the position shown in View C, the direction of the current is reversed. This action is done mechanically in direct current motors. Alternating current motors use the existing 120 direction changes per second in the line current. The current direction change reverses the poles on the armatures as shown view D. Now the entire action is repeated. Like poles repel and turn the armature another one-fourth of a turn. Then like poles attract, turning the armature the final one-forth turn to complete a full revolution of the armature.

INFORMATION PROCESSING MACHINES

Information processing and exchange is as old as human existence. Early results of these actions can be seen on cave walls, Egyptian hieroglyphics, and American Indian petroglyphs. Today the products of information processing are in books and photographs and on tape and film.

Modern information processing is built on a number of important technological advancements. Four of these will be explored here and others will be introduced in Section 6 of this book.

The communication methods basic to all modern communication systems are printing, telecommunications, and/or the computer. Printing is based on movable type, developed in the 1400s. Telecommunications has its roots in Alexander Graham Bell's telephone and Marconi's radio. Computers are based on semiconductor circuits that process data.

PRINTING

We have become accustomed to having printed materials at our fingertips. Books, magazines, newspapers, pamphlets and brochures are everywhere. The abundance of printed materials is of recent history. Until the last 500 years, most people could not read nor had they seen a book. Those books that existed were hand written on parchment paper by monks and scribes. As civilization grew, so did the demand for information. In the 1450s this demand was addressed with a new printing process developed by Johannes Gutenberg, a German goldsmith. It used movable type, as shown in Fig. 7-26.

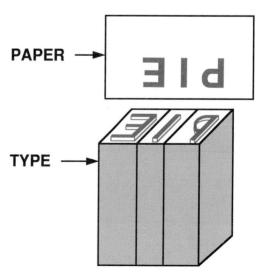

Fig. 7-26. Gutenberg developed movable type that revolutionized book production and started the printing industry.

Before this time the Chinese printed entire pages from blocks of wood with the message carved in them. However, each page required a separate block. Also, the block could not be changed or used again. Gutenberg used his metal-casting knowledge to produce standard sized metal blocks. On the end of each block was a letter of the alphabet in reverse image. These blocks, called type, were all exactly the same height and on a body that was the same thickness. This type became part of the new way to produce a printed page, and is shown in Fig. 7-27.

The type could be assembled in groups to form words, sentences, and paragraphs. The type was then clamped in a frame to make a *page form*. Ink was spread on the type and paper laid on top of the page form. The form was slid into a press that was adapted from a press that squeezed juice from grapes. Pressure was applied by turning a screw that moved a pressure pad downward. The resulting force transferred the ink from the type to the paper. The press was then opened and the printed page was removed from the type form. The process was then ready to be repeated until the desired number of pages were printed. Then, the type form could be taken apart and the individual pieces of type used over again.

The first book Gutenberg printed was the Bible. He and his helpers completed an unknown number of copies. Forty-seven books from this printing still exist.

From this humble start came the printing industry and a number of different printing processes. They will be explored in more depth in Chapter 20.

TELEPHONE

Almost everyone uses a telephone. It allows us to quickly and cheaply communicate information, ideas, and thoughts to another person. It is the simplest form of telecommunication in use today. It is called *hardwired* communication because a separate channel connects each telephone. These channels may be copper wires or fiber-optic strands.

The telephone converts sound into electrical energy. The electrical energy is then transmitted from the sender to the receiver. At the receiver, the electrical energy is converted back into sound waves. In short, the telephone changes information contained in spoken messages into electrical signals.

The first information conversion is done at the telephone microphone (mouthpiece), Fig. 7-28. Carbon granules are placed between two wires in an electric circuit. A diaphragm is placed over the particles. Sound waves will cause the diaphragm to vibrate or move. The movement compresses the carbon and changes its ability to conduct electrical current. Each tone vibrates the diaphragm at a different rate and, therefore, allows a specific amount of electricity to be conducted through

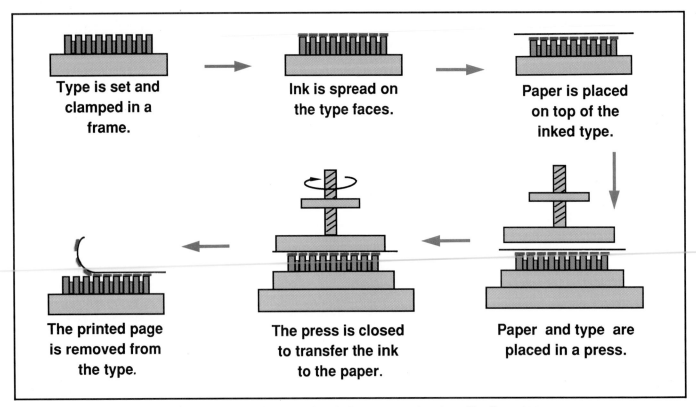

Type is set and clamped in a frame.

Ink is spread on the type faces.

Paper is placed on top of the inked type.

The printed page is removed from the type.

The press is closed to transfer the ink to the paper.

Paper and type are placed in a press.

Fig. 7-27. These are the steps in printing process developed by Gutenberg.

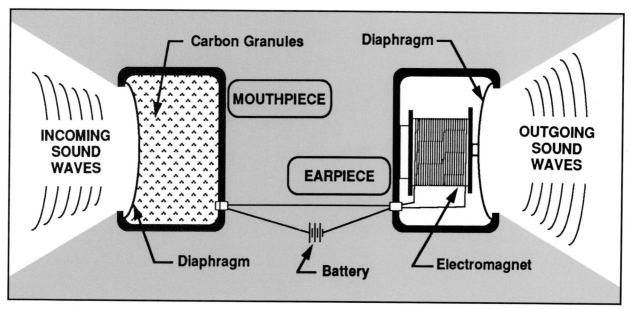

Fig. 7-28. Telephones operate by changing sound waves into electrical impulses and then back into sound waves.

the circuit. The pulse of electricity, which is coded information, is carried to the telephone receiver (ear piece). Here an electromagnet is in the circuit. The electric pulses cause a magnet to vibrate a diaphragm. This vibration produces sound almost identical to the sound that created the electrical impulse in the mouthpiece. The electrical impulse is converted back to information in the form of audible sound.

The use of the telephone has been expanded from a personal communication device to a data transmission instrument. Telephone circuits are now used to transmit coded electronic messages through fax machines and computer modems. The cellular telephone integrates the telephone with the radio, which will be discussed next. It allows people to be connected to the phone system from almost any location.

RADIO

The radio also sends messages from one place to another. Unlike the telephone, which uses wires to connect the sender and receiver, the radio uses radio waves. Radio waves are invisible electromagnetic energy that travels through the atmosphere and space. The radio was originally developed by Guglielmo Marconi for ships to send messages when they were in danger. These *wireless* sets were first used in 1910. Radio is now used to carry information and entertainment to all parts of the world, 7-29.

A radio communication system contains a transmitter to produce the radio waves and a receiver to collect them. The transmitter changes sound (voice and music) into radio waves. It imposes the sound information

Fig. 7-29. Radio communication is part of almost every person's life. In this photo, a Chinese citizen is interviewed by a reporter from a shortwave radio service. (Voice of America)

onto a carrier wave. This action is called *modulation*. Modern transmitters use either frequency modulation (FM) or amplitude modulation (AM) to place the information onto the carrier wave. Frequency modulation changes the rate of the cycles on the carrier wave (or its frequency). Amplitude modulation changes the strength (height or amplitude) of the carrier wave.

Transmitters and Receivers

Look at Fig. 7-30 to see how radio transmitters and receivers work. You will see that the microphone changes sound waves into electrical impulses or signals.

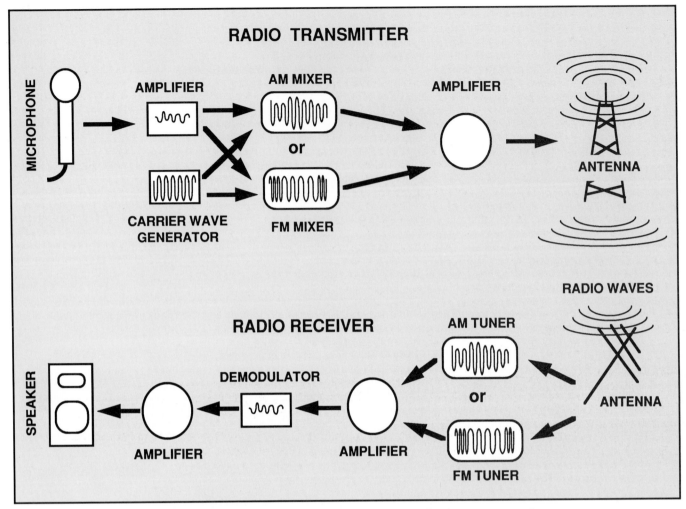

Fig. 7-30. This diagram shows how a radio communication system works.

These signals are carried into the transmitter. A carrier wave generator in the transmitter produces the carrier signal. A mixer merges the sound signals with the carrier wave through frequency or amplitude modulation to make a radio signal. The radio signal is then amplified (made stronger) and sent to a broadcast antenna. From the antenna, radio waves radiate through the atmosphere. A receiving antenna at a distant location collects the waves. A tuner circuit separates the desired radio signal from all other radio signals. The selected signal is then amplified and demodulated (the carrier wave is removed from the signal.) The sound signal remains, and is amplified. Much like a telephone, the amplified signal powers a speaker to produce sound waves that people can hear and understand.

COMPUTERS

Two major information processing machines have changed the way we handle information. They are the calculator and the computer. Both of these machines are different from all other machines in one important way, they can store information. The calculator remembers how to perform math functions. If you enter a **2**, press the + key, enter another **2**, and press the = the calculator will add the two numbers and display a **4**. Its program knows how to add 2 + 2, which equals 4.

Most calculators have fixed programs that cannot be erased or changed. However, **computers** are different. Computers have programs that can be changed. This provides great information processing power.

All computers, from large mainframes to small mini and microcomputers, have five parts. These, as shown in Fig. 7-31, are:

- **Input unit:** Devices used to enter data (numbers, letters) into the system. The input devices may be keyboards, tape drives, disk drives, telephone modems, mice and graphics tablets, or another computer.
- **Processing unit:** The part of the computer, called a central processing unit (CPU), or microprocessor that manipulates the data. It is made up of a microchip or several microchips that may contain millions of microscopic electronic components. The CPU responds to instructions stored in memory.

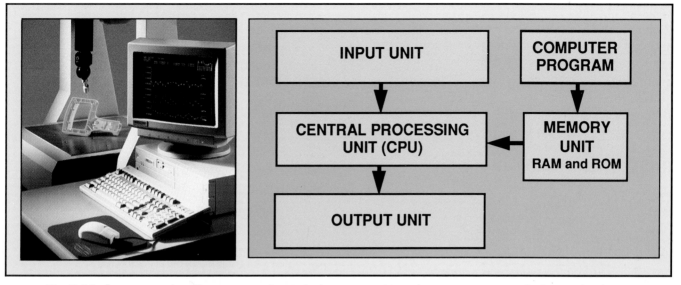

Fig. 7-31. A computer has five parts: an input device, processing unit, memory, output device, and software. (Giddings & Lewis, Inc.)

- **Output unit:** Devices used to display and record to results of the processing unit's actions. These devices include cathode ray tubes (called CRTs or monitors), printers, plotters, disk drives, or tape recorders.
- **Memory unit:** The section of the computer that holds information and instructions. The memory unit has two types of memory. Read-only memory (ROM) contains the instructions that allow the computer to receive and manipulate data. This part of the memory cannot be changed. Random-access memory (RAM) temporarily stores data and feeds it to the central processing unit on command. The RAM is constantly changing as the computer processes data. Also, it usually is erased when the computer is turned off.
- **Program:** The instructions the computer uses to process the data and to produce output. Often the program is called software because it cannot be seen or handled once it is loaded in the computer. The other four units, described above, are hardware. Hardware is the physical parts of the computer system. Typical programs are used to make mathematical calculations, maintain records, prepare drawings and other graphics, and handle text.

Computers are at the heart of many modern information processing systems. They are at every level of business life. They process financial information and prepare financial reports, maintain schedules and ticket records for airlines, operate point-of-purchase units at supermarket check-out stands, guide spacecraft to distant planets, maintain the fuel-air mixture in automobiles, guide washing machines through their wash-rinse-dry cycles, help to prepare layouts for advertising, and control industrial machines.

USING TECHNOLOGY SAFELY

Machines make each of our lives more comfortable. They also injure or kill thousands of people each year. Every day local newspapers report injuries and deaths caused by careless use of automobiles, tools and machines, and home appliances.

If we want technological devices to positively affect our lives, we must use them safely. This includes not just tools and machines that process materials but also devices that process information and energy.

Operators should follow some basic guidelines when they select and use technological devices. These include:
- Select the correct tool for the job.
- Read about the proper operation of the device in owner's manuals and instruction books.
- Seek instruction and advise of the proper use of the device.
- Use the device only in the way described in operation manuals and for applications for which the device was designed.
- Never use a device when personal ability is impaired by medication, lack of rest, or distractions (loud music, people talking to you, etc.)
- Never work alone or where help is not available in case of an accident.

Most people working in business and industry receive proper training on the equipment they use. Their challenge is to follow the safe procedures they are shown and expected to learn. However, many people use technological devices in personal and educational settings. They drive cars, work in home workshops, and school laboratories.

These settings require special attention to safety. Remember safety is both a state of mind and a series of actions. You must be concerned about your safety and the safety of those around you. Likewise, you must complete tasks using safe actions. Working safely requires you to follow the above general rules. The following rules are provided to help you work with laboratory materials and tools safely:

SAFETY WITH PEOPLE

Personal Safety

1. Concentrate on your work. Watching other people or daydreaming can cause accidents.
2. Dress properly. Avoid loose clothing and open shoes. Remove jewelry and watches.
3. Control your hair. Secure any loose hair that may get caught in moving machine parts.
4. Protect your eyes with goggles, safety glasses, or a face shield.
5. Protect your hearing by using hearing protectors when working around loud machines or where high-pitched noises are present.
6. DO NOT use compressed air to blow chips and dirt from machines or benches. Use a brush to gently sweep them away.
7. Think before acting. Always think of what will happen before starting a task.
8. ASK your teacher questions about any operation are unsure of.
9. Seek first aid for any injuries.
10. FOLLOW specific safety practices demonstrated by your teacher.

Safety Around Others

1. Avoid horseplay. What you think of as harmless fun can cause injury to other people.
2. DO NOT talk to anyone who is using a machine, except when necessary. You may distract them and cause an injury.
3. DO NOT cause other people harm by carelessly leaving tools on benches or machines.

SAFETY WITH MATERIALS

1. Handle materials properly and with care.
   a. Be careful when moving long pieces of material. DO NOT hit people and machines with the ends.
   b. Use EXTREME caution when handling sheet metal. The thin, sharp edges can easily cut you and others.
   c. Grip hot materials with pliers or tongs.
   d. Wear gloves when handling hot or sharp materials.
   e. Use extreme caution in handling hot liquids and molten metals.
2. Check all materials for sharp burred edges and pointed ends. Remove these hazards when possible.
3. If a material gives off odors or fumes, place it in a well-ventilated area, fume hood, or spray booth.
4. Lift material properly. Use your legs to do the lifting, not your back. DO NOT overestimate your strength. If you need help, ask for it.
5. Dispose of scrap material properly to avoid accidents.
6. Clean up all spills quickly and correctly.
7. Dispose of hazardous wastes and rags properly.

SAFETY WITH MACHINES

General Rules

1. Use only sharp tools and well-maintained machines.
2. Return all tools and machine accessories to their proper places.
3. Use the RIGHT tool or machine for the RIGHT job.
4. DO NOT use any tool or machine without permission or proper instruction.

Casting and Molding Processes

1. DO NOT try to perform a process that has not been demonstrated to you.
2. Always wear safety glasses.
3. Wear protective clothing, gloves, and face shields when pouring molten metal.
4. DO NOT pour molten material into a mold that is wet or contains any water.
5. Carefully fasten two-part molds together.
6. Perform casting and molding processes in a well-ventilated area.
7. DO NOT leave hot castings or mold parts where other people could be burned by them.
8. Constantly monitor material and machine temperatures during the casting and molding process.

Separating Processes

1. DO NOT try to perform a process that has not been demonstrated to you.
2. Always wear safety glasses.
3. Keep hands away from all moving cutters and blades.

4. Use push sticks to feed all material into wood-cutting machines.
5. Use ALL machine guards.
6. Always stop machines to make measurements and adjustments.
7. DO NOT leave a machine until the cutter has stopped.
8. Whenever possible, clamp all work.
9. Always unplug machines from the electrical outlet before changing blades or cutters.
10. Remove all chuck keys or wrenches before starting the machine.
11. Remove all scraps and tools from the machine before using it.
12. Remove scraps with a push stick.

Forming Processes

1. DO NOT try to perform a process that has not been demonstrated to you.
2. Always wear safety glasses and gloves.
3. Always hold hot materials with pliers or tongs.
4. Place hot parts in a safe place to cool. Keep them away from people and from materials that will burn.
5. Follow correct procedures when lighting torches and furnaces. Use spark lighters, NOT matches.
6. NEVER place your hands or any foreign object between mated dies or rolls.

Finishing Processes

1. DO NOT try to perform a process that has not been demonstrated to you.
2. Always wear safety glasses and a respirator.
3. Apply finishes in properly ventilated areas.
4. NEVER apply finishing materials near an open flame.
5. Always use the right solvent to thin finishes and clean finishing equipment.

Assembling Processes

1. DO NOT try to perform a process that has not been demonstrated to you.
2. Always wear safety glasses.
3. Wear gloves, protective clothing, and goggles for all welding, brazing, and soldering operations.
4. Always light a torch with a spark lighter; never use a match.
5. Handle all hot material with gloves and pliers.
6. Perform welding, brazing, and soldering operations in well-ventilated areas.
7. Use proper tools for all mechanical fastening operations. Be sure screwdrivers, wrenches, and hammers are the proper size, and are in good condition.

SUMMARY

Technological devices are designed to process materials, energy, and information. Tools and machines change the form of materials to make them more useful to people. Technology is used to gather and order data so that we can better understand it. Devices are used to change the form of energy so we can do things like: heat buildings, power vehicles and machines, light up dark areas, transmit information and data, from place to place, and erect structures for shelter and business.

Each one of these processing activities can make our lives better, or it can harm people and the environment. Processing Activities must be carefully selected, properly and safely used, and constantly monitored.

WORDS TO KNOW

All of the following words have been used in this chapter. Do you know their meanings?

Band saw
Chop saw
Circular saw
Computer
Cutting motion
Cutting tool
Cycle
Cylindrical grinding
Drilling
Energy processing
Feed motion
Forstner bits
Input unit
Lathe
Linear motion
Machine tools
Material processing
Memory unit
Multiple-point tool
Output unit
Processing unit
Program
Radial saw
Reciprocating motion
Rotating motion
Sawing machines
Scroll saw
Single-point tool
Spade bits
Stroke
Surface grinding
Table turning saw
Twist drill

TEST YOUR KNOWLEDGE

Write your answers on a seperate sheet of paper. Please do not write in this book.

1. True or false. Cutting and material processing mean the same thing.
2. What is the difference between feed motion and cutting motion?
3. Match the process on the left with the statement on the right that best describes it.

   _____ Cutting along the length of a cylinder to produce a cylindrical shape with a decreasing diameter.
   _____ Reducing the length of a material.
   _____ Producing a straight cylindrical hole.
   _____ Cutting across the end of a piece held in a lathe.
   _____ Reducing the thickness of a material.
   _____ Producing a recess at the top of a hole for a flat head screw.
   _____ Enlarging the diameter of a hole.

   A. Crosscutting.
   B. Facing.
   C. Tapering.
   D. Resawing.
   E. Countersinking.
   F. Drilling.
   G. Reaming.

4. List and describe the five major parts of a computer.
5. The _____ was the first energy converter that did not simply change the form of mechanical energy.
6. _____ is an energy converting device that uses the laws of magnetism and electromagnetism.
7. Match the name of the inventor with the information processing device he developed.

   _____ Alexander Graham Bell.
   _____ Johannes Gutenberg.
   _____ Guglielmo Marconi.

   A. Radio.
   B. Printing with moveable type.
   C. Computer.
   D. Telephone.

APPLYING YOUR KNOWLEDGE

1. Select a product that has been manufactured using material processing technology. List one to three operations that were used to make the products that used each of the following machine tools:

MACHINE TOOLS	PROCESS OR OPERATION
Turning machines	
Sawing machines	
Drilling machines	
Planing or shaping machines	
Grinding machines	

2. Select a major development in material processing, energy processing, or information processing. Develop a two-page report on the development.
3. Select a major development in material processing, energy processing, or information processing. Build a model of the technological device.
4. Produce a simple product that uses material processing technology. Write a summary of the procedure you used to make the product and list the tool or machine you used for each step of the procedure.
5. Communicate data or information using an information processing device. Write a one-page report on the processes you used to complete the task.
6. Select a household device that you use and develop a set of safety rules for its use.

MEASUREMENT TOOLS

After studying this chapter you will be able to:
- *Define measurement.*
- *List and describe the major physical qualities that can be measured.*
- *Compare the U.S. Customary and the metric measurement systems.*
- *List and describe the measurements used to describe the size of a part or product.*
- *Describe the common measurement tools.*
- *Use common measuring tools to measure linear distances, diameters, and angles.*
- *Describe factors in business and industry that are measured.*
- *Relate measurement to quality control.*

Look at the products of technology around you, Fig. 8-1. Most of them are made of several parts that are assembled into a finished item. Each part had to fit correctly. The finished product was, probably, installed in the room in a specific location. For example, the lighting fixtures you see are products with many parts. The fixtures are spaced over various parts of the room. How would you describe the size of the fixtures and their location in the room?

You traveled from your home to school. How far was the trip? How long did you take to make the journey? If you came by car or bus, how much fuel does the vehicle's fuel tank hold?

This book has some physical qualities. How heavy is it? What size are the pages? How thick is it?

Think of how you compare with your classmates. How much do you weigh? How tall are you? How fast can you run the 100-yard dash?

These questions ask you to describe physical qualities. Physical qualities are *characteristics* of an object or event that can be described. If we say a tree is tall, but have nothing to compare it to, saying the tree is tall has no meaning. The tree could be a dwarf apple tree or a giant redwood. If it takes a long time to do something, what do we mean by a long time? Hours? Weeks? Years? Centuries? You must use *measurement* to answer these questions.

Measurement objectively describes the physical qualities of an object. To describe something to someone, you must have a common reference or *standard*. Measurement is the practice of comparing the qualities of an object to a standard. To describe objects using measurement, we must have a system of standards for comparison.

Fig. 8-1. Measurement is vital in describing the physical features of everything shown in this photo. (Merillat Industries, Inc.)

MEASUREMENT SYSTEMS: PAST AND PRESENT

All measurements compare the quality being described against a standard. Early measurement systems used the sizes of human body parts as standards. The Biblical story of Noah records that he built an ark that was 30 cubits long, 50 cubits wide, and 30 cubits high. These are strange measurements to us. But early Egyptians, Romans, and Greeks used these terms. The standard was derived from the human arm, as shown in Fig. 8-2. A cubit was the distance from the tip of the middle finger to the elbow. A shorter measurement was the palm, which was the width of the four fingers. The width of a single finger was called the digit.

However, human arms vary in size. As long as technology was simple, measurements did not have to be highly accurate. In more recent times, measurements have been standardized. Each major physical quality is compared to a standard measurement that has been set by governments and international agreements.

MODERN MEASUREMENT SYSTEMS

Two measurement standards are in use today. These are the **U.S. Customary** system, and the more widely used International System of Units, abbreviated *SI* (from the French name: Le Système International d'Unités). SI is more commonly known as the **metric** system. See Fig. 8-3. The United States is the only industrialized country that has not adopted the metric system for everyday use.

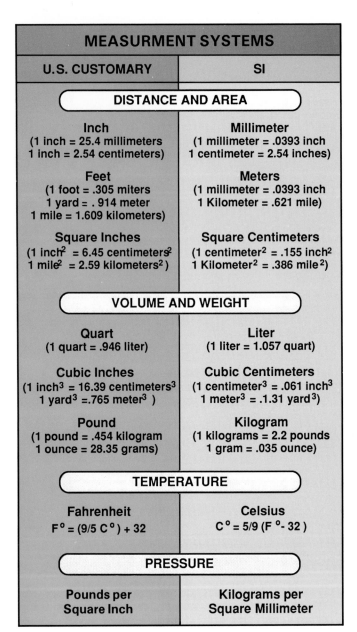

MEASUREMENT SYSTEMS	
U.S. CUSTOMARY	**SI**
DISTANCE AND AREA	
Inch (1 inch = 25.4 millimeters 1 inch = 2.54 centimeters)	**Millimeter** (1 millimeter = .0393 inch 1 centimeter = 2.54 inches)
Feet (1 foot = .305 miters 1 yard = . 914 meter 1 mile = 1.609 kilometers)	**Meters** (1 millimeter = .0393 inch 1 Kilometer = .621 mile)
Square Inches (1 inch^2 = 6.45 centimeters^2 1 mile^2 = 2.59 kilometers^2)	**Square Centimeters** (1 centimeter^2 = .155 inch^2 1 Kilometer^2 = .386 mile^2)
VOLUME AND WEIGHT	
Quart (1 quart = .946 liter)	**Liter** (1 liter = 1.057 quart)
Cubic Inches (1 inch^3 = 16.39 centimeters^3 1 yard^3 = .765 meter^3)	**Cubic Centimeters** (1 centimeter^3 = .061 inch^3 1 meter^3 = .1.31 yard^3)
Pound (1 pound = .454 kilogram 1 ounce = 28.35 grams)	**Kilogram** (1 kilograms = 2.2 pounds 1 gram = .035 ounce)
TEMPERATURE	
Fahrenheit F$^\circ$ = (9/5 C$^\circ$) + 32	**Celsius** C$^\circ$ = 5/9 (F$^\circ$- 32)
PRESSURE	
Pounds per Square Inch	**Kilograms per Square Millimeter**

Fig. 8-3. These common terms are used to describe the sizes of parts and products.

U.S. Customary system

The United States Customary system uses unique units for each quality being measured. Units for the same quality do not follow the same multiples. For example, the system for measuring volume uses ounces, pints, quarts, and gallons. The quart is divided into two pints (each of which is 16 ounces), and it takes 4 quarts to make a gallon. There is not a clear progression from one pint to one quart (two pints) to one gallon (eight pints). Likewise, 12 inches make a foot, three feet make a yard, and 1760 yards equal one mile. The inch is divided into halves, quarters, eighths, sixteenths, thirty-seconds, and sixty-fourths for more accurate measurement. The lack of uniform multiples can make the system confusing.

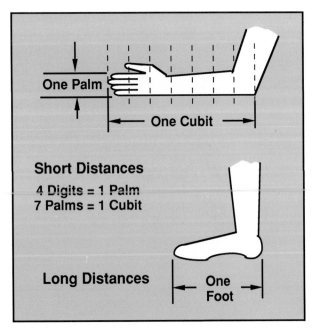

One Palm

One Cubit

Short Distances

4 Digits = 1 Palm
7 Palms = 1 Cubit

Long Distances

One Foot

Fig. 8-2. The basis for early units of measurement were parts of the human body.

Metric System

The metric system is based on ten. There is a logical progression from smaller units to larger ones, since all sizes of units are based on ten. The metric system starts with a base unit. The base unit can be changed to make larger and smaller units. Smaller units are decimal fractions (1/10, 1/100, 1/1000, etc.) of the base unit. Larger units are multiples of ten (10, 100, 1000, etc.) times the base unit.

The metric system uses a prefix to show us how the base unit is being changed. For example, the unit for distance is the meter. For large distances, the *kilo*meter is used. The prefix kilo means 1000, so seven kilometers is equal to 7000 meters. For small distances, the *milli*meter is used. The prefix milli means 1/1000th. Twelve millimeters are 12/1000 of a meter. The metric system lends itself to easy use in mathematical formulas. Look at Fig. 8-4 to see the most common metric prefixes. The metric system uses the same prefixes for all base units.

Another feature of the metric system is that the base units relate to each other in clear ways. A cube that is 10 centimeters on each side has a volume of one liter. A liter of water will have a mass of one kilogram. U.S. Customary units do not relate to each other in such a clear manner.

QUALITIES THAT ARE MEASURED

There are seven physical qualities for which standards have been developed. These qualities are distance, mass, time, temperature, number of particles, electrical current, and light intensity. These qualities are measured as technology is developed and used. However, for this chapter, only four will be discussed. They are:
- Distance–the separation between two points.
- Mass–the amount of matter in an object.
- Temperature–how hot or cold an object or place is.
- Time–how long an event lasts.

DISTANCE

Distance is the separation between two points. The distance unit in the metric system is the meter. The U.S. Customary unit is the foot. Measurements are given for both large distances and short distances. However, the size of the units used to describe the distance between Chicago and Los Angeles are different from those used to describe the distance from the left side to the right side of your desk. Large distances are generally given in a single measurement from Point A to Point B. For example, you might say that it is two miles (just over 3 kilometers) from your home to school.

SI MEASUREMENT PREFIXES			
FACTOR	MAGNITUDE	PREFIX	SYMBOL
10^9	1,000,000,000	giga	G
10^6	1,000,000	mega	M
10^3	1,000	kilo	k
10^2	100	hecto	h
10^1	10	deka	da
10^{-1}	1/10	deci	d
10^{-2}	1/100	centi	c
10^{-3}	1/1000	milli	m
10^{-6}	1/1,000,000	micro	μ
10^{-9}	1/1,000,000, 000	nana	n
10^{-12}	1/1,000,000,000,000	pico	p

Fig. 8-4. These are the common metric prefixes.

For parts or products, however, two or three different distance measurements are often given. Many parts and products are described as having thickness, width, and length. For example, you could describe a block of wood as being 3/4" x 4" x 12". This would mean that the block is three-fourths of an inch thick, four inches wide, and twelve inches long. Round parts are said to have thickness and diameter. A piece of pipe may be described as being 3/4 in. in diameter and 24 in. long. Finished products such as furniture, television sets, and cabinets are described by giving their height, width, and depth. A desk may be 30 inches high by 60 inches wide by 24 inches deep. The examples presented describe the size of objects. These terms and their use are shown in Fig. 8-5.

Volume

A measurement related to distance is *volume*. We give volume in cubic measurement. This describes the size of a cube that takes up the same space as a material or cavity. The metric system uses the liter as the volume unit. A liter is equal to the volume of a cube 10 centimeters on each side. The U.S. Customary system uses the cubic inch as the volume unit. An engine may be described as having a displacement of 2.4 liters. This

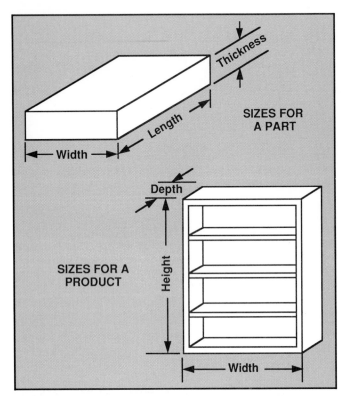

Fig. 8-5. These terms are used to describe the size of objects.

means that the volume of the engine cylinders totals 2.4 liters. A two-liter engine displaces a volume equal to about 122 cubic inches.

MASS

Mass describes the quantity of matter present in an object. Matter is anything that has mass and takes up space. Mass does not vary with location or temperature. For example, if we have two identical parts, one made from copper and the other from lead, the lead part will *weigh* more. Lead has more matter *per unit of volume* than does copper.

Weight

Mass is not the same as weight. Weight describes the gravitational pull exerted on an object. For most purposes, weight and mass are close enough to be used interchangeably. When you buy an apple, you are buying the matter of the apple, not the pull of gravity. We would use a scale to weigh the apple. The scale gives us an idea of the mass of the apple by *measuring the pull of gravity* on the apple. Everywhere on earth the apple will weigh the same. For very accurate measurements (scientific experiments, etc.) balances are used to measure mass. However, for most things, scales are accurate enough. We buy products by the pound (fruit, vegetables, sugar, etc.), by the gram (prescriptions, precious metals, etc.), or by the ton (hay, grain, sand, gravel, etc.).

TEMPERATURE

Temperature is the measurement of how hot or cold a material is. These terms are relative and, therefore, must be compared to a standard. For example, the term hot has different meanings for human comfort and for melting metals. Temperatures that are hot enough to make you uncomfortable would be considered cold for melting steel.

The range of temperature is broken into units. The customary system and the metric system use water as the basis to divide the temperature range into units called degrees. The range from the boiling point of water to the freezing point of water is divided into degrees. The customary *Fahrenheit* scale has 180 degrees between the freezing point and boiling point. The freezing point at sea level is 32°F and the boiling point is 212°F. The metric *Celsius* scale divides this range into 100 degrees, with the freezing point at 0°C and the boiling point at 100°C.

TIME

Time is how long an event lasts. The second is the basic unit of time in both the metric and the U.S. Customary systems. Short-term measurements are given in hours, minutes, seconds, or fractions of a second. These terms measure everything from the time it takes to complete a race (minutes and seconds) to computer speeds (nanoseconds, or billionths of a second). Time may be measured in very long terms such as a millennium (1000 years), a century (100 years), a decade (10 years) or a number of years. These terms are used to describe the age of civilizations, people, buildings, and vehicles.

TYPES OF MEASUREMENT

The measurement of part and product size is important in technological design and production activities. Generally this type of measurement can be divided into two levels of accuracy, as shown in Fig. 8-6:
- Standard measurement.
- Precision measurement.

STANDARD MEASUREMENT

Many production settings do not require close measurements. The length of a house, the width of a playing field, and the angle of the leg on a playground swing set need not be very accurate. If the product is within a fraction of an inch or a degree of angle, it will work fine. These measurements are often given to the

TYPES OF MEASUREMENT

STANDARD	PRECISION

 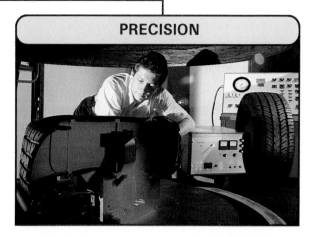

Fig. 8-6. Standard measurement is widely used in the construction industry, as shown on the left. Precision measurement is found in many manufacturing applications, as shown on the right. (Goodyear Tire and Rubber Co.)

foot, inch, or fraction of an inch in the customary system, or to the nearest whole millimeter in the metric system.

The kind of production is only one factor that affects the type of measurement used. The material being measured is also important. For example, wood changes (expands or shrinks) in size with changes in its moisture content and the atmospheric humidity. Measurements closer than 1/32 in. or 1 mm are not useful. Wood can change more than that amount in one day.

Standard measurements are common in cabinet and furniture manufacturing plants, construction industries, and printing companies. The printing industry uses it own system of measurement based on the pica (1/6 in.) and the point (1/72 in.).

PRECISION MEASUREMENT

Standard measurement is not accurate enough for many production applications. Watch parts and engine pistons would be useless if they varied by as much as 1/32 in. (0.8 mm). These parts must be manufactured to an accurate size. For this type of production, precision measurement is required. This type of measurement will measure features to 1/1000 in. (one one-thousandth of an inch) to 1/10,000 in. (one ten-thousandth of an inch) in the customary system. Metric precision measurement will measure features to 0.01 mm (one one-hundredth of a *millimeter*).

Precision measurements are used in manufacturing activities that involve metals, ceramics, plastics, and composites. Laboratory testing, material research, and scientific investigation also use precision measurements.

MEASUREMENT TOOLS

Many types of measurement tools are available to technologists, Fig. 8-7. Among these are tools that the operator manipulates and reads. These are called *direct-reading* measuring tools. In recent years, new measurement tools and machines have been developed. They bring together sensors and computers to automate measurement. These systems are called *indirect-reading* measuring tools.

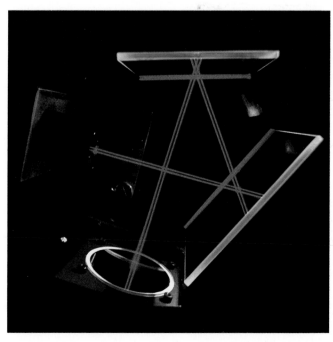

Fig. 8-7. Machine operators use manual measurements for routine work. Indirect measurement systems, such as this laser measurement system, are often built into continuous processing and assembling operations. (GM-Hughes)

Technology: How it Works

LASER: A device that emits a beam of coherent, monochromatic light.

Normal white light, like that from the Sun and electric lamps, is *polychromatic*, meaning it is made up of *many colors, Fig. 1. Each color of light has a certain wavelength.*

A laser is designed to produce light of only one wavelength. The word *laser* stands for *L*ight *A*mplification by *S*timulated *E*mission of *R*adiation. The laser is based on the findings of Albert Einstein and Niels Bohr. When atoms are exposed to an outside source of energy, such as electricity or light, the electrons become excited. This raises the electrons to a higher energy level within the atom. When the electrons fall back to their original energy level they give off light. When this light hits another atom, it causes more light to be given off.

It was discovered that all the atoms of a material will give off light that is the same wavelength. Since all the light has one wavelength, it has one color. Thus, laser light is *monochromatic*. Laser light is *coherent*, meaning that all the wave crests line up. Beams of laser light are very intense and do not to spread out as they pass through the air.

The first practical laser used a ruby rod, Fig. 2. Ruby is an aluminum oxide crystal. Its deep red color is caused by some chromium atoms that replace a few aluminum atoms in the crystal. The ruby rod is polished at each end. The ends are coated with a reflective material. One of the ends has a thinner coating than the other. To start the laser, the chromium atoms are excited by outside energy. Some of the light waves given off by the chromium atoms strike the reflective ends. The light waves that hit the ends bounce back through the ruby rod. The waves strike other atoms, which become excited. The light is *amplified* (made stronger) as it bounces back and forth between the reflective ends of the rod. Once the light becomes strong enough, some will pass through the end of the rod with the thin coating.

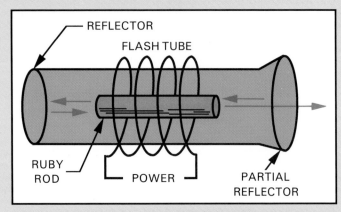

Fig. 2. The first laser used a ruby rod that was excited by a flash tube. Both ends of the rod are reflectors, but one end is a partial reflector. This allows the laser light to escape the rod when it has enough energy.

Many types of lasers are in use today. Lasers are grouped by the *amplifying medium* that they use. For example, the ruby laser used a ruby rod as the amplifying medium. Other lasers use gases, dyes, or semiconductor materials to amplify the light. All lasers operate on the same basic principle as the ruby laser. Each type laser gives off light that has a different wavelength, and therefore, a different color.

Lasers are being used in more applications every day. Lasers can be used for measurement, as shown Fig. 3. Laser are used to send messages through fiber optic cables. Compact disc players and some computer disk drives use lasers to store and retrieve data. Lasers are used for surgery, welding, cutting, and in bar code readers. Lasers are a versatile tool for modern industry.

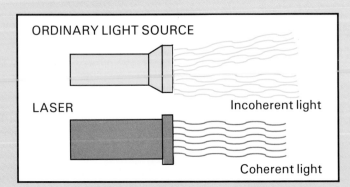

Fig. 1. White light is called *polychromatic* because it is made up of many colors. The light from a laser is one color and is called *monochromatic*.

Fig. 3. This device uses a laser for measurement. (FMC Corp.)

LINEAR	DIAMETER	ANGLE

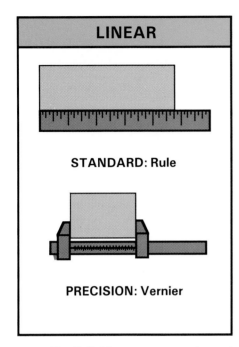

STANDARD: Rule

PRECISION: Vernier

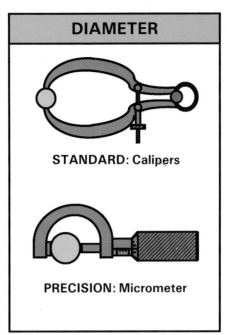

STANDARD: Calipers

PRECISION: Micrometer

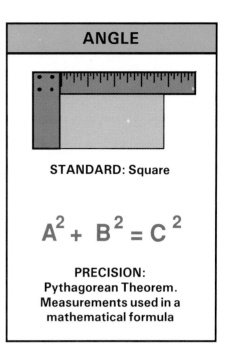

STANDARD: Square

$$A^2 + B^2 = C^2$$

**PRECISION:
Pythagorean Theorem.
Measurements used in a
mathematical formula**

Fig. 8-8. Measurement tools can be used to determine linear sizes, diameters, or the angle of adjacent surfaces.

DIRECT READING MEASUREMENT TOOLS

Three common uses of measurement are finding linear dimensions, diameters, and angles, Fig. 8-8. Each of these three features can be measured using standard or precision devices.

Linear Measurement

The most common linear measurement device is the **rule**, Fig. 8-9. These are rigid or flexible strips of metal, wood, or plastic with measuring marks on their face. Two types of rigid rules are used in linear measuring. These are the machinist's rule, and the woodworker's or bench rule. A bench rule is generally divided into fractions of an inch. The most common divisions are sixteenths (1/16 in.). Metric bench rules are divided into whole millimeters. Machinist's rules are designed for finer measurements. Customary machinist's rules are divided in sixty-fourths (1/64 in.) or into tenths (1/10 in.) and hundredths (1/100 in.). Metric machinist's rules are divided in 0.5 mm increments.

The part is measured with a rule by aligning one end of the part with the end of the rule or with an inch mark. The linear measurement is taken by reading the rule division at the other end of the part.

Flexible rules are often called tape rules. They are used in woodworking and carpentry applications. There is a hook at one end of the rule that is hooked to the end of the board or structure. The tape is pulled out until it reaches the other end of the board or structure. A measurement is then taken.

MAKING LINEAR MEASUREMENTS WITH A RULE

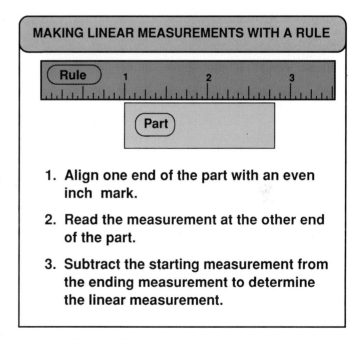

1. **Align one end of the part with an even inch mark.**

2. **Read the measurement at the other end of the part.**

3. **Subtract the starting measurement from the ending measurement to determine the linear measurement.**

Fig. 8-9. Reading a dimension with a rule.

Metric tape rules are divided into 1 mm increments. Commonly, the smallest division on a customary tape rule is one-sixteenth inch. Some tape rules highlight every 16 inches, which is the common spacing of studs in home construction. Tape rules are generally available in lengths from 8 feet to 100 feet or from 2 meters to 30 meters.

Precision linear measurements can be made using a number of machinist's tools. The most common are shown in Fig. 8-10. These are inside and outside micrometers, depth gages, micrometers, and vernier calipers.

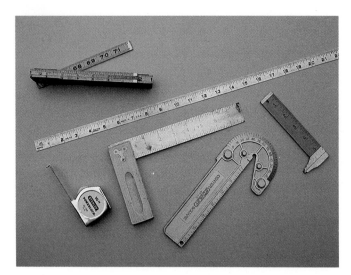

Fig. 8-10. Common linear measurement tools include rigid rules, tape rules, micrometers, and calipers.

Measuring Diameters

A common measurement task involves determining the diameter of round material or parts. A simple, rough measurement may be determined with hole gages or circle templates. These devices have a series of holes into which the stock can be inserted. The smallest hole into which the material will fit establishes the approximate diameter of the item.

More precise diameters can be established using a **micrometer**, as shown in Fig. 8-11. The part to be measured is placed between the anvil (fixed part) and the spindle (movable rod). The spindle is moved forward by turning the barrel. When the spindle and the anvil touch the part, a reading is taken on the barrel. Most customary micrometers will measure to within 1/1000th of an inch, metric micrometers measure to 0.01 (1/100th) of a millimeter.

Measuring Angles

The angle between two adjacent surfaces or intersecting parts is important in many situations. The legs of a desk are generally square (at a 90° angle) with the top. The ends of picture frame parts must be cut at a 45° angle to make a square frame.

Measurements at 90° angles are commonly done with **squares**, Fig. 8-12. These tools have a blade that is at a right angle to the head. The head is placed against one surface of the material. The blade is allowed to rest on an adjacent surface. If the blade touches the surface over its entire length, the part is square. Parts that are not square allow light to pass under the blade.

Some squares have a shoulder on the head that allows the square to be used to measure 45° angles. This angle is important in producing mitered corners on furniture, boxes, and frames.

The most common squares are the rafter or carpenter's square, the machinist's square, the try square, and the combination (90° and 45°) square.

Protractors and sliding T-bevels may be used to measure angles that a square cannot. The protractor allows for direct reading of angles. The protractor is placed over the angle to be measured, and the angle is read. The sliding T-bevel has an adjustable blade. The head is placed on one surface. The blade is clamped along the angle of the second surface. This angle can then be measured by a protractor or by using a mathematical formula.

INDIRECT READING MEASUREMENT TOOLS

In many modern measuring systems, humans no longer take measurements. Sensors gather the measurement data, which is processed by computers or other automatic devices. The final measurement can be

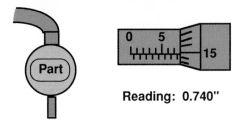

MEASURING WITH A MICROMETER

Reading: 0.740"

1. **Align the part between the spindle and the anvil of the micrometer.**

2. **Gently move the spindle against the part.**

3. **Read the measurement on the barrel.**

   a. **Read to the nearest .050 " -- each mark on the barrel.**

   b. **Add the fraction from the thimble - each mark is .001".**

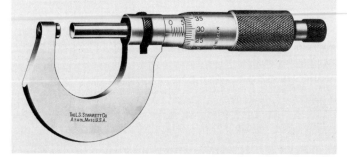

Fig. 8-11. Measuring diameters with a micrometer.

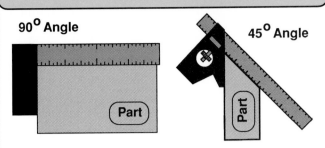

CHECKING ANGLES WITH A SQUARE

90° Angle

45° Angle

Part

Part

1. **Place the head of the square against one surface of the part.**

2. **Hold the part and square up to the light.**

3. **Look for gaps along the blade of the square.**

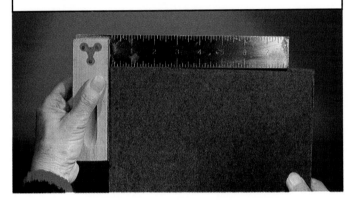

Fig. 8-12. Ninety degree angles can be measured with a square. When an angle is not 90°, the blade of the square does not touch the surface of the work, as shown in the photo.

displayed on a output device such as a digital read-out, computer screen, or printout. These new systems include laser measuring devices, optical comparators, and direct-reading thermometers. If you have weighed yourself on a digital bathroom scale, then you have used and indirect reading measuring device.

MEASUREMENT AND CONTROL

So far you have learned that measurement is used to describe physical qualities. You may measure length, weight, temperature, or other qualities. However, there must be a reason for doing the measuring.

MEASUREMENT AND PRODUCTION PROCESSES

All technological processes produce products or services. These may be goods, buildings, or communication media. Measurement is necessary in designing an artifact. Its size, shape, or other properties are communicated through measurements. Processing equipment is set up and operated using these design measurements. Materials needed to construct the item are ordered using measurement systems. All personal or industrial production is based on measurement systems.

MEASUREMENT AND QUALITY CONTROL

Measurement can also be used to compare the present condition with a desired condition. For example, say you need a 24-inch (610 mm) long shelf for a book case. You would probably cut it from a longer board. First, you would measure and mark the location for the cut. Then, you would saw along the line. If the board is too long, it will not fit into the case. A short board will fail to rest on the shelf supports. Measurement tells you if you have produced a 24-inch-long shelf.

A board of the correct length would meet the intended purpose. It would meet your quality standards. The process of setting standards, measuring features, comparing them to the standards, and making corrective actions is called **quality control**, Fig. 8-13. This process is designed to ensure that products, structures, and services meet our needs.

Quality control involves measuring and analyzing materials entering the system, work in process, and the outputs of the system. It is an on-going process designed to ensure that resources are efficiently used and that customers receive functional products.

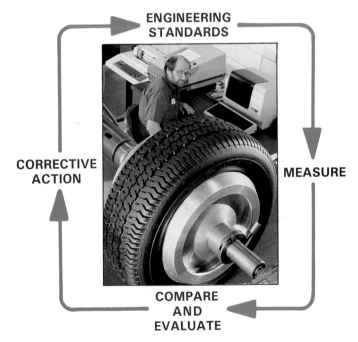

ENGINEERING STANDARDS

CORRECTIVE ACTION

MEASURE

COMPARE AND EVALUATE

Fig. 8-13. The foundation of a quality control system is measurement and analysis. (Goodyear Tire and Rubber Co.)

SUMMARY

Measurement describes distance, mass, time, temperature, number of particles, electrical current, and light intensity. It involves comparing a physical characteristic to an established standard. The common standards are the metric system and the U.S. Customary system. These systems allow people to communicate designs, order materials, set up machines, fabricate products, and control quality.

WORDS TO KNOW

All of the following words have been used in this chapter. Do you know their meanings?

Distance
Mass
Measurement
Metric
Micrometer
Precision measurement
Quality control
Rule
Square
Standard measurement
Temperature
Time
U.S. Customary

TEST YOUR KNOWLEDGE

Write your answers on a seperate sheet of paper. Please do not write in this book.

1. True or false. Measurement is part of a quality control system.
2. Match the physical property with its definition.

    _____ Distance. A. Hotness or coldness.

    _____ Mass. B. Amount of matter.

    _____ Temperature. C. Duration of an event.

    _____ Time. D. Space between two points.

3. List the seven physical qualities that have measurement standards.
4. True or false. A kilometer is longer than a millimeter.
5. Comparing the qualities of an object to a stated standard is called _____.
6. The _____ measurement system expresses distances in feet and inches.
7. The _____ measurement system expresses distances in meters and kilometers.
8. A rigid strip of metal with marks on it that is used to measure linear distances is called a _____.

APPLYING YOUR KNOWLEDGE

1. On a form similar to the one below, write the metric unit and the U.S. Customary unit of measurement for each feature listed.

Measurment	Metric	U.S. Customary
Distance from Los Angeles		
Temperature on a hot day		
Weight of a loaf of bread		
Length of a pencil		
Capacity of a container		

2. Select an object, such as a desk or a book case. Describe its size using both the metric and U.S. Customary measurement systems.
3. Check the squareness of a piece of furniture using a rafter, try, or combination square. Describe what you found and how any out-of-squareness could be corrected.

Precise measurement is required to make the quality products demanded by consumers today. (Sandvik Coromant Co.)

Section Three - Activities

ACTIVITY 3A - DESIGN PROBLEM

Background:

All technology involves a machine or device to process materials, energy or information. These machines are used to change the form of one or more of these resources into a new, more usable form.

Situation:

You have been selected as the public relations director for a local citizens group. Your group is concerned about issues that are not being fully addressed by local politicians and business leaders. Your group also wants to inform the public about these important issues.

Challenge:

Choose an issue that you feel is not being addressed by political and business leaders. Some examples of this type of issue are: the greenhouse effect, fossil fuel dependence, or nuclear power. Use the library for research if necessary. Develop a one-page flyer to communicate your group's position on the issue. If your school has computers available, use one or more computer software programs to help you produce the flyer.

ACTIVITY 3B - FABRICATION PROBLEM

Background:

We live in a material world. All around us are products that have been developed using material processing technology.

Challenge:

Work with a partner to make a game that can be given to a local charity. Fig. 3-A shows the layout for the game.

Materials:

One piece 3/4" x 3 1/2" x 3 1/2" clear pine, redwood, or western red cedar.
Twelve 1/8" diameter x 1" wood pegs, or golf tees.

Procedure:

Carefully watch the demonstration your teacher gives showing the proper use of the tools and machines needed to make the product.

Follow the procedure given on the operation process chart shown in Fig. 3-B to construct the game. Follow

GAME BOARD LAYOUT

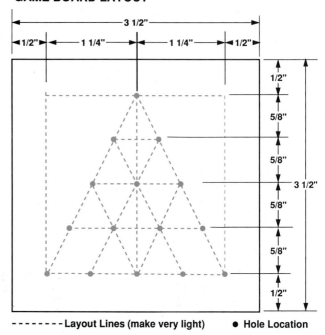

- - - - - - - Layout Lines (make very light)   ● Hole Location

Fig. 3-A. Game board layout.

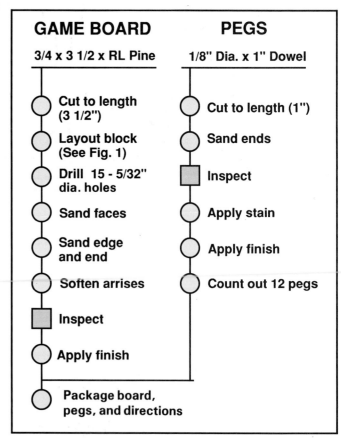

Fig. 3-B.

ALL safety rules discussed by your teacher during the demonstration. Photocopy the directions shown in Fig. 3-C. These will be packaged with the game.

Analysis:

Meet as a class and analyze the material processing activities you completed by answering the following questions.

1. Which material processing tools did you use?
2. Which measuring tools did you use?
3. How could you have increased the speed of the manufacturing process?
4. What changes in the material processing actions would you make to improve the quality of the product?

MIND CHALLENGE

Directions

- **Put a peg in each hole except one.**
- **Select one peg and jump another adjacent peg ending in an empty hole.**
- **Remove the jumped peg.**
- **Continue jumping pegs until only one remains or no more jumps are possible.**

Scoring

✔ **One peg remains - Terrific - 30 points**
✔ **Two pegs remain - Good - 20 points**
✔ **Three pegs remain -   Fair - 10 points**
✔ **Four pegs remain -   Poor -   0 points**

Fig. 3-C.

section 4

Problem Solving and Design in Technology

Complex devices, such as the integrated circuit shown above, requires the teamwork and talents of many people. (Northern Telecom)

THE PROBLEM SOLVING AND DESIGN PROCESS

After studying this chapter you will be able to:
- *Describe the problem-solving process and the design process.*
- *Describe the difference between problem solving and design.*
- *Define and describe innovation.*
- *List and describe the steps followed in designing a product or technological system.*
- *Describe how newly developed technology changes the relationship between people and machines.*
- *Define a technological problem or opportunity.*
- *Describe the types of criteria that are used to focus technological development projects.*
- *List and describe the types of information gathered as a foundation for technological development projects.*
- *Develop criteria and gather information needed in developing solutions to technological problems or opportunities.*

All technology is created for a purpose. It is designed to meet human needs or wants. There is no technology on earth that did not come from someone wanting to solve a problem or address an opportunity. A technological device may have been developed to protect us from the physical environment. It may help us be more informed, or allow us to travel with ease. Technology keeps you warm and dry, lights the space around you, speeds you to distant locations, and allows you to learn about events as they happen around the world.

Most modern technology is not developed by trial and error, nor does it come about by accident. It is the result of focused action by many people. In short, technological devices and systems are designed, developed, engineered, produced, operated, and maintained by people.

This section in your study of technology will present the problem-solving and design activities that are used to develop technology. These actions cut across all types of technology. They can be used to design communication messages and apparatus, buildings and constructed structures, manufactured goods, and transportation devices and systems.

PROBLEM SOLVING

A **problem** exists when people encounter a difficulty, Fig. 9-1. It may be as simple as cracking open a walnut when tools are unavailable. It may be as complex as moving 500 people from Los Angeles to Tokyo in two hours. These problems have two things in common: (1) There is a *goal* (cracking a walnut or moving people) and (2) there is *no clear path from the present state* (an intact walnut or misplaced people) *to the goal*.

Problem solving is a common human activity. Each of us uses a fairly universal process as we approach problems. In short we do the following:
- *Develop an understanding of the problem* through observation and investigation.
- *Devise a plan* for solving the problem.
- *Implement the plan.*
- *Evaluate the plan.*

TECHNOLOGICAL PROBLEM SOLVING

Not all problems are technological. You may have a problem getting along with a classmate or family member. This is a social problem. You may have a problem identifying and describing the weather

Fig. 9-1. Technology is born from a problem or an opportunity, which can be as simple as clearing snow from a road or as complex as landing a person on the moon. (NASA)

produced by some types of clouds. This is a scientific problem. You may be trying to decide whether to keep some money you found on the street. This is an ethical problem.

PROBLEMS LEAD TO SOLUTIONS

However, you face a technological problem if you need to develop tools, machines, or systems to help you do work. Hay production is an example of an activity that has become mechanized. Until World War II, most farmers raised their own hay and stored it in loose stacks. It was used as feed for the animals on the farm. However, in recent years, the self-sufficient family farm has been disappearing. Family farms are being replaced with larger farms that specialize in a specific crop. One group of farmers operate large dairy farms that do not raise hay. Other farmers specialize in raising and selling hay to these dairy farmers and beef cattle feed lots. This crop specialization caused a problem that could be solved by technology. Loose hay can not be easily shipped from the hay farmer to the feed lot.

Labor is Expensive

More recently another agricultural problem has surfaced. Farm labor has become expensive, and many people do not want to do heavy manual labor. The solution to this problem is farm implements. These machines allow one person to do more work, assisted by the machine, Fig. 9-2. Therefore, the process of collecting hay into bales was developed as a solution to this

Fig. 9-2. Technology has changed farming from a labor intensive activity to an equipment intensive activity as shown by these operations. A single farmer can now do the work of many people.

technological problem. This process included the following steps:

1. The field of hay was cut with a mowing machine, and allowed to dry on the ground.
2. A machine called a side delivery rake was used to collect the hay into a narrow pile called a windrow.
3. A tractor pulled the baler along the windrows, where it collected, compacted, and tied the hay into bales.
4. The bales are then pushed out of the machine onto the ground, where farm workers picked them up and loaded the bales onto a truck.

Technology has improved the machines used in harvesting hay. Mowing machines and side delivery rakes have been replaced by a machine called a swather which cuts, conditions, and windrows hay in one operation. The hand loading and hauling of baled hay is being replaced with balewagons that automatically pick-up and stack bales into cubes. Hay can also be gathered into large rolls. These cubes or rolls can be loaded with fork lifts onto trucks that haul the hay to customers. Harvesting hay that was once **labor intensive** has become **equipment intensive** through the use of technological problem solving. The trade-off is that equipment uses energy to do the work that humans used to do. People now operate machines where they once provided a great deal of manual energy to do the job.

TECHNOLOGY CAN IMPROVE EFFICIENCY

Similar examples of technological problem solving can be given for manufacturing operations. Computer-aided design systems have increased the productivity of drafters. The keyboard and the graphics tablet are replacing the T-square, triangles, and drafting board. Drawings are being stored on computer disks instead of in file cabinets. Product specifications and drawings move swiftly between factories on telephone circuits instead of through the mail. Many companies now employ fewer drafters to complete a larger volume of work.

Automation

Robots do many routine manufacturing tasks without human interference or monitoring. Automobile bodies are welded by robots. Likewise, painting and material handling operations are increasingly done by the new "steel collared" workers (the robot).

Machining centers have replaced many individual machine tools and have reduced the need for skilled machinists. These devices and many other manufacturing machines address quality and productivity

Fig. 9-3. Technology allows fewer workers to make large quantities of products. Machines such as this robot are used to do repetitive tasks with high accuracy. (Cincinnati Milicron)

problems facing many companies. They allow fewer workers to produce better products in less time with less scrap, Fig. 9-3.

In transportation, computer-controlled "driverless" people movers speed people between terminals at airports. Guided bus systems, such as the one shown in Fig. 9-4, promise to increase the efficiency of urban transport. Automated equipment on new commercial aircraft have reduced the number of flight deck officers

Fig. 9-4. New transportation systems, such as this guided bus, use fewer drivers to move more people and with greater safety. (Daimler-Benz)

from three to two. Automatic switching, signaling, and train operation systems have replaced the local telegraph operator on railroads and make rapid transit systems more efficient. These devices make transportation systems safer and more efficient.

Opportunity

However, not all technology is developed to solve a problem. There are many conditions that people do not view as a problem. If a "nonproblem" condition can be improved with technology, a technological **opportunity** has been discovered. People may want to use the new device even though an old device was adequate for their needs. The automatic coffeemaker is an example of meeting a technological opportunity. There was no consumer cry for a new way to make coffee. The percolator and drip coffee makers in use were adequate for the job. However, the first Mr. Coffee® automatic drip coffee maker was an almost instant success. Advertising helped many people develop a "need" for the new product. An opportunity was identified and new product was born.

SOLVING TECHNOLOGICAL PROBLEMS AND OPPORTUNITIES

Solving technological problems and meeting technological opportunities require specific action. As shown in Fig. 9-5, this challenge involves four major phases. These and their basic activities are:

- *Identify or recognize a technological problem*: This phase of the technology development includes:
  - ✔ Defining the problem: Explaining the situation that needs a technological solution, and establishing the criteria that the device or system must meet.
  - ✔ Gathering information: Obtaining background information needed to begin developing solutions for the problem or situation.
- *Developing a solution*: This phase of the technology development involves:
  - ✔ Developing possible solutions: Originating a number of different solutions that could solve the problem or meet the opportunity.
  - ✔ Refining the best solutions: Selecting the most promising solutions and integrating, modifying, and improving them.
- *Evaluating the solution*: This phase of the process involves:
  - ✔ Modeling the best solution: Testing and evaluating the proposed solutions through graphic, mathematical, and/or physical modeling techniques.

- ✔ Selecting the best solution: Comparing the design solutions in terms of economic, market, technical, production, and environmental criteria to determine the best solution to the problem.
- *Communicating the solution*: This phase of the process includes:
  - ✔ Interpreting the solution: Communicating the final solution through graphical (engineering drawings), verbal (instructions and reports) and mathematical (formula) means.
  - ✔ Presenting the solution for approval: Obtaining appropriate approval (from management, government, etc.) for implementing the solution.

The remainder of this chapter will explore the first phase of the technology development activity: identifying or recognizing a technological problem. The other phases will be discussed in the following three chapters.

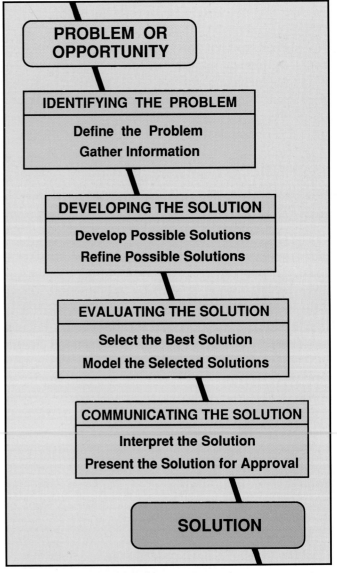

Fig. 9-5. These are the four phases of the technology development process.

Technology: How it Works

PROPELLER: A rotating multi-bladed device that propels a vehicle by moving air or water.

All vehicles have five systems: structure, suspension, propulsion, guidance, and control. The propulsion system causes the vehicle to move. Propulsion systems convert energy into motion.

The propulsion system often uses an engine to create a rotating motion. This motion is transmitted to a device that moves the vehicle. Most land vehicles use wheels to create movement. However, a different system is required for vehicles that travel on water and in air. A common propulsion device is a propeller, Fig. 1. Propellers are used on both airplanes and ships.

A propeller works using two principles of physics. The first principle is: for every action there is an equal and opposite reaction. The propeller forces air or water backwards as it spins. This, in turn, causes the propeller to be pushed forward. Actually, the propeller moves like a screw. The tips of the blades follow a path like the threads of a screw. In theory, one revolution of the propeller would cause the vehicle to move ahead like one turn of a nut would cause it to move on a bolt. However, a boat or a plane does not move that far. Both water and air are "yielding fluids." They compress slightly from the pressure of the spinning propeller. Ships generally move about 60 to 70 percent as much as they would in theory.

The second principle is that as a fluid travels faster its pressure is reduced, Fig. 2. As a propeller spins, the water or air travels faster across the front of the blade than it does on the back. This causes the pressure to decrease on the front. The decrease in pressure draws the propeller forward, adding to its efficiency.

Fig. 1. Propellers are used on airplanes, both old and new. (Piaggio Aviation)

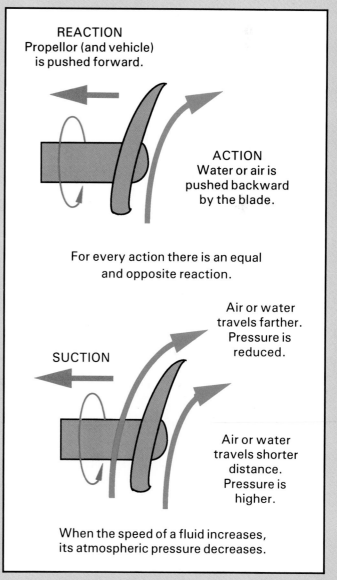

REACTION
Propellor (and vehicle) is pushed forward.

ACTION
Water or air is pushed backward by the blade.

For every action there is an equal and opposite reaction.

SUCTION

Air or water travels farther. Pressure is reduced.

Air or water travels shorter distance. Pressure is higher.

When the speed of a fluid increases, its atmospheric pressure decreases.

Fig. 2. This diagram explains how a propeller works.

The Problem Solving and Design Process 137

IDENTIFYING A TECHNOLOGICAL PROBLEM OR OPPORTUNITY

You have learned that all technology starts with a problem or an opportunity. A person or a group of people must see a goal that is different from the present situation, Fig. 9-6. This goal must be described clearly by the designers of new technological devices. Otherwise, considerable effort and money may be spent without reward.

DEFINING THE PROBLEM

In order to set a goal, the problem must be identified. This may sound simple. However, many problems are not easy to identify. We may know the symptoms without recognizing the problem. For example, we know holes have developed in the ozone layer over the Antarctic and Arctic regions. Also, levels of carbon dioxide (CO_2) and methane gas have increased over the past 30 years. Is the ozone hole and increased CO_2 a problem? It is a problem only in terms of its *consequences*. What happens to the earth because of this hole in the ozone layer? Will increased levels of CO_2 cause the temperature of the earth to rise? Will global warming create deserts out of fertile land? Will ice caps melt causing flooding of areas along sea coasts? Are these situations a problem? We must again look at the consequences. The ozone hole and increased levels of CO_2 are consequences of the original problem. The original problems are caused by our unwise application of technology. It is related to our use of polluting vehicles and power plants, aerosol paint and personal care products, styrofoam food containers, and some refrigerants. These causes present a specific technological problem. Their solutions also rely on technology–energy efficient

machines, better engines, cleaner fuels, and new products that are easy on the environment. The challenge is to do *more work* while using *less energy* and *fewer materials*.

If a designer is to solve a technological problem or opportunity, the situation must be clearly defined. The problem definition should complete two major tasks:
- Explain the situation that needs a technological solution.
- Establish the criteria that the device or system must meet.

Describing the Situation

Defining the problem or opportunity involves recognizing a human need or want that can be met through a new or improved technological device or system. Care must be exercised so that we define the problem or opportunity instead of defining the solution. For example, a book end is a solution, not a problem. The problem is to hold books in an organized manner on a flat surface. Likewise, a chair is a solution. A chair can be described as a device with a seat, back, and four legs. A designer starting with the problem statement "design a chair" would never end up with a bean bag or a wicker basket suspended from a chain. Definitions can restrict creativity or open doors to a variety of solutions.

Look back at our ozone layer "problem." To close the hole is the solution. The problem is *to develop technological devices that emit fewer pollutants into the atmosphere.*

Establishing Criteria

The problem definition leads directly to the next step in developing technological devices and systems: establishing criteria. The designer must know how the effec-

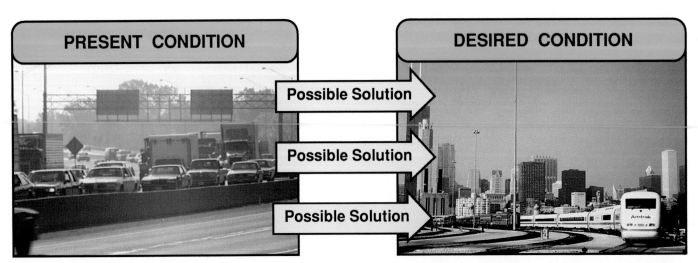

Fig. 9-6. Technology moves people from a present situation to a new situation. Look at the above photos that show a present situation and one possible solution for a desired situation. Describe the present situation and one desired situation. (Siemens)

tiveness of the technology will be evaluated. To do this, we establish **criteria** that the product or system must meet. These, as shown in Fig. 9-7, can be grouped as:

- Technical or engineering criteria describe the operational and safety characteristics that the device or system must meet. These criteria are based on how, where, and by whom the product will be used. An example of a engineering criteria for a new windshield wiper might be: *Must effectively clear the windshield of water when the vehicle is traveling at highway speeds.*
- Production criteria describe the resources available for producing the device or system. These criteria are based on the natural, human, and capital (machine) resources available for the production of the device or system. A production criteria for a product might be: *Must be manufactured using existing equipment in the factory.*
- Market criteria identify the functional, appearance, and value characteristics for the device or system. These criteria are derived from studying what the user expects from the device or system. An example of a market criteria might be: *The product must be compatible with Early American decor.*
- Financial criteria establish the cost-benefit ratio for the device or system. These criteria address the amount of money required to develop, produce, and use the technological device or system and the ultimate benefits from its use. An example of a financial criteria might be: *The product must be priced at five percent less that the major competitor's product.*
- Environmental criteria indicate the intended relationship between the device or system and the natural and social environments. These criteria deal with the impacts of the technological device or system on people, societal institutions, and the environment. An environmental criteria might be: *The device must remove 95 percent of all sulfur dioxide emitted from the electric generating plant smokestacks.*

Gathering Information

The problem definition and criteria provide designers with direction and a focus for their work. The definition and criteria allows the designers to start the next step, gathering information. Designing technological devices and systems requires knowledge. This knowledge is derived from obtaining and studying information. A wide variety of information may be needed, Fig. 9-8. Typically this information includes:

- Historical information about devices and systems that were developed to solve similar problems in the past.

Fig. 9-7. Technological devices and systems must meet both an individual's and society's expectations. These expectations can be communicated through five types of criteria (technical, production, market, financial, and environmental).

- Scientific information about natural laws and principles that must be considered in developing the solution.
- Technological knowledge about information, material, and energy processing techniques that could be used to develop, produce, and operate the device or system.

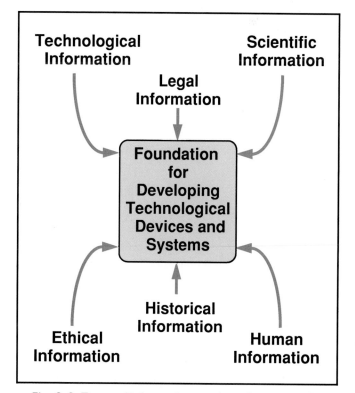

Fig. 9-8. Types of information used as a foundation for technological development activities.

- Human information that will affect the acceptance and use of the device or system. This can be details on ergonomics, body size, consumer preferences, appearance, etc.
- Legal information about the laws and regulations that will control the installation and operation of the device or system.
- Ethical information which describes the values people have toward similar devices and systems.

This information can be obtained through three major activities. First, *historical methods* can be used to gather information already recorded. Books, magazines, and journals housed in libraries may be consulted, Fig. 9-9. Companies can review sales records, customer complaints, and other company files. Judicial codes describe laws that may affect the development project.

Descriptive methods may be used to record observations of present conditions. People may be surveyed to determine product preference, opinions, or goals. The operation of similar devices can be observed and described. Physical qualities, such as size, weather conditions, and weight can be measured and recorded.

Finally, *experimental methods* may be used to compare different conditions. One condition is held constant while the other is varied. This method can be used to determine scientific principles, assess the usefulness of technological processes, or gauge human reactions to situations, Fig. 9-10.

All these methods are used to gather information that can provide the background needed to begin developing solutions for the problem or situation. The procedures for developing these solutions will be the focus of the next chapter in this book.

Fig. 9-10. Laboratory experiments provide valuable information that can be used to solve technological problems. (Goodyear Tire and Rubber Co.)

SUMMARY

Technology is developed by people in response to problems and opportunities. Development activities involve identifying the problem or opportunity, developing solutions for the problem or opportunity, evaluating the chosen solution, and communicating the solution.

The base for all technology development programs is a clear statement of the problem, and a set of criteria that describe the goals for the solution. These two steps are followed by efforts to collect information needed to start developing appropriate devices or systems.

WORDS TO KNOW

All of the following words have been used in this chapter. Do you know their meanings?
Criteria
Equipment intensive
Labor intensive
Opportunity
Problem

TEST YOUR KNOWLEDGE

Write your answers on a seperate sheet of paper. Please do not write in this book.

1. What is the difference between a labor intensive and an equipment intensive process?

Fig. 9-9. Books, magazines, and journals provide a valuable source of information for solving technological problems.

2. True or false. Technology is often developed by trial and error.
3. Match the type of information gathered to the correct information gathering method:
   _____ How other people solved similar problems.
   _____ Which of two methods to process material is best suited for a specific application.
   _____ The number of similar products sold last year by a company.
   _____ Which of two products people would prefer to buy.
   _____ The effects of winter weather on the product.

   A. Descriptive.
   B. Experimental.
   C. Historical.

4. List and define the four major steps in developing a technological device or system.
5. True or false. Technology is designed to meet only people's needs.
6. Statements which outline the expectations for a technological device or system are called _____.
7. List six types of information that can help designers solve technological problems and meet technological opportunities.
8. Describe the difference between technological problems and technological opportunities.

APPLYING YOUR KNOWLEDGE

1. Select a problem or opportunity that technology can address. Use a form similar to the one shown below to help you:
   a. Write a clear definition of the problem or opportunity.
   b. List the criteria that you would consider in solving the problem or meeting the opportunity.

PROBLEM OR OPPORTUNITY:
Engineering criteria:
Production criteria:
Market criteria:
Financial criteria:
Environmental criteria:

2. Suppose you were given the following problem to solve: Design a device that would allow you to generate enough electricity to power a television set while riding an exercise bicycle. Where would you go to gather information to start the design process?

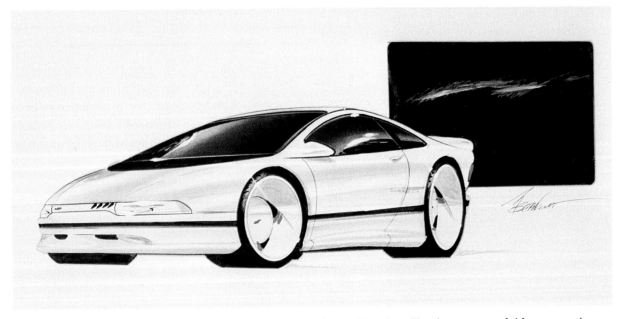

Designers use drawings to visualize and develop design ideas. Drawings like these are useful for presenting ideas to other members of a design team. (Ford Motor Co.)

10
CHAPTER

DEVELOPING DESIGN SOLUTIONS

After studying this chapter you will be able to:
- *List and describe the two types of technological design.*
- *List and describe the three stages in developing technological designs.*
- *Describe ways that designers, engineers, and architects can use sketches to develop initial design ideas.*
- *List and describe the three types of pictorial sketches used in product design.*
- *Prepare rough sketches of design ideas.*
- *Refine rough sketches to show more detail and improve the design.*
- *Prepare dimensioned (detailed) sketches of a technological design.*

Technological devices are designed to meet identified problems and opportunities. These situations, as shown in Fig. 10-1, can be divided into two major types:
- System design problems and opportunities.
- Product design problems and opportunities.

SYSTEM DESIGN

System design deals with the arrangement of components to produce a desired result. For example, automotive braking systems are a result of system design efforts. Look at the drum brake system shown in Fig. 10-2. This design brings together mechanical and hydraulic components into a speed reduction system. The brake pedal unit is a mechanical linkage. When the pedal is depressed, a plunger in the master cylinder is moved. This motion causes the fluid to move in the hydraulic system that connects the master cylinder to the wheel cylinders. The fluid movement pushes the pistons outward in the wheel cylinders. These pistons are attached to the brake shoes. The piston movement causes the shoe to be forced against the brake drum. This mechanical action creates friction between the shoe and the drum which slows the automobile.

System design can be used in all technological areas. It is an important part of construction technology, Fig. 10-3. Electrical, heating and cooling, plumbing, and communication systems are designed for buildings. In

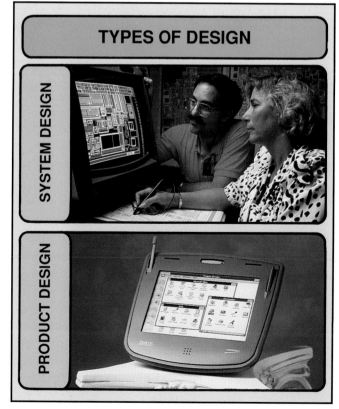

Fig. 10-1. Designers create both systems and products. (Harris Corp., Zenith Data Systems)

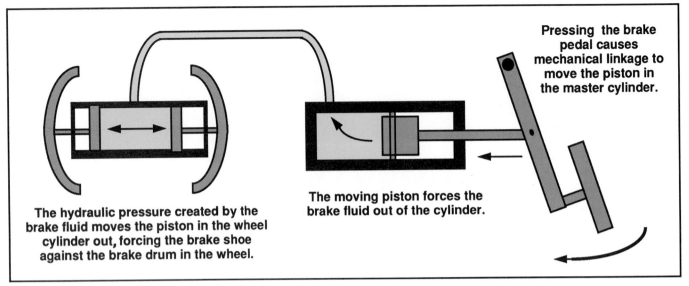

The moving piston forces the brake fluid out of the cylinder.

Pressing the brake pedal causes mechanical linkage to move the piston in the master cylinder.

The hydraulic pressure created by the brake fluid moves the piston in the wheel cylinder out, forcing the brake shoe against the brake drum in the wheel.

Fig. 10-2. This brake system was the result of system design efforts.

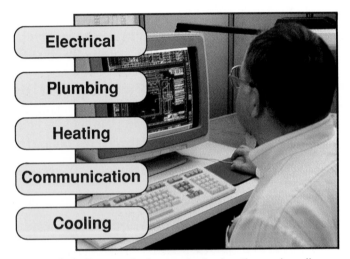

Electrical

Plumbing

Heating

Communication

Cooling

Fig. 10-3. System designers devise heating and cooling, plumbing, communication, and electrical systems for buildings. (General Electric Co.)

manufacturing, the methods of production, warehousing, and material handling must be designed. Messages are carried over fiber optic and microwave communication systems. Transportation systems combine manufactured vehicles and other components to move goods and passengers from place to place.

PRODUCT DESIGN

Product design deals with the development of manufactured products and architecture deals with constructed structures, Fig. 10-4. The goal of both activities is to develop a product or structure that meets the customer's needs. This task means that the product must function well, operate safely and efficiently, be easily maintained and repaired, have a pleasant appearance, and deliver good value.

In addition, products and structures must be designed so that they can be produced economically and efficiently. They must also be sold in a competitive environment. In short, the product must be designed for:
- Function–easy and efficient to operate and maintain.
- Production–easy to manufacture or construct.
- Marketing–appealing to the end user.

DEVELOPING DESIGN SOLUTIONS

System and product designs start with a clear definition of the situation or opportunity. Procedures for developing the definition were discussed in the previous chapter. This problem definition leads to the next step product design–developing design solutions. These solutions often evolve through three steps, as shown in Fig. 10-5.
- Developing preliminary solutions.
- Refining the best solution.
- Detailing the best solution.

This process can be described as "imagineering." First, the designer's *imagination* is used to develop a number of unique solutions or designs. These solutions are then *engineered* back to reality through design refinement and detailing activities. This step starts with broad thinking. This kind of thinking is called **divergent thinking**. It seeks to think of as many different (divergent) solutions as possible. The most promising solutions are then refined and reduced until one "best" answer is found. The refinement of ideas requires **convergent thinking**. The goal is to narrow and focus (converge) the ideas until the most feasible solution is found.

Keep in mind that the *best* solution may not be the one that works best or is the least expensive. Trade-offs

TYPES OF PRODUCT DESIGN

MANUFACTURED PRODUCT

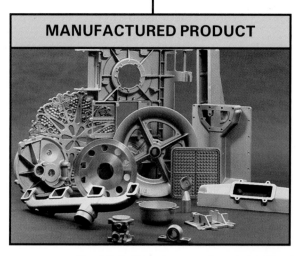

CONSTRUCTED STRUCTURE

Fig. 10-4. Structures and manufactured parts are designed with product design procedures.

between appearance, function and cost are often made. Many of us cannot always afford the very best answer. Our budget and the length of time we expect to keep the product enters into our product choices. For example, it may be unwise to purchase a $900 racing bicycle to occasionally ride to the corner store. However, if you regularly ride a bicycle to work or go on cycling vacations, the expense may be justified. Likewise, a camera with a high-quality lens, automatic film advance, and complex shutter controls is beyond the needs of the "snapshot" photographer. It would be the camera of

choice for a professional photographer. In fact, a "use once and throw away" camera may be the best solution for the person who takes very few photographs.

Product design activities produce a wide range of products. This allows the consumer to select one that meets his or her performance needs and financial resources.

DEVELOPING PRELIMINARY SOLUTIONS

Designs start in the minds of designers, engineers, or architects. There are a number of ways that ideas can be stimulated. Three popular techniques are brainstorming, classification, and "what if" scenarios.

Brainstorming

Brainstorming is a process that requires at least two people. However, groups of three or more participants are better. **Brainstorming** involves *seeking creative solutions to an identified problem*. Members of the group offer individual solutions to that they think will work. Proposed solutions will often cause other members of the group to think of more ideas. The strategy uses a concept called synergism. It builds on the individual contributions of the participants to make a larger whole. The number of ideas generated by the group is more than the number they could develop if everyone worked alone.

Brainstorming activities work best when the group accepts some basic rules. These include:

- *Encourage wild, far out ideas.* There are no "bad" or "stupid" ideas. Wild, but promising, ideas can always be engineered back to reality.

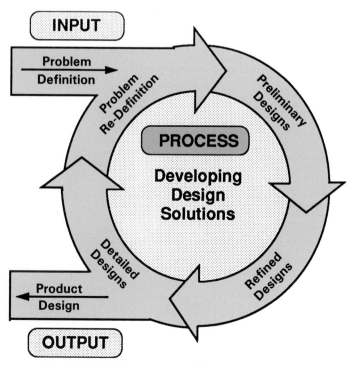

Fig. 10-5. This is a model of the product design process.

- *Record the ideas without reacting to them.* Criticizing ideas will cause people to not offer ideas. They will provide only those that the group will like in order to avoid criticism.
- *Seek quantity, not quality.* The chances of good ideas emerging are increased as the number of ideas increase.
- *Keep up a rapid pace.* A rapidly paced session will keep the mind alert and reduce the chance of judging the ideas.

Classification

Classification can be done by one person, or a group of people. **Classification** involves dividing the problem into major segments. Then each segment is reduced into smaller parts. A classification of buildings could be homes, business and commercial, and industrial. Homes may be further classified as houses, apartments, condominiums, mobile homes, etc. A house can be classified by its major features: foundations, floors, walls, ceilings, roofs, doors, windows, etc. Foundations can then be classified as poured concrete, concrete block, wood posts, timber, etc. This processes may result in a classification chart. This type of chart is often developed as a tree chart with each level having a number of branches below it, as shown in Fig. 10-6. It would end up looking much like a family tree that people use to trace their ancestors.

''What If'' Scenarios

"What if" scenarios start with a wild proposal. Then its good and bad points are investigated. The good points can be used to develop solutions. For example, peeling paint is a problem for house painters. They must remove the old paint from a house before repainting. A wild solution would suggest mixing an explosive material with the paint before it is applied. Whenever the building is ready to be repainted the old paint could be blown off the building. Obviously, exploding house paint is ridiculous. It could, however, lead to a solution. Paint sticks to a house through the adhesion between the paint and the siding. Maybe a material could be mixed with the paint that would cause it to lose adhesion when a special chemical is applied. At repainting time, the chemical could be sprayed on to loosen the paint. Then the paint could the be easily removed from the siding.

ROUGH SKETCHING

Once the designer has conceived a number of ideas in her or his mind they must be recorded. The most common recording method is to develop **rough sketches** of the products, structures, or system components, as shown in Fig. 10-7. These sketches are as much a part of the thinking processes as they are a communication media. The designer is forced to think through concepts such as size, shape, balance, and appearance. The sketches then become a library of ideas for later design efforts.

The term "rough" is not used to describe the quality of the drawing. Rough sketches are not necessarily crude. They often represent good sketching techniques. The term rough describes the state of the design ideas. It suggests that the designs are incomplete and unrefined.

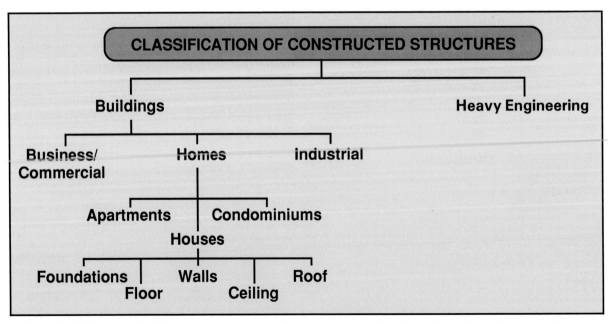

Fig. 10-6. This is a classification chart of constructed structures. This type of chart is called a tree chart.

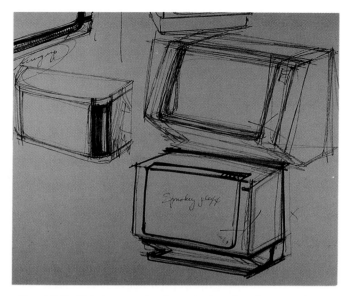

Fig. 10-7. This is a group of rough sketches for a product.

REFINING DESIGN SOLUTIONS

The rough sketches allow designers to capture a wide variety of solutions for the design problem or opportunity. The sketches are like books in a library: they contain a number of different thoughts, views, and ideas. They can be selected, refined, grouped together, or broken apart.

Refining original designs in the "library of ideas" is the second step in developing a design solution, Fig. 10-8. Promising ideas are studied and improved. This process may involve working with one or more good rough sketch. The size and shape of the product or structure may be changed and improved. Details may be added and the shape may be worked on. In short, the design is becoming refined as problems are worked out and the proportions become more balanced.

Refined design ideas may also be developed by merging ideas from two or more rough sketches into a **refined sketch**. The overall shape may come from one sketch while specific details come from others. This approach is one of integration, which blends the different ideas into a unified whole. The new idea may not look anything like the original rough sketches.

DETAILING DESIGN SOLUTIONS

Rough and refined sketches do not tell the whole story. Look back at the sketches shown in Fig. 10-8. What size is the product in the sketch? You can't tell. The sketches communicate shape and proportion. They do not communicate size. For this task, a third type of sketch is needed, called a **detailed sketch**. It communicates the information needed to build a model of the product or structure. Detailed sketches can also be used as a guide to prepare engineering drawings for manufactured products and architectural drawings for constructed structures. Engineering and architectural drawings will be discussed in Chapter 12.

Detailed sketches are helpful when models of products or structures are made. Building models requires three major types of information. These are shown on the drawing in Fig. 10-9 and are:

- *Size information* explains the overall dimensions of the object, or the size of features on an object. This information might include the thickness, width, and length of a part, or the diameter and depth of a hole, or the width and depth of a groove.
- *Location information* gives the position of features within the object. This information may establish the location of the center of a hole, the edge of a groove, or the position of a taper.

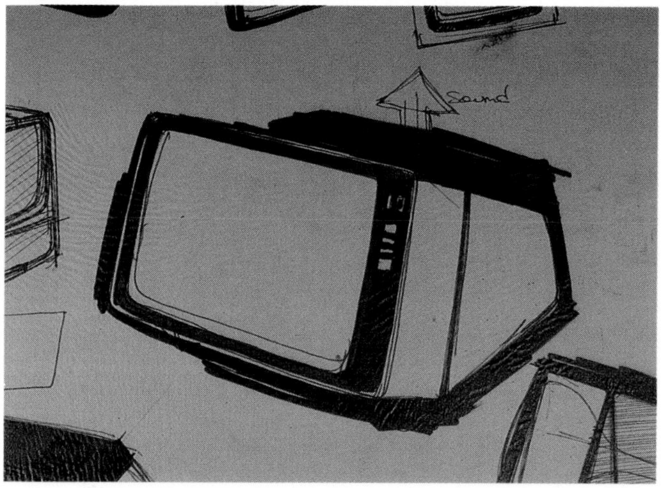

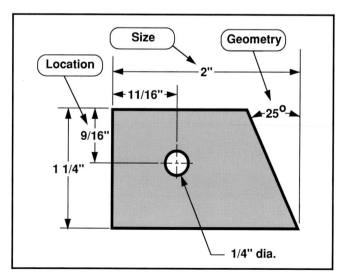

Fig. 10-8. This is a refined product sketch. Notice how the shape has been worked out and that shading has been added to improve the looks of the sketch. (RCA)

Fig. 10-9. The types of information provided on detailed sketches are location, size, and geometry.

- *Geometry information* describes the geometric shape or relationship of features on the object. This information could communicate the relationship of intersecting surfaces (square, 45° angle, etc.), or the shapes of holes (rectangular, round, etc.) or the shapes of other features.

DEVELOPING PICTORIAL SKETCHES

Designers often use pictorial sketching techniques to capture and further refine product design ideas. These techniques try to show the artifact much like the human eye would see it. Therefore, a single view is used to show how the front, sides, and top would appear.

Three basic techniques are used to develop pictorial sketches. These are:
- Oblique sketches.
- Isometric sketches.
- Perspective sketches.

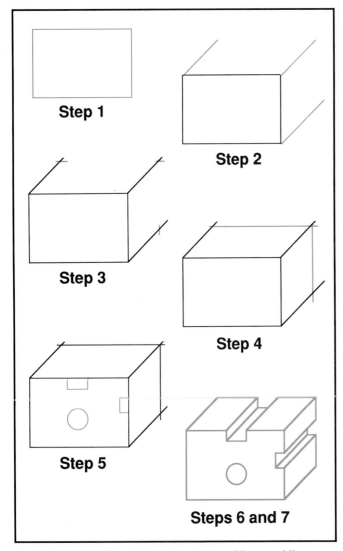

Fig. 10-10. These are the steps in making an oblique sketch.

Developing Oblique Sketches

Oblique sketches are the easiest pictorial sketch to produce. An **oblique sketch** shows the front view as if a person was looking directly at it. The sides and top extend back from the front view. They are shown with parallel lines that are generally drawn at 45° to the front view.

To produce an oblique sketch, the designer completes steps like those shown in Fig. 10-10:
1. Lightly draw a rectangle that is the overall width and height of the object.
2. Lightly extend parallel lines from each corner of the box back at 45°.
3. Lightly mark the extension lines at a point equal to the depth of the object.
4. Lightly connect the depth lines to form a box.
5. Add any details such as holes, notches, and grooves onto the front view.
6. Extend the details the depth of the object.
7. Complete the sketch by darkening in the object and detail outlines.

The procedure listed above will produce a cavalier oblique drawing. This type of drawing causes the sides and top to look deeper than they are. To compensate for this appearance cabinet oblique drawings are often used, as shown in Fig. 10-11. This type of drawing shorten the lines that project back from the front to one-half their original lengths.

Developing Isometric Sketches

Isometric sketches are the second type of pictorial drawings used to produce refined sketches. The word isometric means equal measure. **Isometric sketches** get their name because the angles formed by the lines at the upper right corner are equal–each is 120°.

Isometric sketching is used when the top, sides, and front are equally important. The object is shown as if it were viewed from one corner.

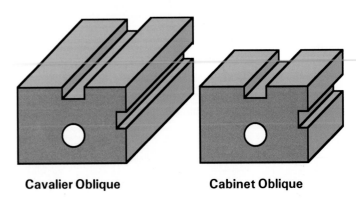

Cavalier Oblique   **Cabinet Oblique**

Fig. 10-11. The two types of oblique drawings are cavalier and cabinet. Cabinet drawings use one-half the depth of the object for a more natural appearance.

Technology: How it Works

ELEGANT SOLUTION: A product that meets a human need in the simplest, most direct way.

We are surrounded by complex products and devices. Many of these have all kinds of frills and little "add-ons" that designers thought were necessary. However, many of these add-ons complicate the design and make products hard to use.

The opposite of this is the product that meets a need in the simplest and most effective way. Engineers call these *elegant solutions*. They should be the goal of every designer of technological artifacts.

Think of an adhesive bandage, commonly known as the Band-Aid®. It is designed to hold gauze over a small cut or scratch. Can you think of a better solution? Over the years no one has thought of one and, therefore, the Band-Aid is an elegant solution.

Another commonplace product is the paper clip, Fig. 1. It temporarily holds sheets of papers together. Everyone uses paper clips and no one has improved on them. Again, this is an elegant solution.

Think of a number of other devices that work so well that improving them is a challenge. These devices include the zipper, pin hinge, paper stapler, ballpoint pen, pipe cleaner, and thong sandal. Small details may change, such as the types of materials used, but the basic design remains the same.

However, elegant solutions do not have to be simple products. There are many examples of complex technological products that elegantly solve problems.

Consider the Apollo spacecraft, Fig. 2. It carried three men into space and back with relative ease. Its complex systems propelled and navigated the capsule. The spacecraft also provided heat, light, and fresh air for the three astronauts inside. The capsule also protected the astronauts from the outside environment.

Elegant solutions are everywhere. Have you ever tried to read and record large volumes of numerical data? It is time-consuming process and prone to error. A bar code

Fig. 2. The Apollo spacecraft is an example of an elegant solution. Complex devices can also be elegant solutions.

reader, such as the one shown in Fig. 3, makes this task quick and accurate. The unit shown is the size of a hand-held calculator. Despite its small size, the device efficiently reads and processes the information coded in the bars. Bar code readers are widely used to take inventory in retail stores.

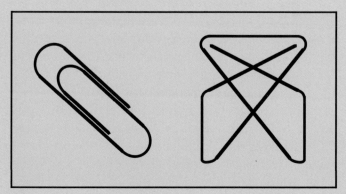

Fig. 1. The paper clip is an example of an elegant solution.

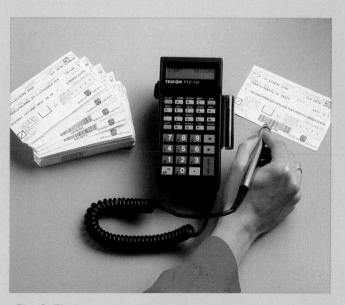

Fig. 3. The bar code reader is still another elegant solution to a modern problem.

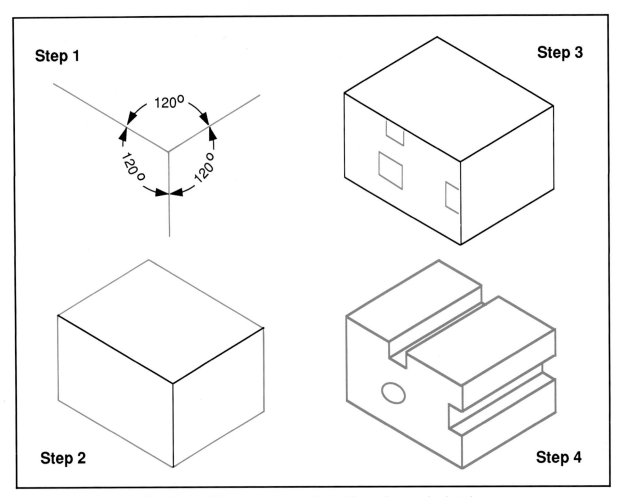

Fig. 10-12. These are the steps in making an isometric sketch.

To produce an isometric drawing, four major steps are required. These, as shown in Fig. 10-12, include:

1. Lightly draw the upper right corner of an isometric box that will hold the object.
2. Complete the box by lightly drawing lines parallel to the three original lines.
3. Locate the major features such as notches, tapers, and holes.
4. Complete the drawing by darkening the features and darkening the object outline.

Developing Perspective Sketches

Perspective sketches show the object as the human eye or a camera would see it. This realism is obtained by having parallel lines meet at a distant vantage point. When you look down a railroad track you will see a similar effect. The rails remain the same distance apart, yet your eye sees them converge (come together) in the distance.

Developing the perspective, or "human eye" view is more difficult to draw than oblique or isometric views. However, perspectives are the most realistic of the three pictorial sketches.

Types of Perspective Views. There are three major types of perspective views: one-point, two-point, and three-point. The difference between these types is determined by the number of vanishing points used, Fig. 10-13.

One-point perspective shows an object as if you were directly in front of it. All the lines extending away from the viewing plane converge at one point. The one-point perspective is like an oblique drawing with tapered sides and top.

Two point perspective shows how an object would appear if you stood at one corner. It is constructed much like an isometric drawing. Again, the sides are tapered as the lines extend toward the vanishing points.

A three-point perspective shows how the eye sees the length, width and height of a object. All lines in this drawing extend toward a vanishing point.

The appearance of a perspective drawing will change as the horizon changes, Fig. 10-14. Changing the position of the horizon line can cause the object to be seen as if the observer was looking down on the object (aerial view), directly at it (general view) or up at it (ground view). The designer must decide which of these views best suits the object and the audience that will see the sketch.

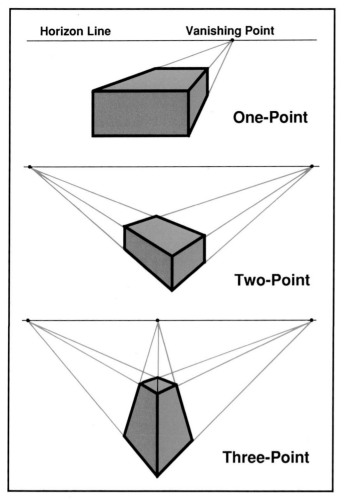

Fig. 10-13. The types of perspective drawings are one-point, two-point, and three-point.

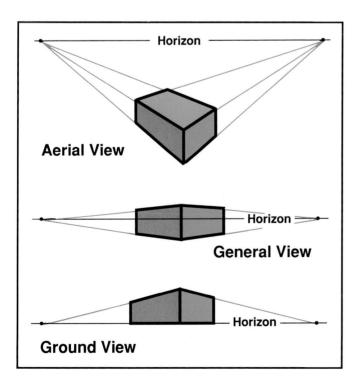

Fig. 10-14. Changing the location of the horizon will change the appearance of a perspective drawing.

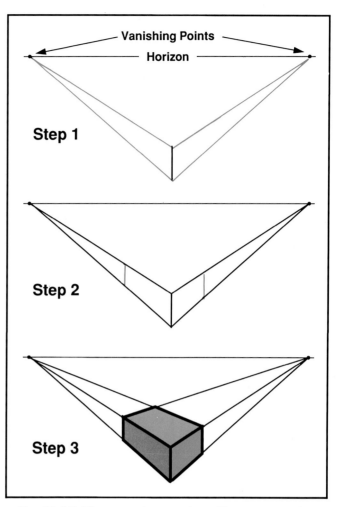

Fig. 10-15. These are the steps in making a perspective sketch.

Developing the basic structure for one, two, or three-point perspective sketches follow the same basic steps. These, as shown in Fig. 10-15, are:

1. Establish the horizon line, vanishing points and the front of the object. Connect the front line(s) to the vanishing point(s).
2. Establish the depth of the objects along the lines that extend to the vanishing point(s).
3. Connect the depth lines to the vanishing point(s). Darken in the object.

Details are then added to complete the sketch. Often perspective sketches are shaded to add to their communication value.

SUMMARY

Design solutions are the result of the work of designers and architects. They must first study the definition of the problem or opportunity. Then, they must generate a number of possible solutions. This is often done using rough sketches. These sketches become a library of design ideas. From this library, specific ideas are selected and refined to bring the solution into focus. Finally, the refined ideas are described through detailed sketches. These sketches are the foundation for making models of the designs and evaluating them. The evaluation of designs will be the subject of the next chapter in this book.

WORDS TO KNOW

All of the following words have been used in this chapter. Do you know their meanings?

Brainstorming
Classification
Convergent thinking
Detailed sketch
Divergent thinking
Isometric sketch
Oblique sketch
Perspective sketch
Refined sketch
Rough sketch
"What if" scenario

TEST YOUR KNOWLEDGE

Write your answers on a seperate sheet of paper. Please do not write in this book.

1. List the three types of pictorial sketches that can be used to design a product.
2. True or false. An architect uses product design principles to design a waste water system for a home.
3. Match the types of drawings on the right to the correct statement(s) on the left:

    _____ An equal measure drawing.   A. Isometric.
       B. Oblique.
    _____ Shows the front view as if you were looking directly at it.   C. One-point perspective.
    _____ Sides extend back at 45°.   D. Two-point perspective.

_____ upper right corner of a cube forms 120° angles.
_____ has two vanishing points.
_____ shows a railroad track as you would see it if you were standing in its middle.

4. _____ sketches form the library of ideas during the design process.
5. True or false. Products should be designed with function, production, and marketing in mind.
6. _____ perspective drawings will best show a skyscraper.
7. List the three types of perspective drawings.

APPLYING YOUR KNOWLEDGE

1. Develop a set of rough sketches for the following design definition:

> Problem or opportunity: The director of the school cafeteria would like a holder that would contain a salt shaker, a pepper shaker, 20 rectangular (1 x 1 1/2″) packages of sugar, and a bottle of ketchup. The holder should be easily removed from the table at the end of the lunch period.

2. Refine the best sketch that you produced for the lunchroom table organizer.
3. Develop a detailed sketch for your lunchroom table organizer.
4. Select a device in the technology laboratory and develop a perspective sketch of it.

11
CHAPTER

EVALUATING DESIGN SOLUTIONS

After studying this chapter you will be able to:
- *Define a model.*
- *Describe how models are used in the design of technological devices and systems.*
- *List the three major types of models.*
- *Describe, construct, and use graphic models.*
- *Describe, construct, and use mathematical models.*
- *Describe, construct, and use physical models.*
- *List and discuss the types of analysis used to evaluate a design.*
- *Analyze a product design.*

We use the products of technology every day. Each product started with a problem or an opportunity that could be defined. Then, designers were challenged to solve this problem. A number of different ideas were explored. These solutions started in the designers' mind and were recorded on paper. Rough, refined, and detailed sketches were developed. These sketches contained design ideas, not product plans. At some point, two-dimensional drawings had to be made into three-dimensional models that could be evaluated, Fig. 11-1. This conversion from drawings to models is the focus of this chapter and includes three basic activities:
- Modeling the solution.
- Evaluating the model.
- Redesigning the product or structure.

MODELING DESIGN SOLUTIONS

Everyone is familiar with models. Children play with dolls, model cars, and toy trains. They build towns out of wooden blocks, convert cardboard boxes into frontier forts, and have grand prix races with their tricycles. They pretend that they are dealing with real-life situations.

SKETCHES

MODEL

Fig. 11-1. The design process changes product sketches (left) into models (right) that can be evaluated. (Pontiac)

Likewise, museums use models to show how things work or the events of the past, Fig. 11-2. These models show a slice of historical life so we can better understand the present day.

This activity of imitating reality is widely used in product and system design. People simulate expected conditions to test their design ideas. This process is called **modeling** or simulation.

Simply stated, modeling is the activity of *simulating actual events, structures, or conditions*. For example, architects may build a model of a building to show clients how it will look. They may use structural models to test a building's ability to withstand an earthquake and the forces of the wind. Economists may devise a model to predict how the economy will react to certain conditions. Weather forecasters use models to show the public the location and movement of storms. They also use sophisticated computer models to predict the intensity and movement of hurricanes, Fig. 11-3.

A model allows us to reduce complex mechanisms and events into an easily understood form. They allow us to focus on important parts of the total problem. This focus permits us to build understanding one part at a time. For example, an automobile is a very complex device. It is almost impossible to study and understand it as a whole. However, if you look at and understand

Fig. 11-3. This forecaster is viewing a computer model of a storm. (Harris Corp.)

the systems that make up the automobile one at a time, the whole becomes clear. You can model and study its power train, cooling system, electrical system, lubricating system, or its suspension system, Fig. 11-4.

There are literally thousands of models used each day. Each of them starts as a *conception*, an idea in the human mind. This idea is often communicated through verbal descriptions. These written or oral descriptions provide the foundations for the development of models. Each model is designed to help people understand what a device looks like and how it functions, or how various events take place.

All models, as shown in Fig. 11-5, can be grouped into three types:
- Graphic models.
- Mathematical models.
- Physical models.

Fig. 11-2. This museum model shows how yarn was spun from wool in the colonial period.

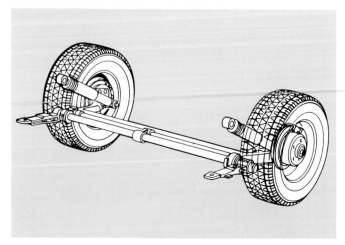

Fig. 11-4. This is a graphic model of a automotive torsion bar rear suspension system. (Chrysler Corp.)

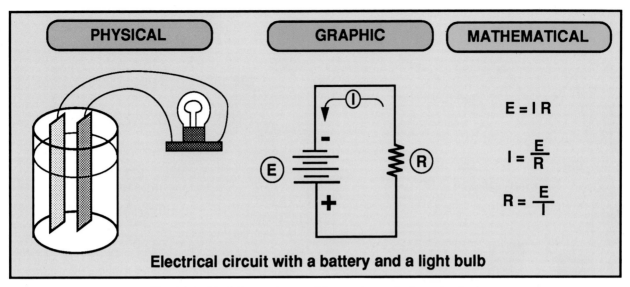

$$E = I R$$

$$I = \frac{E}{R}$$

$$R = \frac{E}{I}$$

Electrical circuit with a battery and a light bulb

Fig. 11-5. Models may be graphic, mathematical, or physical.

GRAPHIC MODELS

Designers cannot make physical models early in the product and structure development process. They do not have enough information about the design to construct a physical model. However, they must explore ideas for components and systems. One way to do this is to use a **graphic model**. Typical graphic models are conceptual drawings, graphs, charts, and diagrams. Each of these graphic models serve a specific purpose.

Conceptual models capture the designer's ideas for specific structures and products. They show a general view of the components and their relationships, Fig. 11-6. Conceptual models are often the first step in evaluating a design solution. Relationships and working parameters of systems and components can be studied, modified, and improved using conceptual models. The refined and detailed sketches discussed in the previous chapter could serve as conceptual models.

For example, conceptual models could be developed for a toy train. These models would explore ways to connect the cars together, fabricate wheel and axle assemblies, and attach the car bodies to chassis assemblies.

Graphs allow designers to organize and plot data. **Graphs** display numerical information that can be used to design products and assess testing results.

For example, a graph can be developed that shows vehicle speed and braking distance for different types of brakes. The data can be charted on a line graph, Fig. 11-7. This information will help designers select the type of braking system to be used in a specific vehicle. Likewise, plotting data on the colors of shirts purchased during a specific period can help designer select colors for next year's products. This type of data can be shown of either a bar graph or a pie graph.

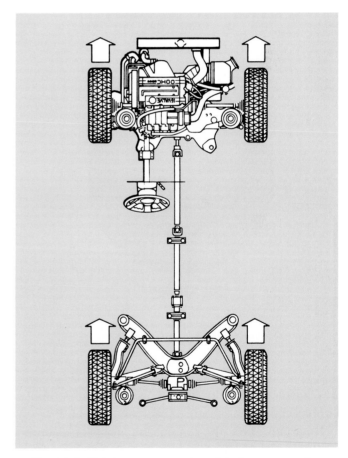

Fig. 11-6. This is a conceptual model of an automotive four-wheel drive system. (Chrysler Corp.)

Charts show the relationship between people, actions, or operations. They are useful in selecting and sequencing tasks needed to complete a job. A number of different charts are used for specific tasks. Flow process charts help computer programmers write logical programs. Flow charts help the manufacturing engi-

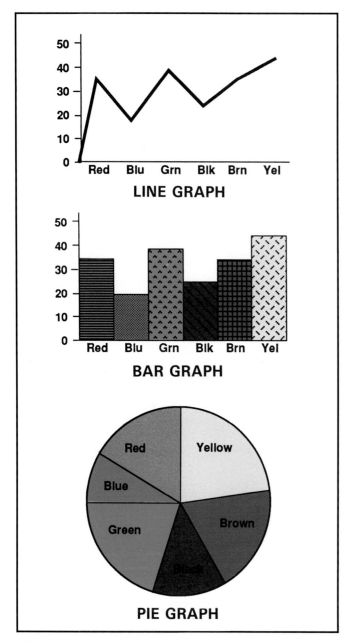

LINE GRAPH

BAR GRAPH

PIE GRAPH

Fig. 11-7. Graphs are used to show the relationship among mathematical data gathered about specific factors.

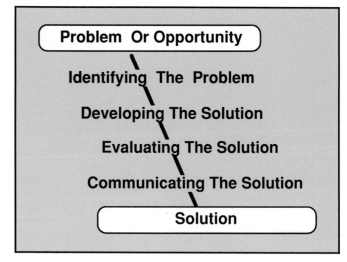

Fig. 11-8. Charts can be used to show the sequence of operations or the structure of organizations.

Another type of diagram that many people see is a play diagram. It is developed by a football coach to show how the players (components) should interact during an offensive or defensive play (system). Analysts on television often use this type of diagram to show what happened during the game. The play diagram summarizes the purpose of all diagrams: show how something is designed to happen or how an event took place.

MATHEMATICAL MODELS

Mathematical models show relationships in terms of formulas. For example the relationship between voltage, amperage, and resistance in an electrical circuit is shown by the formula $E = IR$, where E = electromotive force measured in volts, I = electrical current measured in amperes, and R = electrical resistance measured in ohms.

neer develop efficient manufacturing plants. Organization charts show the flow of authority and responsibility within a company, school, or business, Fig. 11-8.

Diagrams show the relationship between components in a system. A schematic diagram can be used to indicate the components in electrical, mechanical, or fluidic (hydraulic or pneumatic) systems, Fig. 11-9. Schematics do not show specific location of the parts. Their position in the system is the important information that is communicated.

Flow diagrams show how parts move through a manufacturing facility. Lines and arrows indicate the path the material takes as it moves from operation to operation.

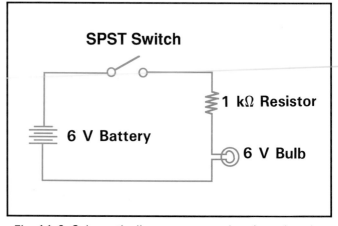

Fig. 11-9. Schematic diagrams are used to show the relationship among components in electrical, mechanical, and fluidic systems.

In a similar way, the relationship between the force needed to move an object and the distance it is lifted in shown in the formula for work: Work = Force x Distance.

The formulas used to explain chemical reactions are also mathematical models. The formula below shows the chemical reactions that take place as an automotive lead-acid storage battery is charged and discharged.

$$Pb_2O + Pb + H_2SO_4 \xrightleftharpoons[\text{charge}]{\text{discharge}} 2PbSO_4 + 2H_2$$

The examples given so far are for simple mathematical models. More complex models employing thousands of formulas are used to predict the results of complex relationships. These formula can be part of economic models which predict the economic growth for a period of time. Also, complex mathematical models track storms and space flights, predict ocean currents and land erosion, and help scientists conduct complex experiments.

PHYSICAL MODELS

A **physical model** is a three-dimensional representation of reality. There are two types of physical models: mock-ups and prototypes.

Mock-ups

The first type of physical model is designed to show people how a product or structure will look. This type

Fig. 11-10. These designers are fabricating a clay mock-up of a new automobile. (Daimler-Benz)

of model is an appearance model or **mock-up**. It is used to *evaluate the styling, balance, color, or other aesthetic feature* of a technological artifact, Fig. 11-10.

Mock-ups are generally constructed of materials that are easy to work with. Commonly these material include wood, clay, styrofoam, paper, and paperboard (cardboard, posterboard, etc.), Fig. 11-11.

Fig. 11-11. These designers are developing mock-ups of new toys. (Hasbro Inc.)

Technology: How it Works

RADAR: A system that can detect, locate, and determine the speed of distant objects with the use of radio waves.

Radar is an acronym for **ra**dio **d**etection **a**nd **r**anging. Practical radar systems were developed in the 1930s as a means of detecting enemy aircraft and ships. They were used widely during World War II. Radar systems contain three separate subsystems. First, a transmitter that produces and radiates radio signals. Second, a receiver that captures the radio signals. Finally, a display system that presents the data gathered.

The transmitter produces and sends out short bursts of powerful radio signals from an antenna, Fig. 1. These waves travel off into space unless they strike an object. When the waves strike an object, some of the signal is reflected back in the direction of the antenna. There, the signal is received and amplified.

The radar's signal is emitted in bursts. This allows a single antenna to both send and receive the signal. The radar system uses the elapsed time between the sending and receiving of the signal to determine distances of objects. Radio waves travel at the speed of light (186,000 miles per second). This allows elapsed time to be used as an accurate means to determine distances.

There are several different types of radar. The intended use of the system determines the area that the beam covers. Detection radar broadcasts a wide beam of radio waves. Ships use a flat arc that sweeps an area just above the water in order to avoid other objects in the water. An aircraft uses a radio beam that sweeps both up and down as well as right and left.

Location radar uses a narrow, flashlight-type beam. The beam is focused on an object so that accurate elevation, distance, and speed data can be obtained. This type of radar has many applications. Most people are familiar with its use by police officers to enforce speed limits.

Most radar antennas rotate to provide large viewing areas. Ground-based systems generally rotate in a 360° circle. On-board radar systems often view only the areas in the direction of the vehicle's travel.

Once the radar beam has been reflected and received, it must be displayed. Most radar systems display their information on a cathode-ray (television-type) tube. These screens may show an object's location in terms of elevation and/or distance. More complex systems show the object on an ''electronic map'' of the area, which is superimposed on the display unit.

Many commercial aviation systems use a second radar system. This system receives a signal that is broadcast from the aircraft. This signal identifies the height and identity of the aircraft.

Fig. 1. Pictured is a massive radio antenna.

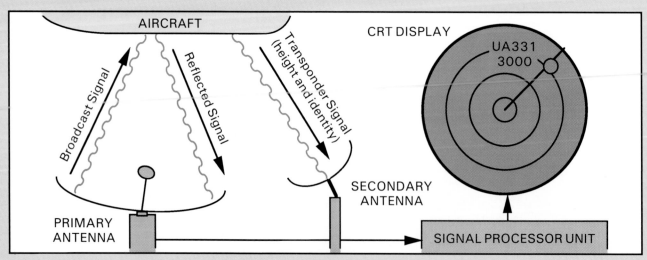

Fig. 2. This is a typical radar system used for commercial aviation.

Prototype

The second physical model is a prototype. This model is a working model of a system, assembly, or a product. **Prototypes** are built to test the operation, maintenance, and/or safety of the item, Fig. 11-12. They are generally built of the same material as the final product. However, in some cases, substitute materials are used. Some automobile manufactures have found that a specific plastic reacts to external forces in the same manner as steel does. Since the plastic prototypes are easier to fabricate, they are used in place of steel ones.

Both types of physical models may be built full-size or to scale. *Full-size* models are needed to test the product's operation. For example, a full-sized model is needed for people to evaluate to the comfort of a new bus seat.

In other cases full-sized models are impractical. Building a full-size model of a new skyscraper would be a waste of money. A scale model is used when the product or structure is too large to construct in full-size just for a test. A scale model is proportional to actual size. This means that the model's size is related to actual size by a ratio. A ratio of 4 to 1 (written as 4:1) means that four units in actual size are equal to one unit on the model. A scale model of a new building is used to show a client how the structure will look, how it fits on the site, and how it will be landscaped.

COMPUTER MODELING AND SIMULATION

The advent of the computer has made modeling much easier. In many cases, physical models are no longer needed. A computer can develop and analyze the structure or product.

Computers can be used to develop three types of three-dimensional models: wireframe, surface, and solid. Two of these are shown in Fig. 11-13. The top view is a **wireframe model.** It is developed by connecting all the edges of the object. The process produces a structure made up of straight and curved lines.

The bottom view is a **surface model**. It can be thought of as a wireframe with a sheet of plastic drawn over it. Surface models show how the product will appear to an observer. Surface models can be colored to test the effects of color on the product's appearance and acceptance.

Surface models are widely used in developing sheet metal products. The surface model is first developed. Then the computer will unfold the model to produce a cutting pattern for the metal.

Solid models are the most complex. They look like surface models except the computer "thinks" of them as

Fig. 11-12. This model is being used to test the aerodynamics of an aircraft. (Daimler-Benz)

solids. This allows the designer to direct the computer to cut-away parts, insert bolts and valves, and rotate moving parts. Solid models can be used to establish fits for mating parts and the procedures needed to assemble parts into products.

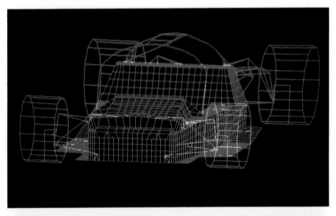

Fig. 11-13. Two types of computer models are the wireframe (above) and surface models (below). They are used here in the design of a race car. (Daimler-Benz)

Fig. 11-14. This computer program is plotting the stress placed on a structure.

Other computer simulations allow designers to test strengths of materials and structures, Fig. 11-14. They design the structure on the computer, then apply stress. These may simulate to forces created by weight, operating conditions, or outside conditions such as wind or earthquakes.

Finally, computer models may be used to observe the product during normal operation. For example, computer programs simulate the flow of metals and plastics into molds. Other programs, like the one shown in Fig. 11-15, test air circulation in buildings and vehicles. This information is essential for the designers of heating and cooling systems.

DESIGN ANALYSIS

Modeling helps designers and decision makers enter an important design activity: design analysis. This stage requires people to carefully evaluate a design in terms of:
- Function.
- Specifications.
- Human factors.
- Market acceptance.
- Economics.

FUNCTIONAL ANALYSIS

Every product, structure, or technological system is designed to meet human needs or wants. Functional analysis evaluates the degree to which the product meets its goal. It answers the basic question: "Will the device operate effectively under the conditions for which it was designed?" These conditions may relate to the outside environment (weather, terrain, water, etc.), operating conditions (stress, heat, gases, etc., produced during use), human use and abuse, and normal wear (material fatigue, part distortion, etc.).

SPECIFICATION ANALYSIS

Every product must meet certain specifications. These specifications may be given in terms of size, weight, speed, accuracy, strength, or a number of other factors. The specifications for all new products and structures must be analyzed. It is important that they are adequate for the function of the product. However, excessive specifications will add to the cost of the product. Holding bicycle handle bar diameters to 1/1000" (.025 mm) is foolish. That level of precision is not needed for the part to function. However, close tolerances are required for spark plug threads. The goal is to produce an economical, efficient, and durable product that will operate properly and safely.

Specifications must also relate to the material and manufacturing processes to be used. For example, holding wood parts to tolerances smaller than 1/64" (.397 mm) is impossible. Normal expansion and contraction due to changes in humidity can cause this much change. Likewise, specifying aluminum for the

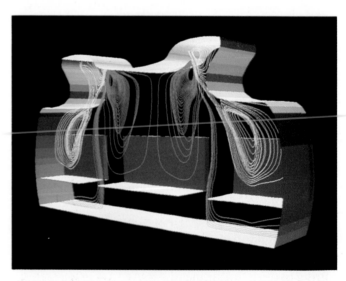

Fig. 11-15. This is a computer simulation of the air circulation in the interior of an aircraft. Note the seats (in purple) and the shape of the overhead bins. (Boeing)

internal parts of a jet engine is unwise. The temperatures inside the engine will melt aluminum parts. Similarly, specifying green sand casting as a process to produce precision parts is a mistake. Green sand casting produces low-cost parts, but they cannot be held to close tolerances.

HUMAN FACTORS ANALYSIS

All products are designed to meet human needs. Devices and structures must be designed for the people who will use them, travel in them, or live and work in them. Designing products and structures around the people who use them is the focus of *human factors analysis*, or more commonly known as **ergonomics**. This science considers the size and movement of the human body, mental attitudes and abilities, and senses such as hearing, sight, taste, and touch.

Ergonomics also considers the type of surroundings that are the most pleasing and help people to become more productive. A good example of matching the environment to humans is an aircraft flight deck, Fig. 11-16. All the controls are within easy reach. Dials and indicator lights are within the pilot's field of vision. Windows are located so that the pilots have a clear view of the sky ahead and above them.

MARKET ANALYSIS

Most products of technology are sold to customers. These customers may be the general public, government agencies, or businesses. During design activities the market for the product must be studied. Then the designs must be analyzed in terms of that market. Market analysis will include finding customer expectations for the product's appearance, function, and cost. It will also include studying present and anticipated competition. Often market analysis is done by taking surveys of potential customers and analyzing competing products that are available on the market.

ECONOMIC ANALYSIS

As we said earlier, most technological devices and structures are developed by private companies. They risk money to develop, produce, and market the items. In turn, they hope to make a profit as a reward for their risk-taking, or *entrepreneurship*. To increase their chances of success, a financial analysis for the new products is made. The product is studied in terms of the costs of development, production, and marketing. These data are compared with expected sales income to determine the financial wisdom for producing the product. Often the product is judged on its return on investment (abbreviated ROI). This figure indicates the percentage of return based on the money invested in developing, producing, and selling the product. The higher the ROI, the better the anticipated financial returns to the company.

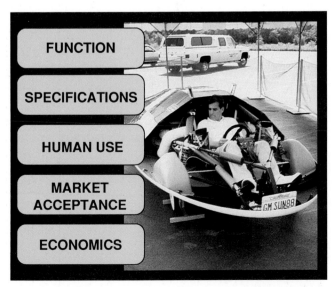

Fig. 11-16. Product designs must be analyzed in each of the five areas shown above. (GM-Rochester)

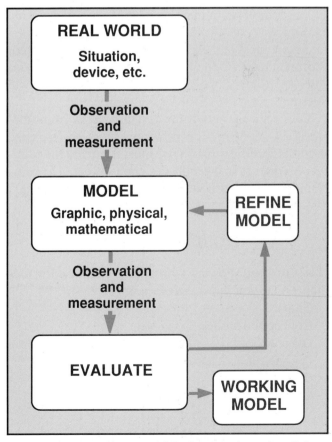

Fig. 11-17. Models are developed and evaluated until they simulate reality, or meet stated design criteria.

PRODUCT OR STRUCTURE REDESIGN

The final goal of all product and structure development activities is to design an artifact that will help people control or modify the environment. This goal requires a continuing processes. Needs are identified and defined. Solutions are designed and modeled. The designs are analyzed and flaws are identified. Redesign is often required, Fig. 11-17. Problems in the original design must be solved. New designs may be developed or the original designs may be altered. New models will be built and evaluated until an acceptable product or structure emerges.

However, this is not the end. Products have a life expectancy. New technologies and changes in people's needs and attitudes make products obsolete. Old products will need redesigning and new product definitions will appear. Product and structure design is a never-ending processes.

SUMMARY

Designers develop product ideas to meet human needs and wants. These designs must be evaluated. This requires that graphic, mathematical, or physical models be built. The models allow designers to study and refine specific mechanisms and the total product.

The completed model is subject to careful analysis. Its function, specifications, ergonomic qualities, market acceptance, and economic qualities are evaluated. Shortcomings in the design are addressed through redesign activities. Finally, an acceptable design will emerge.

This is not the end of the design process. Manufacturing and construction personnel must be informed about its specifications. They must fabricate the product or structure to meet the designer's intent. The communication of specifications is the subject of the next chapter.

WORDS TO KNOW

All of the following words have been used in this chapter. Do you know their meanings?

Chart
Conceptual model
Diagram
Ergonomics
Graph
Graphic model
Mathematical model
Mock-up
Modeling
Physical model
Prototype
Solid model
Surface model
Wireframe model

TEST YOUR KNOWLEDGE

Write your answers on a seperate sheet of paper. Please do not write in this book.

1. List three ways models are used in technological design.
2. True or false. Models can be used to study part of a total problem or mechanism.
3. Match the statements on the left with the appropriate type of model on the right:

   _____ Appearance model.   A. Mock-up.
   _____ Test the operation of a   B. Prototype.
   product.
   _____ Functional model.
   _____ Test the safety of the product.
   _____ Shows only styling features.

4. List the three types of models.
5. True or false. Product redesign is a common activity.
6. Matching products to the sizes and abilities of humans uses the science of _____.
7. List five major types of analysis to which product models are subjected.
8. List and describe the three types of computer models.

APPLYING YOUR KNOWLEDGE

1. Develop a graphic model, and build a physical model of a microwave communication tower. Assume the toothpick represents a piece of steel that is 15 feet long. Build the tower of that it is 10″ square at the bottom and 60″ high.
2. Select a simple game that is on the market. Analyze it in terms of:
   - its specifications by developing a set of manufacturing specifications.
   - its function and market acceptance by playing it with classmates. Write a brief report that summarizes your analysis.

12
CHAPTER

COMMUNICATING DESIGN SOLUTIONS

After studying this chapter you will be able to:
- Discuss the importance of communicating design solutions accurately.
- List the three ways that design solutions are communicated.
- Describe the three types of engineering drawings.
- Describe and construct a detailed drawing using the multi-view method.
- Describe and construct a pictorial assembly drawing.
- Describe and construct a systems drawing.
- Describe the types of information contained on a specification sheet.
- Describe and prepare a bill of materials for a simple product.
- List and describe the types of reports used to gain approval for designed products.

In the 1790s, Eli Whitney revolutionized the way products are made. He developed a system for mass producing muskets for the Army. The foundation of his new system was the concept of *interchangeable parts*. A part made for one gun would fit all other guns of the same make and model.

Today, we take interchangeable parts for granted. If you break a part of a product you expect to be able to buy a replacement. This ability requires a well-developed communication system between those who develop and engineer the product and those who make the parts and assemble the product, Fig. 12-1.

TYPES OF PRODUCT DESIGN COMMUNICATION

The workers who make the parts and assemble them into products must be well informed. They must have knowledge of manufacturing processes. They must be able to set up and operate machines, apply finishing

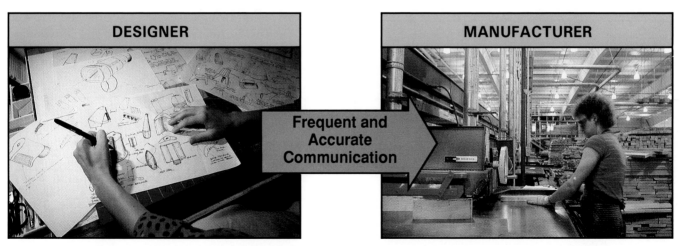

Fig. 12-1. The design for the product must be communicated from the designers to the manufacturer.
(Ohio Art Company and American Woodmark)

materials, and perform assembly operations. However, this is only part of the manufacturing knowledge they need. They must "know" the product. This means they have knowledge about the materials to be used in its manufacture. The workers must also know the size and shape of each part. Finally, they must know how the product is assembled from parts and fasteners (bolts, screws, rivets, etc.).

This knowledge of the product is delivered through three basic kinds of documents. They are, as shown in Fig. 12-2:

- Engineering drawings.
- Bills of materials.
- Specification sheets.

Let's look at how each of these are developed and the information each one communicates.

ENGINEERING DRAWINGS

Engineering drawings communicate basic information needed to construct the product or structure. In manufacturing they are called engineering, or working drawings. However, in construction they are called architectural drawings. This chapter will focus on manufacturing drawings, realizing that architectural drawings are quite different. Nevertheless, the basic principles used to prepare both are very similar.

There are three common types of engineering drawing used to communicate product information. These are, as shown in Fig. 12-3:

- Detail Drawings: Drawings that show specific information needed to produce a part.
- Assembly Drawings: Drawings that show how parts go together to make a sub-assembly or product.
- Systems Drawings: Drawings that show the relationship between electrical, hydraulic, or pneumatic components.

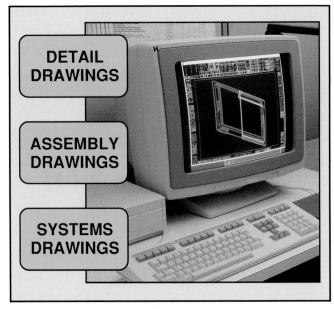

Fig. 12-3. These are the types of engineering drawings. These drawings can be generated by computer or by hand.

Detail Drawings

Most products are made up of several parts. Each of these parts must be manufactured to meet the designer's specifications. These specifications are often communicated on **detail drawings**. Commonly detail drawings contain *all the information needed to manufacture one part*. Therefore, there are generally a number of different detail drawings for a complete product.

Most detail drawings are prepared using the *multiview* method. This drawing method places one or more views of the object in one drawing, as shown in Fig. 12-4. The number of views will depend on how complex the part is. The most common multi-view drawings, as shown in Fig. 12-5, are:

- **One-view drawings** are used to show the layout of flat, sheet metal parts. The thickness of the material is listed on the drawing.
- **Two-view drawings** are used to show the size and shape of cylindrical parts. The front and top view are generally identical. Therefore, only one is needed. The two views shown are the front view, which shows the features along the length of the part, and the end view.
- **Three-view drawings** are used to show the size and shape of rectangular and complex parts. Generally a top, right side, and end view are shown. This arrangement is called third-angle projection and is used in the United State, Britain, and Canada. Other countries use first-angle projection which shows the top, front, and left side views. For very complex parts additional views, called *auxiliary views*, may be needed.

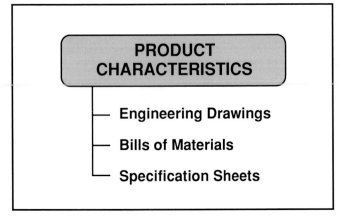

Fig. 12-2. These are the types of documents that communicate product characteristics from the designers to the manufacturing staff.

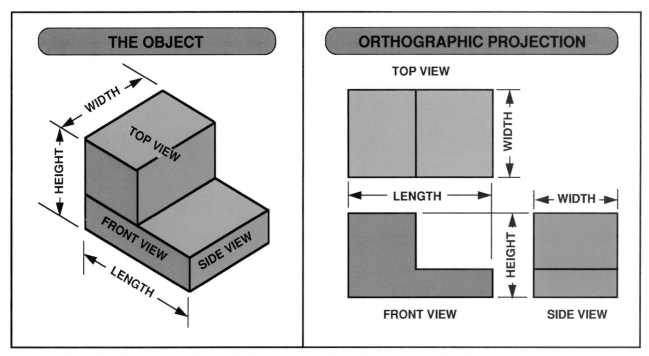

Fig. 12-4. The orthographic projection shown on the right represents the object shown on the left.

In all cases, the least number of views are used. Creating unnecessary views costs time and money. Each drawing must meet the "acid test" of all engineering drawings. It should communicate all of the information needed to make the part. The designer and the drafter will not be on the manufacturing floor to answer questions or supply missing information.

Preparing Multi-View Drawings. Multi-view drawings use **orthographic projection** to project informa-

tion at right angles to new views. Drafters complete several steps as they prepare a three-view drawing. The most important of these, as shown in Fig. 12-6 are:

1. The object is studied to determine the best way to show it. The surface that has the most detail is chosen to be shown in the front view. The goal is to have as few hidden details as possible.
2. The front view is drawn in the lower left quadrant of the paper. An accepted practice is to construct a

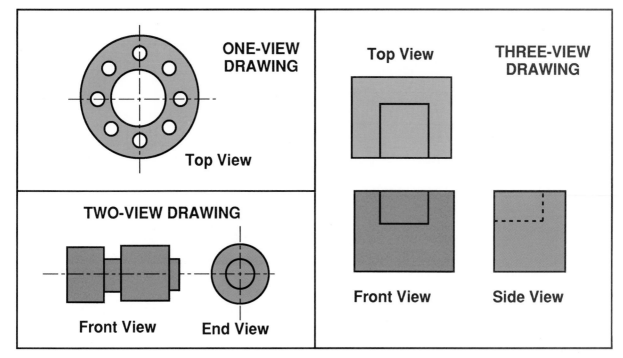

Fig. 12-5. These are the types of multi-view drawings.

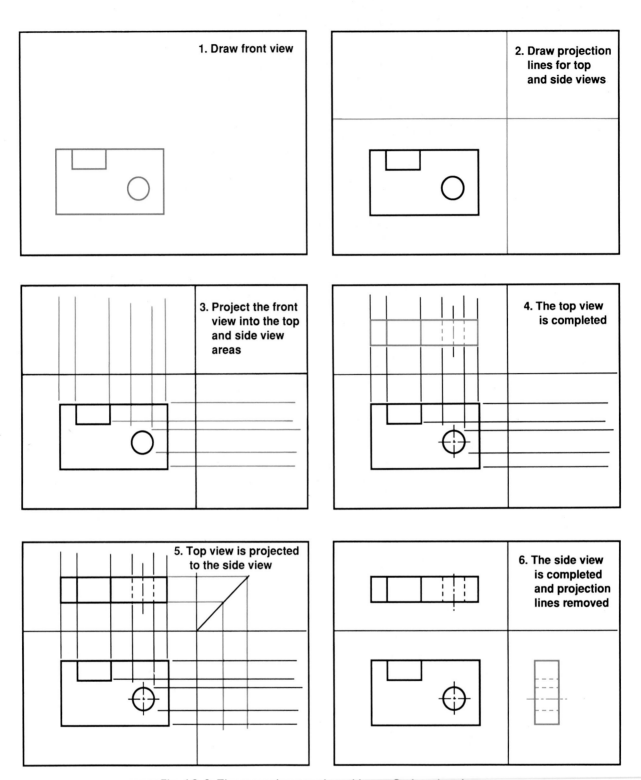

Fig. 12-6. These are the steps in making an 3-view drawing.

box that will enclose the object using light construction lines. Then details on the object are located and lightly drawn.

3. Projection lines are lightly drawn above and to the right of the front view. These lines are used to project the size of the object to the other views.

4. At the same distance above and to the right of the front view, lines are lightly drawn (perpendicular to the projection lines) as the place to start the right side and top views.

5. The outline and details of the top view are lightly drawn.

6. A 45° line is drawn in the upper right quadrant. The outline and features of the top view are projected to the side view. This is done by projecting sizes and features to the 45° line, and then down to

cross the projection lines drawn earlier from the front view. The side view will appear where these projection lines cross. Like those in step 4, these lines should be very light so they can be erased.

7. The side view is completed by constructing the overall shape from the front and top projections. Details are located and dark object lines in all the views are made.

8. Dimension and extension lines, which will be discussed later, are now added. Finally, the projection lines are erased to complete the drawing.

Dimensioning Drawings. The size and shape of an object are communicated by detail drawings. To accomplish this task, dimensions are included. The basics of dimensioning were presented in Chapter 10. At that time you were introduced to three important types of dimensions:

• **Size dimensions:** Indicates the size of the object (length, width, and length) and its major features (diameter and depth of holes, width and depth of notches, etc.).

• **Location dimensions:** Indicates the position of features on the object, such as center points for holes, edges of grooves, starting points for arcs, etc.

• **Geometry dimensions:** Indicates the shape of features and the angle at which surfaces meet (round holes, square corners, etc.)

All these dimensions must be included on the drawings. One technique that ensures they are present suggests:

• Dimension the size of the object first followed by that of all major features.

• Dimension the location of all features next.

• Indicate any necessary geometric dimension last. (Angles not indicated are assumed to be 90°.)

Dimensioning uses two kinds of lines. First **extension lines** indicate the points from which the measurements are taken. Between the extension lines are **dimension lines**. These have arrows pointing to the extension lines that indicate the range of the dimension. The actual size of the dimension is shown near the center of the dimension line.

These dimensions may be given in fractions or decimals of an inch. Additionally, a **tolerance** may be included with the dimension. This number indicates the amount of deviation in the dimension that will be allowed. For example a +/- (plus or minus) 1/64 in.(0.4 mm) after the dimension indicates that the size may be 1/64 in. larger or 1/64 in. smaller that the dimension and still be acceptable.

Alphabet of Lines for Drawings. Drawings should be easily read by the user. Therefore, a set of **drafting standards** have been developed so that all drawings

communicate well. These rules are much like the rules of grammar that a writer uses. They allow each reader to interpret the prose (using grammar rules), or the drawing (using drafting standards) in a similar manner.

One set of essential drafting standards deals with lines and line weights, Fig 12-7. The shape of the object is of primary importance. Therefore, the lines that outline the object and its major details must stand out. These solid lines are called **object lines** and are the darkest on all drawings.

Some details are hidden in one or more of the views. However, their shape and location are important to understanding the drawing. Therefore, they are shown but with lighter, dotted lines called **hidden lines.**

A third type of line locates holes in the part. These lines pass through the center of the hole and are thus called **center lines**. These are constructed of a series of light long and short dashes.

Dimension and extension lines, which were introduced earlier, are the same weight as hidden lines. They are important, but should not dominate the drawing. Remember, at first glance the outline of the object and its details are the dominant feature that should "jump out" at the viewer.

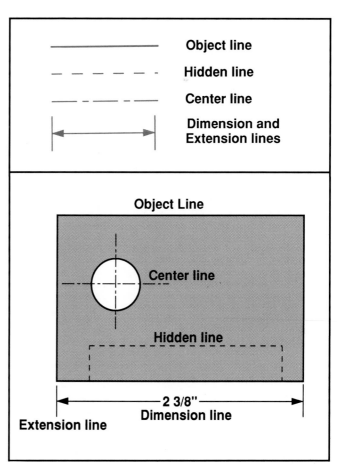

Fig. 12-7. This "alphabet of lines" shows the types and weights of lines used in drawings.

Technology: How it Works

COMPUTER-AIDED DESIGN SYSTEMS (CAD): Systems that use computers to create, change, test, and store drawings that communicate design information.

Computers are used throughout industry to increase productivity and to improve product quality. One area in which computer systems have made vast inroads is the drawing room. Computer-generated drawings and designs are commonplace in many industries. The systems used to develop these designs are known as CAD (computer-aided design) systems, Fig. 1.

Data can be entered into the CAD system using keyboard commands, a mouse, or a menu pad. A menu pad has a number of common drawing commands on its surface. The drafter simply points to the commands with a wand. This enters the command into the computer. These controls allow drafters to produce standard two-dimensional detail and assembly drawings. Additional features and text may then be placed on any engineering drawing, Fig. 2.

A growing use for CAD systems is in producing solid models (three-dimensional representations) of objects. This process uses a concept similar to weaving a rug. Strands are intertwined to produce a frame for the object. This produces a special drawing called a *wireframe representation,* Fig. 3. This frame is the skeleton that is required to produce the object. Such a skeleton can be covered to produce a solid representation, Fig. 4. This three-dimensional drawing is often used when considering styling and appearance. Solid models can also show how parts fit together in an assembly.

In addition to their display functions, CAD systems can be used to test and examine parts. Many CAD systems can calculate the mass, volume, reactions to stress, and other attributes of a designed part. CAD systems have become an important tool for the design engineer.

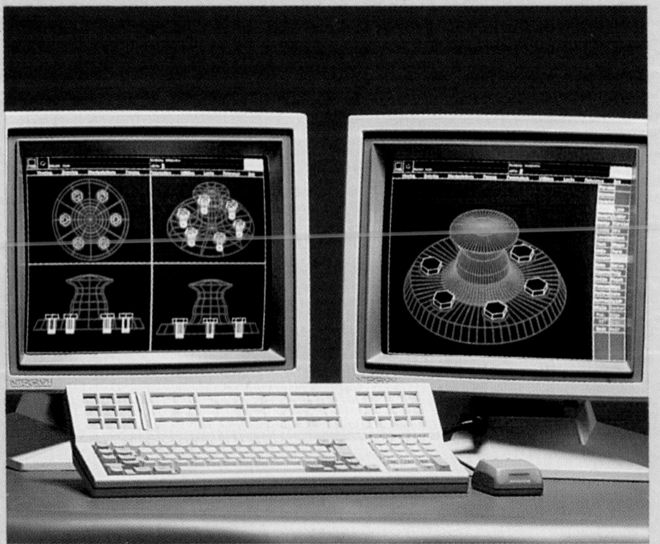

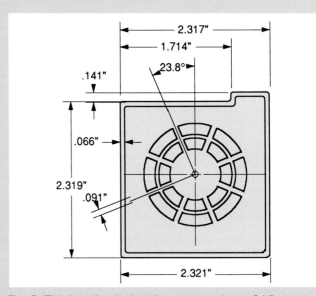

Fig. 2. This is a simple drawing produced on a CAD system.

Fig. 3. Wireframe representations allow you to see through an object.

Fig. 1. This is a typical CAD station.

Fig. 4. Shown is a solid representation of a turbine.

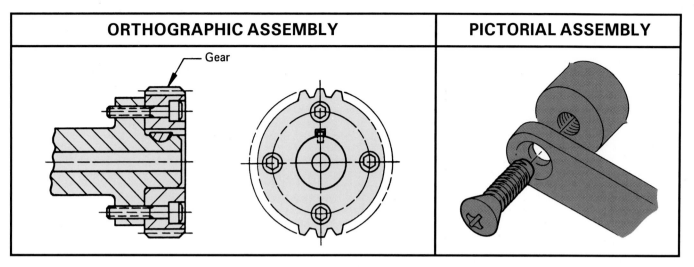

ORTHOGRAPHIC ASSEMBLY	PICTORIAL ASSEMBLY

Fig. 12-8. Assembly drawings can be pictorial or orthographic. (General Motors Corp.)

ASSEMBLY DRAWINGS

A second type of engineering drawing is the assembly drawing. **Assembly drawings** show how parts fit together to make *assemblies*, which are put together to make products. There are two types of assembly drawings, as shown in Fig. 12-8. They are orthographic assembly drawings and pictorial assembly drawings.

Orthographic assembly drawings use a single view to show the mating of the parts. **Pictorial assembly drawings** show the assembly using oblique, isometric, or perspective views like those discussed in Chapter 10.

Either type of drawing may be a standard view or an exploded view, Fig. 12-9. Standard views are constructed using the normal techniques for constructing orthographic or pictorial drawings. *Standard views* show the product in one piece, as it would be after it is assembled.

Exploded views show the parts that make up a product as if it were taken apart. The parts are arranged in the proper relationship to each other on the drawing. This type of drawing is often found in owner's manuals and parts books. They are used to show the parts that comprise a product. Each part on the drawing generally has a code that allows the owners or repair person to order a replacement part.

Most assembly drawings do not have dimensions. The exception would be when the assembler must manually position the parts for assembly. However, this practice is avoided whenever possible. Bolt holes, keyways, joints, shoulders, and other features that position the parts for assembly are preferred.

SYSTEMS DRAWINGS

Systems drawings are used to show how parts in a system relate to each other and work together. They are used for electrical, hydraulic (fluid), and pneumatic (gas) systems. They are often called *schematic drawings*. They do not attempt to show the actual position of the parts in a product. Assembly drawings would do this. They are designed to show the connections for wires, pipes, and tubes.

Systems drawings use symbols to represent the components. Standard symbols have been developed for electrical, pneumatic, and hydraulic parts. Fig. 12-10 includes some common symbols for electrical and electronic components.

Systems drawings are developed by first arranging the major components on the sheet. Then connecting

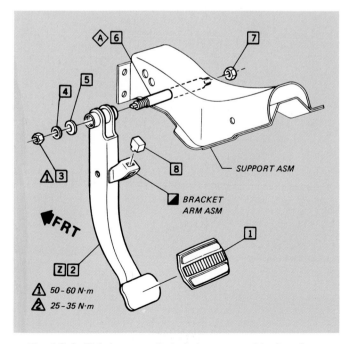

Fig. 12-9. This is an exploded view assembly drawing. (General Motors Corp.)

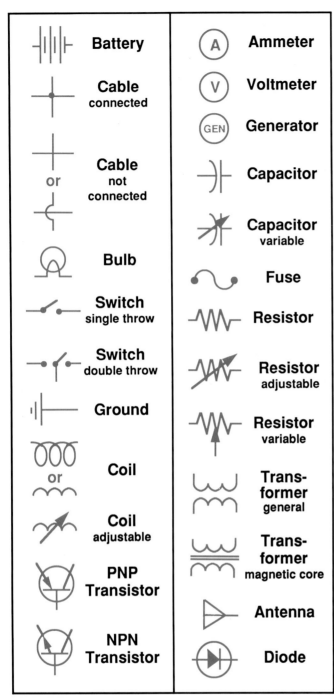

⊣⊢⊢⊢	**Battery**
—•—	**Cable** connected
—┼— or —╧—	**Cable** not connected
⊘	**Bulb**
—•⁄—	**Switch** single throw
—•⁄•—	**Switch** double throw
⊣⊢	**Ground**
ᓮᓮᓮ or ᔫᔫ	**Coil**
⤳	**Coil** adjustable
⊗	**PNP Transistor**
⊗	**NPN Transistor**
Ⓐ	**Ammeter**
Ⓥ	**Voltmeter**
⒢	**Generator**
⊣⊢	**Capacitor**
⤳	**Capacitor** variable
⌒⌒	**Fuse**
—ᏢᏙᏙ—	**Resistor**
—ᏢᏙᏙ—	**Resistor** adjustable
—ᏢᏙᏙ—	**Resistor** variable
ᓵᓵᓵ	**Transformer** general
ᓵᓵᓵ	**Transformer** magnetic core
▷	**Antenna**
⊕	**Diode**

Fig. 12-10. These are some common symbols used on electrical and electronic systems drawings. Systems drawings are also called schematic drawings.

wires, pipes, and tubes are indicated. Special drawing techniques are used to indicate when the lines connect or simply cross each other.

COMPUTER-AIDED DRAFTING

Computers are a valuable technological tool. Computer systems can be applied to a number of tasks. A common industrial application for computer systems is in preparing drawings and models. This application uses a computer, a plotter or printer, input devices (keyboard, mouse, graphics tablet, etc.) and **CAD** (Computer-Aided Drafting) software.

There is nothing magical about this system. It allows an operator to complete the steps of laying out and producing drawings following the methods that a drafter uses. However, CAD systems do the job more quickly and uniformly. Also, computer drawings are easier to correct, store, and communicate. Computer drafters can send their drawings across the country or around the world in seconds using telephone circuits, Fig. 12-11.

BILL OF MATERIALS

Not all information needed to produce a product can be contained on detail, assembly, and system drawings. Additional documents are needed to provide complete production information. One important document is a **bill of materials**. The name of this document causes some confusion. We all pay bills for things we buy and use. A bill of materials does not contain cost information. Instead, it is a list of the materials needed to make *one* complete product. An example is shown in Fig. 12-12.

Most bills of material contains the following information for each part on the product:

- A *part number* which can be used on assembly drawings and for ordering repair parts.

Fig. 12-11. This drafter is developing a electronic circuit drawing for a microprocessor chip on a CAD terminal. (GM-Hughes)

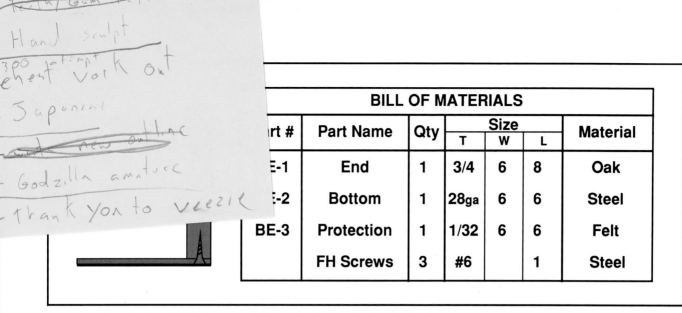

BILL OF MATERIALS						
Part #	Part Name	Qty	Size			Material
			T	W	L	
E-1	End	1	3/4	6	8	Oak
E-2	Bottom	1	28ga	6	6	Steel
BE-3	Protection	1	1/32	6	6	Felt
	FH Screws	3	#6		1	Steel

Fig. 12-12. A bill of materials for a simple bookend tells the materials needed to make one product.

- A descriptive *name* for the part.
- The number or *quantity* (abbreviated qty.) of parts needed to manufacture one product.
- The *size of the part*, indicating either its thickness, width, and length (for rectangular parts), or its diameter and length (for round parts). Sizes are given it the order shown: T x W x L or Dia. x L.
- The *material* out of which the part is to be made.

The items on a bill of material are listed in a priority order. Manufactured parts are listed first. Parts that are purchased ready to use and fasteners are listed after the manufactured parts.

SPECIFICATION SHEETS

Not all materials can be shown on a drawing. Can you make a drawing of engine oil, an adhesive, or sandpaper? If you did make a drawing of any of these items, it would be of little value. These and thousands of other items are not chosen for their size and shape. Other properties are important in their selection.

For example, some important factors in selecting adhesives are the working time (time between application and clamping), clamping time (time the work must be held together for the glue to set), and shear strength. Window glass must be transparent. Insulating materials must stop heat from passing through them.

The important properties that a material must possess for a specific application are communicated by **specification sheets**, Fig. 12-13. These properties, which were introduced in Chapter 4, may include:
- Physical properties: moisture content, density, porosity, surface condition.
- Mechanical properties: strength, hardness, ductility, and elasticity.
- Chemical properties: corrosion resistance.

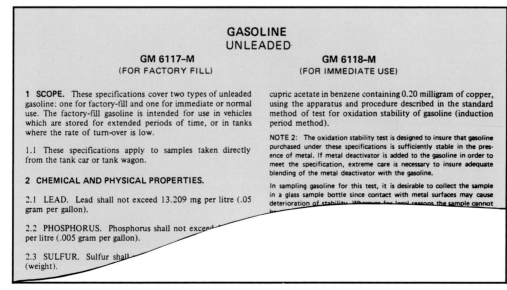

Fig. 12-13. The characteristics of materials are described on specification sheets.

- Thermal properties: resistance to thermal shock, thermal conductivity, heat resistance.
- Electrical properties: resistance, conductivity.
- Magnetic properties: permeability.
- Acoustical properties: sound absorption, sound conductivity.
- Optical properties: color, transparency, optical reflectivity.

Specifications are included on two types of sheets. **Technical data sheets** are prepared by manufacturers to communicate the specifications for products that they have on the market. These kinds of products are often called standard, or "off-the-shelf" materials and components. They are generally kept in stock by the manufacturer and are often listed in a supplier's catalog.

For example, you could write to a manufacturer and tell them about your need for an adhesive. You would probably receive technical data sheets on several adhesives the manufacturer makes that meet your needs. You would study the specifications for each product and choose the one that meets your needs.

Large organizations may prepare their own specifications for materials and products they need. They send them to suppliers, who then compete to supply a specific item. An example is the Military Specification system, also known as the MILSPEC system, that are prepared by the government. Large manufacturing companies also have specification systems.

PRESENTING DESIGNS FOR APPROVAL

So far we have discussed methods of communicating designs for manufacture. However, before anything can be built, it must be approved by someone in charge. This "someone" may be company management, government agencies, or the customer.

Generally, the approval process requires two types of communication. First, written reports must be prepared. They may include need statements, proposed design solutions, cost estimates, marketing strategies, economic forecasts, and environmental impact statements.

The written reports are supported by oral reports to those who will approval to the project. The highlights of the written report are presented. Graphs, illustrations, and other visual media are used to help communicate the design data, Fig. 12-14. This presentation is designed to make the final "sale" of the design idea. The result is official approval to proceed with the manufacturing or construction project. The actual practices used to produce the product or construct the structure will be the focus of the next section of this book.

SUMMARY

Products or structures are seldom built by the designer. The exception is the home hobbyist who constructs products for themselves, builds their own home or a storage shed. However, most technological artifacts are built by specialists. The designer must communicate vital information to these manufacturing personnel. They do this through engineering drawings which communicate the details of each part, the way parts are assembled into products, and the arrangement of system components.

Additionally, they prepare bills of material to list the parts needed to construct on product. Finally, they must communicate the specifications for materials and supplies used in making the product.

Fig. 12-14. Graphic devices are used to help communicate the product and its importance to decision makers. (American Greeting Co.)

WORDS TO KNOW

All of the following words have been used in this chapter. Do you know their meanings?

Assembly drawings
Bill of materials
Center line
Detail drawings
Dimension line
Drafting standards
Engineering Drawings
Exploded view
Extension line
Geometry dimensions
Hidden line
Location dimensions
Object line
One-view drawings
Orthographic assembly drawings
Orthographic projection
Pictorial assembly drawings
Size dimensions
Specification sheets
Systems drawings
Technical data sheets
Three-view drawings
Tolerance
Two-view drawings

TEST YOUR KNOWLEDGE

Write your answers on a seperate sheet of paper. Please do not write in this book.

1. List the three documents that are used to communicate design solutions.

2. True or false. Detail drawings show how parts go together to make products.

3. Match the statements on the left with the appropriate type of drawing on the right:

    _____ Sometimes called schematics.

    _____ One is prepared for each part.

    _____ The two types are pictorial and orthographic.

    _____ Can be normal or exploded.

    _____ Generally are shown in orthographic projection.

    _____ Often used to show electrical circuits.

    A. Detail drawings.
    B. Assembly drawings.
    C. Systems drawings.

4. _____-view orthographic projections are used to show the size and shape of flat sheet metal parts.

5. True or false. Extension lines are used to show how far a hole extends beyond the end of a part.

6. A _____ is the document that lists all the parts needed to produce one product.

7. List the information provided for each part that is shown on a bill of materials.

APPLYING YOUR KNOWLEDGE

1. Disassemble a simple product such as a ballpoint pen or a flashlight. Prepare:
    A. A detail drawing for one part.
    B. An assembly drawing for the product.
    C. A bill of materials for the product.

2. Select a simple game that is on the market. List the materials and parts that:
    A. Would need a drawing to produce.
    B. Would need specification sheets prepared for them.

Section Four - Activities

ACTIVITY 4A - DESIGN PROBLEM

Background:

All technological devices are the result of design efforts by people. People define problems and opportunities, think up many solutions, and model the solutions that are selected. Finally, the designer communicates the design to production personnel.

Situation:

Road Games Inc. is a company that specializes in designing small, compact games that people can take with them on trips. You have been recently employed in the creative concepts department of the company.

Challenge:

Design a travel game that uses pegs and a 3/4" x 3 1/2" x 3 1/2" wood board. Dice can be used in the game, but are not required. Your boss expects you to produce rough sketches for five different ideas, a refined sketch for the best idea, a prototype, and a detail drawing for the game board. Note: Activity 3B is an example of a game that fits this criteria.

ACTIVITY 4B - FABRICATION PROBLEM

Background:

Product designers often work to improve an existing product. They also develop new and improved products to meet the changing demands and requirements of the customer.

Challenge:

You are a product designer for the Acme Book end Company. The company makes bookends for different markets. Each market gets unique graphics and special shapes for the wood member of the bookend. Your boss has asked you to modify an existing product to meet these new criteria:
- A new shape of the wood member.
- New graphics that will appeal to high school students.

Equipment:

Sketch paper
Drawing paper
Pencils, magic markers, etc.
Drafting ruler, T-square
and triangles.

Procedure:

Look at Fig. 4-A. Redesign the bookend to meet the new criteria.

DESIGNING THE SHAPE

1. Photocopy the book end layout sheet, Fig 4-B.
2. Sketch four new shapes for the book end wood member.
3. Select the best shape and circle it with a colored marker.

DECORATING THE PRODUCT

1. Make four layouts using the shape you chose for the wood member.
2. Sketch four new graphic designs for the wood member. Make sure the design will appeal to high school students.
3. Select the best design and circle it with a colored marker.

COMMUNICATING THE DESIGN

1. Obtain a piece of drawing paper.
2. Draw a border that is 1/2" in from all edges of the paper.
3. Draw a title box like the one in Fig. 4-C.
4. Produce a dimensioned two-view orthographic drawing of your design for the new book end.

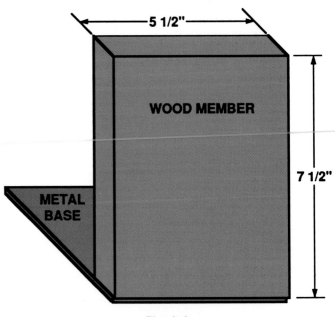

Fig. 4-A.

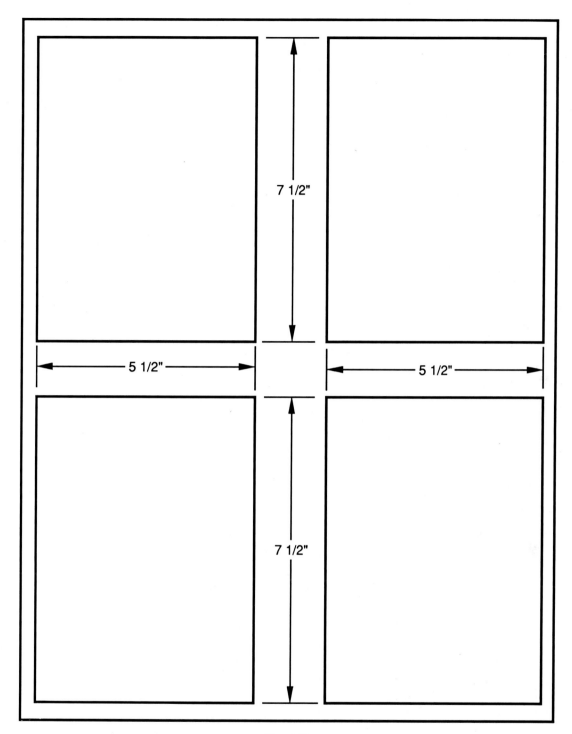

Fig. 4-B.

School:	Part Name:	Drawn By:
		Date:
Class:	Part Number:	Checked By:
		Date:

Fig. 4-C.

section 5

Applying Technology: Producing Products and Structures

13. **Using Technology to Produce Artifacts**

14. **Obtaining Material Resources**

15. **Processing Resources**

16. **Manufacturing Products**

17. **Constructing Structures**

18. **Using and Servicing Products and Structures**

Section 5 Activities

TEKTRONIC **177**

Industrial materials are the result of resource processing.

13
CHAPTER

USING TECHNOLOGY TO PRODUCE ARTIFACTS

After studying this chapter you will be able to:
- Define production.
- List and describe the three types of activities that make up production.
- Tell the difference between primary processing, manufacturing, and construction systems.
- Describe the inputs and outputs of primary processing activities.
- Describe the inputs and outputs of secondary manufacturing (processing) activities.
- Describe the inputs and outputs of construction activities.
- Describe how maintenance and servicing are used with products and structures.

Some of our earliest technology was developed for production systems, Fig. 13-1. These systems produce what is called **form utility**. They change the form of materials to make them more valuable. Look at Fig. 13-2. Which is more valuable to people: the lumber, shown in the top view, or the home that is built from it shown in the bottom view? You probably answered the home. Wood, in the form of a home, is more valuable to you than in the form of lumber. Therefore, the actions that change lumber into houses provide form utility.

PRODUCTION ACTIVITIES

Production activities have two major goals: to produce a product or produce a structure. Production activities that make products are called **manufacturing** activities. **Construction** activities produce a structure such as a building or a roadway.

Fig. 13-1. These people are practicing early production activities using technology of the colonial period.

Using Technology to Produce Artifacts   179

Fig. 13-2. The lumber (top) may be changed in form to become a home (bottom). (Weyerhauser Timber Co.)

There are literally thousands of different production activities that fall under the categories of manufacturing and construction. However, these activities can be grouped into three major types:

- Resource processing (Primary manufacturing).
- Product manufacturing (Secondary manufacturing).
- Structure construction (Construction).

Each of these systems plays a unique role in converting a material resource into a product or structure to meet human needs and wants.

RESOURCE PROCESSING SYSTEMS

Few materials occur in nature in a usable state. Generally, they must be converted into new forms before products and structures can be made. This conversion uses **primary processing** technology.

Typically, these processing systems involve two actions. First the material must be located and obtained from the earth. This may involve growing and harvesting trees, crops, and domesticated animals. It may involve searching for minerals or hydrocarbons (petroleum and coal). Drilling or mining may be required to extract these resources from the earth.

Once the natural resource has been obtained, it must be transported to a processing mill. There the natural resource is changed into **industrial materials**. These materials are the inputs to secondary manufacturing activities. For example, iron ore, limestone, and coke may be changed into steel, Fig. 13-3. The steel is then processed into bars, rods, sheets, or pipes. This standard material becomes the raw inputs for systems that make products for industrial companies and retail consumers. It may end up as part of a lathe, automobile, broadcast tower, airline terminal, or one of thousands of other products and structures.

Specialized processing systems are used to manufacture food products, medicines, and chemicals. Other systems process petroleum into fuels and lubricants, natural gas into plastics, and coal into coke, Fig. 13-4.

Fig. 13-3. Primary processing activities change natural resources into industrial goods like the steel rods shown above. (Inland Steel Co.)

Fig. 13-4. These and other products are produced using process systems. (Goodyear Tire and Rubber Co.)

More detailed information on obtaining material resources and processing them into industrial materials is contained in Chapters 14 and 15 of this book.

PRODUCT MANUFACTURING SYSTEMS

Most standard materials have limited use to the average person. Suppose someone gave you ten sheets of plywood. You might feel that this is a good gift. However, what are you going to use them for? You will probably have to ask the question: "What should I build out of the plywood?" You might build a bookcase or a storage locker. This action would employ **secondary processing activities**.

Secondary processing systems change industrial materials into products, Fig. 13-5. They cause the material to take on a desired size and shape. This may involve casting, applying force for forming, or machining using tools or other cutting devices. Secondary processes also assemble parts into products by welding, fastening, or gluing them together. Other secondary processes apply coatings to the material, part, or product. These coatings protect the item from the environment and improve its appearance. Finally, secondary

processing activities may change the properties of the material. This change may cause the material to be harder, stronger or more resistant to fatigue.

The outputs of secondary manufacturing processing are either industrial or consumer products. **Industrial products** are items that are used by companies in conducting their businesses. For example, a computer terminal is an industrial product that may be used in the accounting activities of a company. Likewise, a furnace is an industrial product that becomes part of a building.

Consumer products are outputs that are developed for the end user in the product cycle–people such as home owners, athletes, or students, just to name a few. Examples of consumer products are soap powder, television sets, lawn mowers, baseball bats, or furniture.

The range of secondary manufacturing processes will be introduced and discussed in Chapter 16.

STRUCTURE CONSTRUCTION SYSTEMS

The three basic physical needs of all people are food, clothing, and shelter. Food comes to us from processing industries. Clothing comes from manufacturing industries. Shelter is the output of construction activities.

Construction may be described as producing a structure on the site it will be used. It is different from manufacturing, in that a manufactured product is produced in a factory and shipped to its point of use.

Constructed works can be grouped into the categories of buildings and civil structures. Homes, factories, stores, and offices are typical buildings. Common civil structures are roads, railways, canals, dams, power transmission lines, communication towers, and pipelines.

Fig 13-5. Secondary manufacturing processes change industrial materials into products like this finishing sander. (Makita)

Technology: How it Works

HIGH-RISE BUILDING: A multi-story residential or commercial building that has a skeleton frame.

A construction marvel, even today, are the Great Pyramids at Gizeh in Egypt. However, this type of tall structure was not common. Later, the great cathedrals were built in Europe. The cathedrals have high, thick stone walls. This type of wall is called *load-bearing*, because it holds all the weight (load) of the roof.

This type of construction has height limits. Load-bearing walls are not practical in buildings over five stories tall. The walls become too thick and heavy.

As cities became larger, taller buildings were needed. It took the development of the elevator and the skeleton frame to make tall buildings practical. The elevator allowed people to move between floors without walking up of stairs. A skeleton frame carries the weight of the building, much like the human skeleton carries the body. These developments made high-rise buildings, or skyscrapers, possible.

The first high-rise building was built in Chicago in 1885. Built for the Home Insurance Company, it was ten stories tall. The exterior walls provided protection from the weather, but did not carry any load. This type of wall is called a *curtain wall*, because it merely "hangs" from the frame.

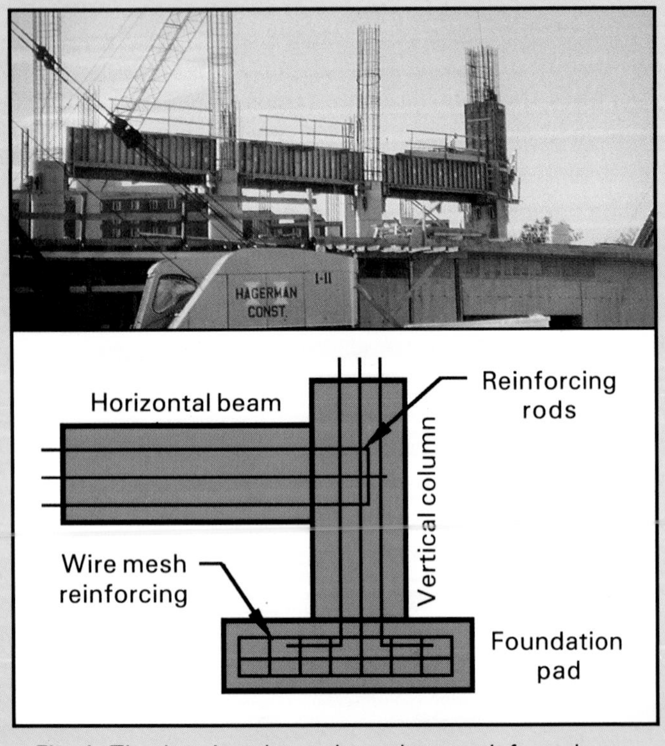

Today high-rise buildings use two types of framework. The first type is reinforced concrete, Fig. 1. This type of framework is cast on site. Forms are erected around a network of steel rods. The concrete is poured inside the forms. After the concrete has cured, the forms are removed.

The weight of reinforced concrete limits its use to buildings of moderate height. Taller buildings use steel skeleton. Steel is fabricated into angles and "I" beams. Steel columns and beams are bolted or riveted together, Fig. 2. This allows the beams to expand and contract uniformly with temperature changes.

Once the frame is erected the floor and roof are installed. The exterior walls are then put in place, Fig. 3. These walls can be steel or aluminum panels, brick, concrete block, sheets of glass, or other materials.

Fig. 2. This is a steel-framed building. Notice the different shapes of the steel members.

Fig. 1. The drawing above shows how a reinforced concrete structure is put together. The photograph shows a building under construction.

Horizontal beam

Reinforcing rods

Vertical column

Wire mesh reinforcing

Foundation pad

Fig. 3. The building in this picture is having its exterior walls finished. The brick is in place and the windows are installed.

Each one of these structures are developed to meet a specific need, Fig. 13-6. Their development involves preparing the site, building foundations and superstructures, installing utilities, enclosing the superstructure, and landscaping the site.

Construction activities will be presented in detail in Chapter 17.

SERVICING PRODUCTS AND STRUCTURES

All products and structures are subject to wear and tear. They can become damaged, worn, or outdated over time. Also, some products need attention during operation. All these conditions call for service and repair.

Servicing, or maintenance, is the scheduled adjustment, lubrication, or cleaning required to keep the product or structure operating properly. Automobiles require oil changes, and engine tune-ups periodically, Fig. 13-7. Buildings must be cleaned and floors must be waxed. Machine need adjustment and cleaning. These are servicing acts and are usually performed at a specific point in time. The goal of servicing is to keep the product or structure in good working order.

Repair, on the other hand, involves fixing a broken, damaged, or defective product or building. It is designed to return a disabled product or structure to working condition. Repair requires diagnosing (determining) the problem and fixing or replacing defective parts or materials. Repairing a building may include replacing damaged ceiling tiles, applying new wall coverings, or fixing a hole in a wall. Product repair can

Fig. 13-7. This person is selecting engine tune-up parts to service an automobile. (GM-Rochester)

involve replacing worn-out or broken parts and adjusting mechanisms.

The actions involved in servicing products and structures are presented in Chapter 18.

SUMMARY

The remaining chapters in this section of your textbook will explore ways people apply technology to produce products and structures, Fig. 13-8. You will see how material resources are obtained by growing, harvesting and extracting. Then, you will explore the ways natural resources are converted into industrial

Fig. 13-6. Structures like these grain storage bins are the result of construction activities.

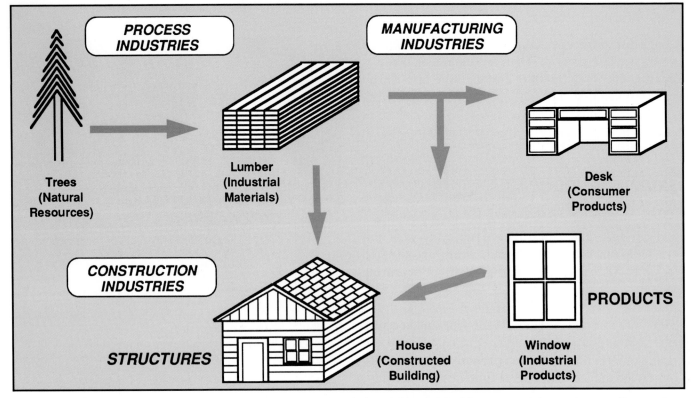

Fig. 13-8. Production systems convert natural resources into industrial materials. These materials may be manufactured into industrial or consumer products. Materials and some manufactured goods are used in constructing structures.

materials. Later you will study how consumer and industrial products are manufactured. Following that you will examine the ways materials and manufactured products are combined to make structures. Finally, you will survey the area of servicing and maintaining products and structures.

WORDS TO KNOW

All of the following words have been used in this chapter. Do you know their meanings?

Construction
Consumer product
Form utility
Industrial material
Industrial product
Manufacturing
Primary processing
Repair
Secondary processing
Servicing

TEST YOUR KNOWLEDGE

Write your answers on a seperate sheet of paper. Please do not write in this book.
1. Describe form utility.
2. True or false. Manufacturing activities are designed to produce products.
3. True or false. Logs from the forest are an industrial material.
4. Buildings are the outputs of _____ activities
5. Changing industrial materials into products uses _____ manufacturing processes.
6. True or false. Maintenance puts a broken product back into working order.
7. Food is the output of _____ systems.

APPLYING YOUR KNOWLEDGE

Select three products or structures that you see around you. Complete a chart like the one that follows for each product.

PRODUCT			
Primary natural resource(s)			
Industrial material(s) used			
Product of manufacturing or construction			
Service or maintenance required			

14

OBTAINING MATERIAL RESOURCES

After studying this chapter you will be able to:
- *Describe natural resources as material inputs to production systems.*
- *List and describe the types of natural resources that are used as inputs to production systems.*
- *List and describe the types of genetic materials that are used in production systems.*
- *List and describe the types of fossil fuel materials that are used in production systems.*
- *List and describe the types of minerals that are used in production systems.*
- *Describe how genetic materials are obtained for use in production systems.*
- *Describe how fossil fuel materials are located and obtained for use in production systems.*
- *Describe how minerals are located and obtained for use in production systems.*

You have learned that technology involves people designing and using tools and artifacts. This action extends human abilities to control or modify their environment. When there is no object or technical means, there is no technology present. All technological objects (human-made objects, called artifacts) are made of materials. The materials that each object is made from can be traced back to one or more natural resources. Plastic materials are made from oil and natural gas. Glass is made from silica sand and soda ash. Cotton is grown and harvested from plants. Plywood comes from trees in the forest. Steel is made from iron ore, limestone, and coal. Each material comes from natural resources. These materials form the foundation for all production activities. Without material resources, production is not possible.

TYPES OF NATURAL MATERIAL RESOURCES

Production technology uses materials and energy as inputs and makes products and structures as outputs. There are three types of natural resources which can become the inputs to production systems. These materials are shown in Fig. 14-1 and include:
- Genetic materials.
- Fossil fuel materials.
- Minerals.

GENETIC MATERIALS

Many resources come from living things. These resources, as you learned in Chapter 4, are called organic materials. Some organic materials are from organisms that have been dead for hundreds of years. This type of organic material includes fossil fuels that will be discussed later in this chapter. Other organic materials are obtained during the normal life cycle of plants or animals. These materials may be called **genetic materials**.

Genetic materials are obtained by three activities: farming, fishing, and forestry. Each of these activities works directly with nature as plants and animals move through the stages of life, shown in Fig. 14-2.

Typical genetic materials that are used in production systems are grains (wheat, oats, barley, corn, etc.), vegetable fibers (wood, flax, cotton, etc.), and animals or fish (meat, hides, wool, etc.)

The origin of all genetic materials is in birth or **germination**. The appearance of animal life is called birth. Plant life generally starts with the germination of seeds or spores.

Young plants and animals grow rapidly early in their life cycle. Their growth generally slows down as they reach older age. This period is called **maturity**. The

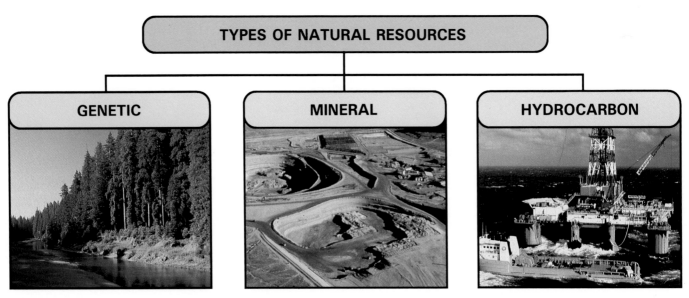

TYPES OF NATURAL RESOURCES

GENETIC

MINERAL

HYDROCARBON

Fig. 14-1. There are three major types of natural resources.

organism is still healthy at maturity but it stays about the same size. The length of this maturity stage may be a matter of days, as with mushrooms, or centuries, as is the case with redwood trees. However, all organic life ends at some point. The plant or animal dies from old age or disease.

FOSSIL FUEL MATERIALS

Fossil fuels are mixtures of carbon and hydrogen. They are called *hydrocarbons*. Hydrocarbons include a

vast number of products in use today, from fuels to medicines. However, most hydrocarbon products come from only three fossil fuel resources. These are:
- Petroleum.
- Natural gas.
- Coal.

Petroleum is an oily, flammable mixture of hydrocarbons that has no specific composition. Instead, it is a mixture of a number of different solid and liquid hydrocarbons. The composition of petroleum will vary with where it is found on earth. Petroleum is the princi-

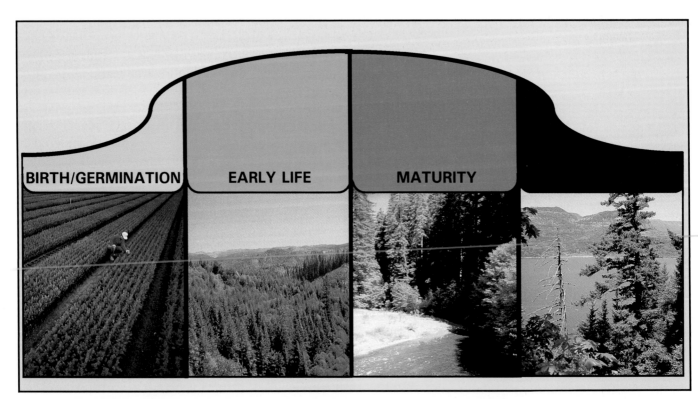

BIRTH/GERMINATION   EARLY LIFE   MATURITY

Fig. 14-2. All genetic materials have a life cycle starting with birth, or germination, and ending with death.

pal source of the liquids we use as gasoline, diesel fuel, heating oil, and lubricants, Fig 14-3. Lighter hydrocarbon products that come from petroleum are gases. They include methane, ethane, butane, and propane. These gases are widely used in producing plastic resins such as polypropylene and polyethylene.

Petroleum is the fuel most used by transportation vehicles. However, exhaust from gasoline and diesel engines produces harmful air pollutants including nitrous oxides and carbon monoxide.

Natural gas is a combustible gas that occurs in porous rock. It is composed of light hydrocarbons. Typically, natural gas is about 85 percent methane. The rest is made up of propane and butane. Natural gas is used as a fuel for homes and industry. Natural gas is also used to make plastics, chemicals, and fertilizers.

Natural gas burns cleanly compared to other fossil fuels. It requires complex pipeline networks to distribute it to potential users. In some cases it is compressed and used as a fuel for vehicles. In this form it is called *compressed natural gas*, or CNG.

Coal is a combustible solid that is composed mostly of carbon. It started as plant matter thousands of years ago. In moist areas the plant matter did not decay easily and layered up to make *peat*. Peat is brownish-black plant matter that looks like decayed wood. When dry, it will burn but gives off a great deal of smoke. The peat was buried by sediment, and pressure and heat changed the peat into coal.

The principal types of coal are:
- Lignite: A soft and porous material made from peat that has been pressed by natural action. It gives off more heat than peat and is used in electrical generating plants and for industrial heating.

- Bituminous coal: The most commonly found coal, which is harder than lignite. It is sometimes called soft coal because it can be easily broken into various sizes. Bituminous coal is widely used for power generation, and heating. It can also be used for coal gasification and chemical processes. See Fig. 14-4.
- Anthracite coal: The hardest coal is a shiny black material that burns without smoke. Anthracite has the highest carbon content of all the types of coal. Anthracite coal is used for heating and to produce coke for steel making.

Coal does not burn cleanly, and its sulfur content is a source of chemicals that make acid rain. Also, its bulk makes it costly to ship.

MINERALS

Minerals are any substance with a specific chemical composition that occur naturally. This is different from fossil fuel resources, which are all chemical mixtures. Typical minerals are iron ore, bauxite (aluminum ore), and sulfur.

There are a number of ways to classify minerals. One way is to group them by their chemical composition. This grouping would include native elements (elements that occur naturally in a pure form), oxides, sulfides, nitrates, carbonates, borates, and phosphates.

A more useful way to group minerals is available. This method groups the minerals that have economic value into families with similar features. These groups include:
- **Ores:** minerals that have a metal chemically combined with other elements. Ores can be processed to separate the metal from other elements.

Fig. 14-3. Petroleum is a vital natural resource that provides the fuels to power vehicles and other machines. (AMOCO)

Fig. 14-4. Production materials make up products and structures.

- Nonmetallic minerals: substances that *do not* have metallic qualities, such as sulfur.
- Ceramic minerals: Fine-grained minerals that are formable when wet and become hard when dried or fired.
- Gems: Stones that are cut and polished and prized for their beauty and hardness.

LOCATING AND OBTAINING NATURAL RESOURCES

All natural materials are found on the earth. However, they are not all visible as we travel across the land and water or through the air. In fact, two major actions are required to obtain natural resources. First, they must be *located* and then *extracted* or *gathered*.

OBTAINING GENETIC MATERIALS

Most genetic materials are easy to find. Trees and farm crops are on easily used plots of land. Domesticated animals and the fish raised on fish farms are contained in specific locations. Only commercial fishermen must seek genetic resources that are sometimes hard to find.

The challenge for people dealing with genetic resources is to harvest the plant or animal at the proper stage of growth. This will vary with the growth cycle and growing habits of the organism. Many trees are harvested during their mature phase. However, some young trees will not grow in the shade of older trees. This may require that all the trees in a single plot be harvested at one time. Then new trees can be planted and they will grow in the cleared area.

Likewise, most farm crops are planted at one time in fields. The plants may be fertilized and irrigated to stimulate their growth, Fig. 14-5. Most of the crop will mature at the same time. Therefore, the entire crop can be harvested at one time.

Let's look at harvesting a genetic material–trees. The forest management process requires that each stand of trees be evaluated and designated for a specific use. **Wilderness areas** are set aside. No roads or logging is permitted in a wilderness. **National parks**, and many state and provincial parks, protect scenic beauty. Roads are allowed in the parks but logging is not. **National**, state, and provincial **forests**, generally, are multi-use lands. Lakes, hiking trails, and camping areas are set aside for recreational use. Logging is permitted in selected areas to harvest mature trees. **Private forests** are generally managed to intensely produce trees.

Fig. 14-5. Intensive farm management, like this irrigation process, can increase crop yields.

The logging process requires both planning and action. Planning for removing trees involves several steps. First, the forest is studied to determine whether it is ready for harvesting. This involves a process called **timber cruising.** Teams of two or three foresters measure the diameter and height of the trees. Their task is to find stands of trees that can be economically harvested. They also prepare topographical maps showing the location and elevations of the features on the potential logging site.

Forest engineers must then plan the proper way to harvest the trees. Logging roads and loading sites are planned. The type of logging must also be selected to match the terrain and the type of forest. There are three logging methods:
- **Clear-cutting:** All trees, regardless of species or size, are removed from a plot of land which is generally less that 1000 acres, Fig 14-6. This process allows for

Fig. 14-6. Clear cutting, like shown above, increases the yield of Douglas fir in a managed forest.

replanting the area with trees that cannot grow in competition with mature trees. Also, the number of tree species can be controlled.

- **Seed-tree cutting**: All trees, regardless of species, are removed from a large area except three or four per acre. These trees are used to reseed the area. Again, the number of species that will be reseeded is controlled by the type of seed trees left.
- **Selective cutting**: Mature trees of a desired species are selected and cut from the plot of land. This technique is used in many pine forests where tree density is limited.

These steps are the prelude for the main activity of **logging**, Fig. 14-7. Equipment to remove the trees is moved into the forest. The **feller** uses a chain saw to cut down (fell) the appropriate trees. Smaller trees can be harvested by a machine which shears the trees off at ground level. Care is taken to drop the tree into a clear area so it is not damaged or causes damage to other trees.

A **bucker** removes the limbs and top. These parts of the tree are called slash. The slash is piled for later burning or chipping so that nutrients are returned to the soil. The trunk is then cut into lengths called logs.

The logs are gathered in a central location called the landing. This process is called **yarding** and can be done in several ways. First chokers (cables) are used to bind the logs into bundles. Then cables can be used to drag logs using high-lead and skyline yarding, as shown in Fig. 14-8. These systems use a metal spar (pole), cables and an engine. High-lead yarding drags the logs to the landing. Skyline yarding lifts and carries the logs over rough or broken terrain.

On gentle terrain, ground yarding is used. This system uses tractors and an implement called an arch. They drag the logs, bound together with a choker, to the landing. Very steep terrain may require helicopter yarding, as shown in Fig. 14-7.

Once logs arrive at the landing they are loaded on trucks and moved to the processing plant. This may be a lumber, plywood, particleboard, hardboard, or paper *mill*. The logs are often stored at the mill in ponds or stacked up and sprayed with water to prevent cracking and insect damage.

Harvesting a genetic material such as trees is much different than finding fossil fuels. We discuss fossil fuels next.

Fig. 14-7. Logging starts with selecting and cutting trees. The felled trees are collected using yarding. Then, the trees are hauled to a mill for storage and further processing. (Boise Cascade)

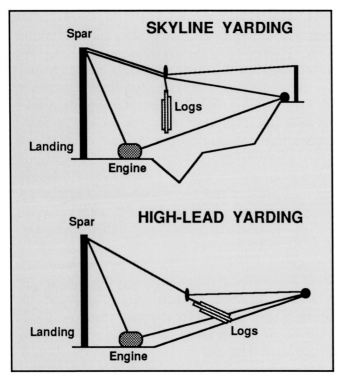

Fig. 14-8. Two ways used to move logs to the landing are high-lead and skyline yarding.

Fig. 14-9. This geologist is seeking deposits of petroleum. (AMOCO)

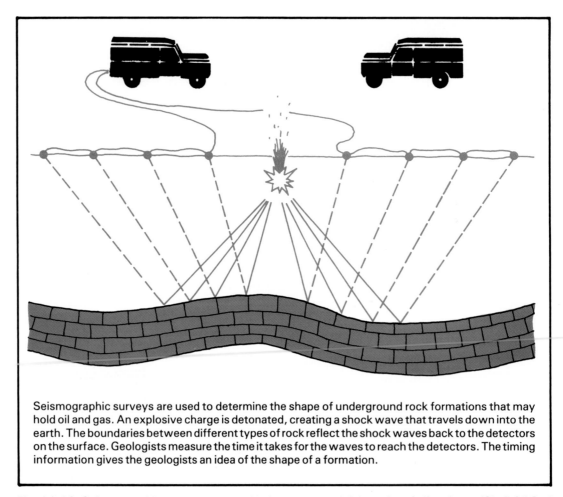

Seismographic surveys are used to determine the shape of underground rock formations that may hold oil and gas. An explosive charge is detonated, creating a shock wave that travels down into the earth. The boundaries between different types of rock reflect the shock waves back to the detectors on the surface. Geologists measure the time it takes for the waves to reach the detectors. The timing information gives the geologists an idea of the shape of a formation.

Fig. 14-10. Seismographic surveys are used to locate potential deposits of oil and gas. (Shell Oil Co.)

OBTAINING FOSSIL FUEL RESOURCES

Most fossil fuel resources are buried under the surface of the earth. Fossil fuels can be pools of petroleum, pockets of natural gas, or veins of coal. Locating and extracting these resources is a major challenge. The techniques used for natural gas and petroleum are much different from those used for coal. Therefore, let's look at them separately.

Obtaining Petroleum and Natural Gas

People do not look for oil (petroleum) and gas. They look for rock formations that may contain deposits of oil and gas. Finding these rock formations is the job of geologists and geophysicists, Fig. 14-9.

It is thought that petroleum comes from decayed plant and animal matter. This organic matter was covered by layers of sediment from rivers. The layers built up and created great pressures and heat. Over millions of years the pressure and heat turned the organic matter into oil. Oil and gas are generally found in porous rock under a layer of impervious (dense) rock. The oil and gas collects under the dense rock. These deposits may be under oceans, mountains, deserts, or swamps. They may be near the surface, as they are in the Middle East, or several miles beneath the land or sea.

There are a number of ways to explore for petroleum and natural gas. One the most accurate ways is **seismographic study**, Fig. 14-10. This technique uses shock waves like those in an earthquake. A small explosive charge is detonated in a shallow hole. The shock waves from the explosion travel into the earth. When the waves hit a rock layer they reflect back to the surface. Seismographic equipment uses two listening posts to measure the shock waves that bounce off various rock

Fig. 14-11. This scientist is studying satellite photographs to determine where to conduct seismographic surveys. (American Petroleum Institute)

layers. Measuring the time it takes the waves to go down to the layers and reflect back allows the geologist to construct a map of the rock formations.

Other methods use geological mapping, which measures the strength of magnetic forces, fossil study, and core samples from drilling to search for the deposits, Fig. 14-11.

The geological study helps people select a good site for exploration. If an area has never produced oil or gas it is called a **potential field**. Producing fields are called **proven reserves**.

A drilling rig is brought to promising sites. This rig may be either a land rig or and off-shore drilling platform, Fig. 14-12.

Fig. 14-12. Oil and gas rigs allow people to drill wells on land and under lakes and seas. (AMOCO, Gulf Oil Co.)

Technology: How it Works

COMBINE: An machine that separates the grain from the straw in crops such as wheat, oats, barley, or rye.

Fig. 1. A modern combine.

Modern agriculture is based on labor-saving machines. These machines paved the way for efficient production of food for a growing world population. Machines also freed labor from farming to work in the industrial and service sectors instead.

An early farm machine was the reaper, which cut and bundled the grain. Another machine was the thresher, which separated the seed from the straw. These devices were put together to make the modern combine—a combination of a reaper or header, and a thresher, Fig. 1.

Fig. 2 shows how the parts of a combine work. The grain is drawn into a moving cutter bar and cut off near the ground level (1). The cut stalks with their grain heads are moved into the separator section on a conveyor (2). At the top of the conveyor is a cylinder that beats the stalks to separate the grain from the straw (3). The mixture of grain and straw moves onto a set of plates with holes in them (4). The straw and grain are vibrated or "walked" toward the back of the machine. The grain and chaff (very small pieces of straw) fall through the holes. In the area below the straw walkers, air is blown across the grain and chaff. Small, immature grain and chaff are blown away and exit out the back of the combine. The heavier, mature grain and remaining chaff are further screened and fall to the bottom of the machine. Heavier pieces of chaff and unseparated grain on the plates are returned to the cylinder area at the front of the machine (5). The separated grain is conveyed to a storage hopper (6). When the hopper is full the grain is unloaded onto trucks to transport it to storage or to a processing plant.

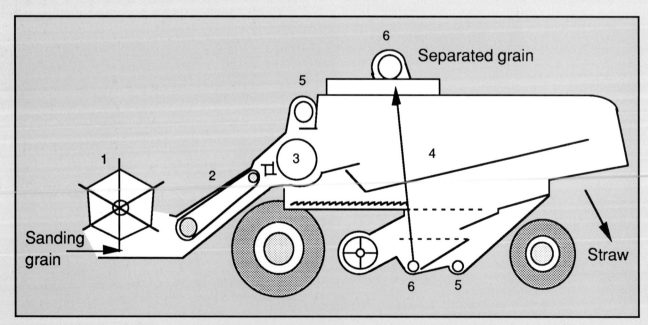

Fig. 2. This drawing shows the parts of a combine and the path of the grain and straw inside.

Drilling involves rotating a drill bit on the end of a drill pipe. Lengths of pipe are added to the drill pipe as the hole gets deeper, Fig. 14-13.

During drilling, a mixture of water, clay and chemicals is pumped down the drill pipe. This mixture is called **mud**. It flows out through holes in the drill bit to cool and lubricate. The mud also picks up the ground-up rock and carries it to the surface. Finally, it seals off porous rock and maintains pressure on the rock. Pressure is maintained to prevent a blowout. A **blowout** occurs when oil surges out of the well. Blowouts are very dangerous and waste large quantities of oil and gas.

Early oil wells were drilled straight down or at a specific angle. Modern techniques allow the well to be drilled along a curve to reach deposits that cannot be otherwise tapped, Fig. 14-14.

Fig. 14-14. Modern techniques allow wells to be drilled along curved lines to tap difficult to reach deposits of oil and gas. (AMOCO)

Once an oil or gas deposit is found, the drilling rig is replaced with a system of valves and pumps, Fig 14-15. The recovered resource flows through pipes into storage tanks. From the well the petroleum is transported to refineries. Natural gas is compressed and sent to petrochemical plants. Pipeline companies sell natural gas to home heating and electric power customers.

Obtaining Coal

Coal is the most abundant fossil fuel and is found on every continent. However, most of the known reserves are in the northern hemisphere. These reserves are

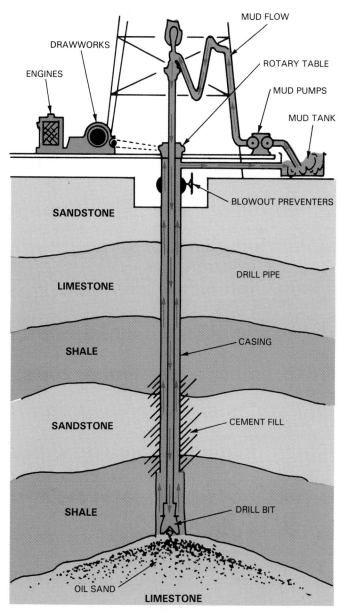

Fig. 14-13. This is a typical drilling process. (Mobil Oil Co.)

Fig. 14-15. A valve system, called a Christmas tree, on top of an oil well. (AMAX)

generally recovered through mining operations. There are three types of mining: Surface or open pit, underground, and fluid.

The first two are used to mine coal. **Open pit mining** is used when the coal vein is not very deep underground. Surface mining of coal generally involves four steps:

1. The surface layer of soil and rock are stripped from above the coal. This material is called over-burden. It is saved for later use in reclaiming the land.
2. The coal is dug up with giant shovels, Fig. 14-16.
3. The coal is loaded on trucks or railcars to be transported to a processing plant.
4. The site is reclaimed by replacing the topsoil and replanting the area.

Underground mining requires shafts in the earth to reach the coal deposits. The three types of underground mining are shaft, drift, and slope mines, Fig. 14-17. **Shaft mining** requires a vertical shaft to reach the coal deposit. Then, horizontal shafts are dug to remove the coal, Fig. 14-18.

Slope mining is used when the vein is not too deep under the ground. A sloping shaft is dug to reach the coal. Then a horizontal shaft is dug to follow and remove the coal vein.

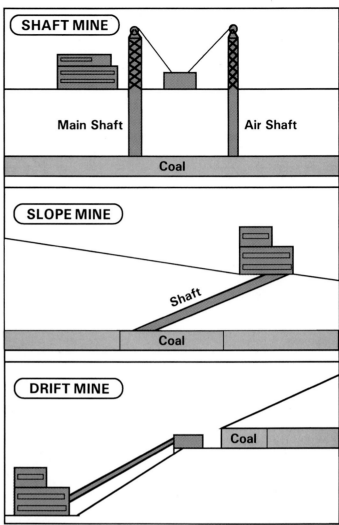

Fig. 14-17. These are the three types of underground mines.

Fig. 14-16. Surface mining uses giant shovels to dig up and load the natural resource. (American Electric Power)

Fig. 14-18. This long-wall mining machine is recovering coal from an underground mine. (FMC Corp.)

Drift mining is used when the coal vein extends to the surface of the earth. Then a horizontal shaft is dug directly into the vein.

Underground mining uses elevators to remove the mined coal. People and equipment are also moved in and out of the mine with the same elevators.

Slope and drift mines often use coal cars or conveyors to remove the coal, Fig. 14-19.

OBTAINING MINERALS

There are a number of ways that minerals can be extracted from the earth and oceans. One way is through **evaporation**. Sea water or water from salt lakes can be pumped into basins. Solar energy can be used to cause the water to evaporate. The mineral resource will be left behind. A number of minerals are recovered in this manner from the Great Salt Lake in Utah.

Mining can also be used. These are the same techniques that were described in the section on recovering coal. However, open pit mines for minerals are generally much deeper. They can extend several thousand feet into the earth. They will appear as a giant inverted

Fig. 14-20. This is an open pit mineral mine. (Brush-Wellman)

cone with ridges around the edge. The spiral ridge is the road used to move equipment into the mine and minerals out of it, Fig. 14-20.

An additional mining method can be used. It is called **fluid mining**. This techniques uses two wells that extend into the mineral deposit. Hot water is pumped down one of the wells. The water dissolves the mineral and is forced up the other well. This mining process is often called the Frasch process. It is widely used to mine sulfur that is found in the limestone rocks that cover salt domes.

SUMMARY

All production systems change the form of materials. Common production materials come from genetic, fossil fuel, and mineral resources. These resources must be located and extracted. This may involve harvesting trees, farm crops, and animal and fish life. They may be obtained from drilling oil and gas wells. Finally, material resources may be extracted from the earth using mining techniques. They may be removed from open pit or underground mines.

WORDS TO KNOW

All of the following words have been used in this chapter. Do you know their meanings?

Blowout
Bucker
Clear cutting
Coal
Drift mining

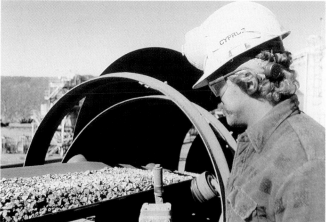

Fig. 14-19. Coal is removed from slope and drift mines on tractors (top) or conveyors (bottom). (AMAX)

Evaporation
Feller
Fluid mining
Fossil fuels
Gems
Genetic materials
Germination
Logging
Maturity
Minerals
Mud
National forests
National parks
Natural gas
Open pit mining
Ores
Petroleum
Potential field
Private forest
Proven reserve
Seed-tree cutting
Seismographic study
Selective cutting
Shaft mining
Slope mining
Timber cruising
Underground mining
Wilderness areas
Yarding

TEST YOUR KNOWLEDGE

Write your answers on a seperate sheet of paper. Please do not write in this book.

1. List the three types of natural resources.
2. True or false. Clear cutting is always the wrong way to harvest trees.
3. Match the correct type of mining on the right with the statements on the left.

   _____ Uses shafts to reach a mineral deposit.   A. Surface.
   _____ Follows a mineral vein that is exposed at the surface.   B. Underground.
                                                                              C. Fluid.
   _____ Uses hot water pumped down a well to bring the mineral to the surface.   D. Shaft.
                                                                              E. Slope.
   _____ Uses an angled shaft to reach the mineral deposit.   F. Drift.

   _____ Is often called open pit mining.
   _____ Includes shaft, drift, and slope mining.
   _____ Uses vertical shafts to reach the mineral and horizontal shafts to recover it.
   _____ Is called the Frasch process when it is used to mine sulfur.

4. Moving logs to a central site in the forest so they can be loaded on trucks is called _____.
5. True or false. Coal is often recovered using fluid mining.
6. Using sound waves to locate potential deposits of petroleum or natural gas is called _____ study.
7. Materials that have life cycles are called _____ materials.
8. List the three types of fossil fuel resources used in production systems.

APPLYING YOUR KNOWLEDGE

1. Select a resource from the categories of genetic, fossil fuels, and minerals. Then research how they are located (exploration) and recovered from the earth (production). Use a chart similar to the one below to record your findings.

Resource:
Exploration processes:
Extraction processes:

2. Build a model to show a selected mining or drilling technique.
3. Read an article and develop a report on a controversial resource recovery issue. This may deal with a number of issues including protecting the environment, wildlife, or the supply of the resource.

15
CHAPTER

PROCESSING RESOURCES

After studying this chapter you will be able to:
- *Define primary processing as a production action.*
- *Diagram how primary processing relates to other processing actions used to convert materials into products and structures.*
- *Describe mechanical processing.*
- *Describe how trees are processed into industrial materials.*
- *Describe thermal processing.*
- *Describe how steel is manufactured from raw materials.*
- *Describe how petroleum is processed into fuels and lubricants.*
- *Describe how glass is produced and formed into sheet materials.*
- *Describe chemical and electrochemical processing.*
- *Describe how aluminum ore is processed into pure aluminum.*

You have learned that natural resources are used to produce products and structures. You also learned that there are several steps in the production process. One of these steps comes between obtaining the resource and manufacturing the product or constructing the structure. This step can be called **primary processing**, Fig. 15-1. The goal of primary processing is to convert material resources into industrial materials. Industrial materials are often called standard stock.

For example, primary processing converts wheat into flour; aluminum ore into aluminum sheets, bars and rods; logs into lumber, plywood, particleboard, hardboard, and paper; natural gas into plastic pellets,

film, and sheets; and silica sand and soda ash into glass. These industrial materials are used as inputs to further manufacturing or construction activities.

TYPES OF PRIMARY PROCESSING

Primary processing uses many production actions. These actions can be grouped in a number of ways. One

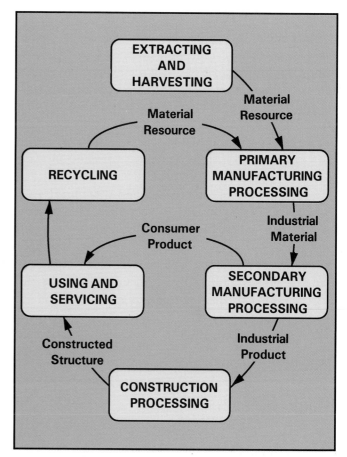

Fig. 15-1. This is a model of the production process. Note how primary processing fits into the cycle.

useful way to group processes is by the type of energy used. This grouping, as shown in Fig. 15-2, includes:
- Mechanical processes.
- Thermal (heat) processes.
- Chemical and electrochemical processes.

This system provides a general overview of most primary processing activities. However, many materials are produced using more than one type of process. For example, steel is made from iron ore, coke, and limestone using a *thermal* process. Some steel that is produced is formed into bars, rods, and sheets using *mechanical* processes. Some sheets of steel are coated with zinc to produce galvanized steel. This process uses an *electrochemical* process.

In this chapter you will be introduced to the manufacture of several materials that you come into contact with daily. These materials include lumber, plywood, steel, glass, petroleum products, and aluminum.

The primary processes used to produce these materials are viewed in terms of the first process used to change them from a raw material to an industrial material. The other processes used after the initial action will be briefly covered during each specific discussion.

MECHANICAL PROCESSES

Mechanical processes use mechanical forces to change the form of natural resources. They may use compression (pressure) to crush the material to reduce its size or change its texture. Other processes use shearing forces to cut and fracture the material. Still other processes run the material over screens to sort it by size.

A number of natural resources are first processed using mechanical means. The most common of these produces forest products from trees.

As you learned in Chapter 4, wood is a natural composite. Wood is cellulose fibers held together by a natural adhesive called lignin. The processing of forest products uses two methods. The material may be cut or sheared into new shapes. These mechanical actions are used to produce lumber, plywood, and particleboard.

Other wood processing techniques use chemical action. They digest the wood so that the lignin releases the cellulose fibers. The cellulose fibers are then processed into new materials. Chemical processes are used to produce hardboard, cardboard, and paper.

Producing Lumber

Trees were one of the first natural resources used by humans. Trees provided the raw material for shelters and crude tools. Wood was also used as the primary fuel for cooking and heating. In fact, even today, wood

Fig. 15-2. The types of primary processes are mechanical, thermal, chemical, and electrochemical. This photo shows the beginning of a mechanical process—the turning of logs into lumber.

is a major energy source in developing countries around the world.

By the time of ancient Egypt, wood had become a basic material for carpentry and boat building. This use of wood, a natural resource, continues in today's modern civilization.

One form of wood that is widely used is lumber. A piece of **lumber** is a flat strip or slab of wood. Lumber is available in two types:
- **Softwood lumber** is produced from *needle-bearing* trees such as pine, cedar, and fir. Softwood lumber is used by the construction industry, for shipping containers and crates, and for railroad ties. Softwood

Fig. 15-3. Logs are shipped to the mill on trucks or floated down rivers. Lumber production starts when the logs arrive at the mill. (Boise Cascade)

lumber is produced in specific sizes called nominal sizes. Typical sizes range from 1 by 4 (3/4 in. x 3 1/2 in.) and the common 2 by 4 (1 1/2 in. x 3 1/2 in.) to as large as 4 by 12 (3 1/2 in. x 11 1/2 in.). These materials are available in standard lengths in one-foot increments. Generally, they are available in lengths from 6 ft. to 16 ft. You may want to study a building supply catalog to see the many standard sizes of lumber that are available.

- **Hardwood lumber** is produced from deciduous, or *leaf bearing*, trees that lose their leaves at the end of each growing season. Hardwoods are widely used in the cabinet and furniture industry, for making shipping pallets, and in manufacturing household decorations and utensils. Hardwood lumber is produced in standard thicknesses. These range from 5/8 in. thick rough to one-inch thick rough boards, known as 4/4 (pronounced "four-quarter"), to as large as 4 in. thick rough boards, or 16/4. The boards are available in random widths and lengths. Hardwoods are usually not cut to specific widths and lengths like softwood lumber.

The largest quantity of lumber is produced from softwood trees. Therefore, let's look at how lumber is manufactured from logs.

In the previous chapter, you learned how trees are harvested and shipped to the mill, Fig. 15-3. At the mill, logs are stored in ponds to prevent checking (cracking) and to protect them from insect damage. These logs are the material input for lumber manufacturing.

The logs follow some basic steps as they are changed from a natural resource into an industrial material: lumber. These steps, as shown in Fig. 15-4, include:

1. The log is removed from the pond and cut to a standard length. This length is established to give the mill a uniform input and maximum yield from the log.
2. The log is debarked. This process may use mechanical trimmers or high pressure water jets to remove the bark. The bark is a by-product that can be used as fuel for the mill or sold as landscaping mulch.
3. The log is cut into boards and cants at the **head rig**. A head rig is a very large band saw that cuts narrow slabs from the log. When the square center section (called a *cant*) remains, a decision is made. The cant may remain at the head rig to be cut into thick boards, or it may move to the next step.
4. The cant is cut into thin boards at a **resaw**. This machine is a group of circular or scroll saw-type blades evenly spaced to cut many boards at once. Small logs often move directly from the debarker to the gang saw, bypassing the head rig.

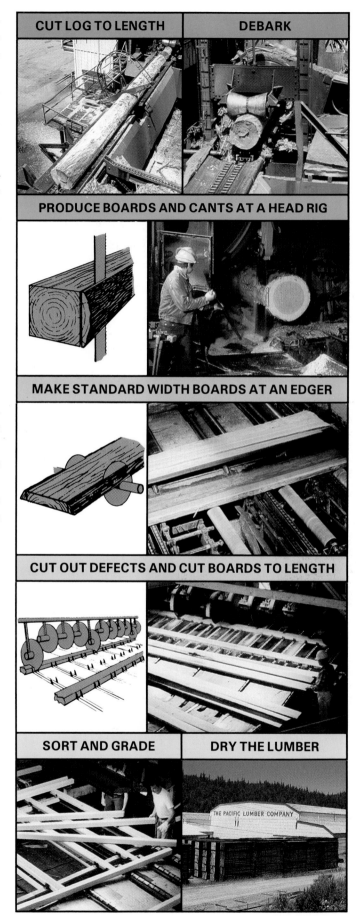

Fig. 15-4. These are the steps in lumber manufacturing.

5. The boards are cut to standard widths at an **edger saw**. This machine has a number of blades on a shaft. The blades can be adjusted at various locations to produce standard widths from 2 to over 12 inches.

6. The edged boards are cut to standard lengths at a **trim saw**. This machine has a series of blades spaced two feet apart. The operator can actuate any or all the blades. This allows for cutting out defects and producing standard lengths of lumber. The boards may be 6, 8, 10, 12, 14, or 16 ft. long. All the blades cut low-quality boards. This cuts the board into two-foot long scrap pieces. The scrap from the edgers and trim saws is used as fuel for the sawmill or becomes the raw material for hardboard, particleboard, waferboard, oriented strand board, and paper.

The processed lumber then moves onto the **green chain**. The boards move down a conveyor and are inspected and sorted by quality. The lumber is dried to make it a more stable product. The two kinds of drying are air and kiln (oven) drying.

The dried lumber is shipped as an industrial material. Some lumber receives special processing. Short boards can be processed into longer boards, Fig. 15-5. Special finger joints are cut in the ends of the boards. The boards are then glued together to form a continuous ribbon of lumber. The ribbon is cut into standard lengths as it leaves a glue curing machine.

Producing Plywood

Another common forest product is plywood. Plywood is a composite material made up of several layers, Fig. 15-6. Plywood is more stable than solid lumber, because cross-grained layers in the plywood reduce warping and expansion.

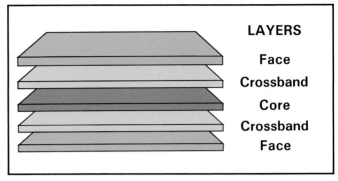

Fig. 15-6. Plywood is made of layers of wood that are glued together under pressure. The number and the thickness of the layers are varied to make different types of plywood.

The outside layers are called **faces**. Between the faces are layers called **crossbands**. The grain of the crossbands is at a right angle (90 degrees) to the face grain. The layer in the center is called the **core**. Its grain is parallel with the face grain. Plywood with only three layers does not have crossbands. Three-layer plywood has a core with its grain running at a right angle to the face layers.

There are three types of cores used for plywood. Fig. 15-7. The most common is **veneer core plywood**. A **veneer** is a thin sheet of wood that is sliced, sawed, or peeled from a log. Plywood used for cabinet work and furniture usually has a lumber core or a particleboard core. **Lumber core** plywood has a core made from pieces of solid lumber that have been glued up to form a sheet. **Particleboard core** plywood has a core made of particleboard. Particleboard is made up of wood chips that are glued together under heat and pressure.

Typically plywood is available in four ft. by eight ft. sheets. Thicknesses from 1/8 in. to 3/4 in. are available.

Veneer core plywood is produced in two stages. The first stage makes the veneer. The veneer is sliced or

Fig. 15-5. These operations are changing short lengths of redwood lumber into long boards. Later, they are made into molding for homes.

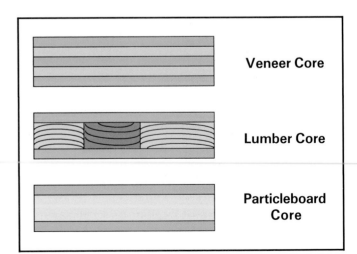

Fig. 15-7. Plywood is available with three types of cores: veneer, lumber, and particleboard.

peeled from the log and moves through a dryer. The dried veneer is sheared into workable size pieces. Defects are cut out or are patched.

Now the veneer is ready for the second stage: plywood production. Glue is applied between the layers. The layers of veneer are stacked up. Then, the sheet is placed in a heated press. The press is closed and the glue cures under heat and pressure.

After pressing, the sheet is removed, trimmed to size and sanded. The completed sheets are inspected and loaded for shipment. Look at Fig. 15-8. This series of photos show the steps used in manufacturing common plywood.

Log is loaded in a veneer lathe.

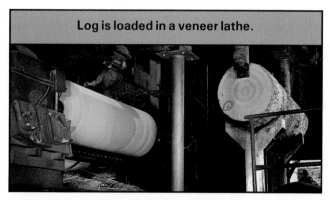

Ribbon of veneer is cut from the log.

Veneer is cut and spliced.

Defects are patched.

Veneer is glued and pressed into a sheet.

Sheets are cut to size and sanded to thickness.

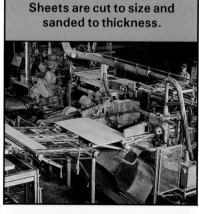

Finished sheets are shipped to customers.

Fig. 15-8. These are the steps in plywood manufacture.

FLOWLINE OF STEELMAKING

From iron ore, limestone, and coal in the earth's crust to space-age steels—this fundamental flowline shows only major steps in an intricate progression of processes with their many options.

Fig. 15-9. This illustration shows the steps it takes to turn iron ore into steel products. The three main stages are smelting, steelmaking, and steel finishing. (American Iron and Steel Institute).

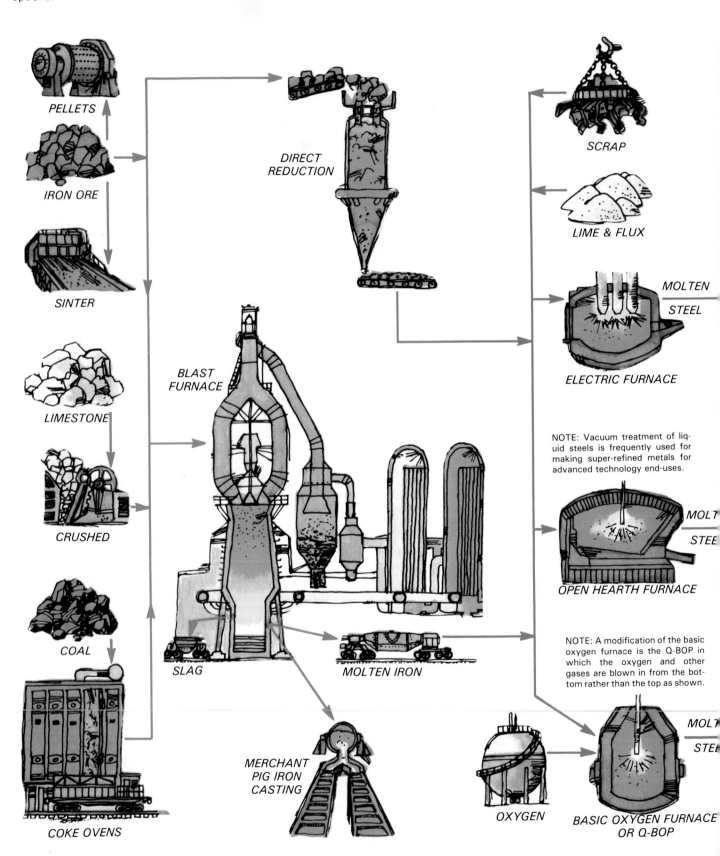

PELLETS

IRON ORE

SINTER

LIMESTONE

CRUSHED

COAL

COKE OVENS

DIRECT REDUCTION

BLAST FURNACE

SLAG

MERCHANT PIG IRON CASTING

MOLTEN IRON

OXYGEN

SCRAP

LIME & FLUX

ELECTRIC FURNACE

MOLTEN STEEL

NOTE: Vacuum treatment of liquid steels is frequently used for making super-refined metals for advanced technology end-uses.

OPEN HEARTH FURNACE

MOLT STEE

NOTE: A modification of the basic oxygen furnace is the Q-BOP in which the oxygen and other gases are blown in from the bottom rather than the top as shown.

BASIC OXYGEN FURNACE OR Q-BOP

MOLT STEE

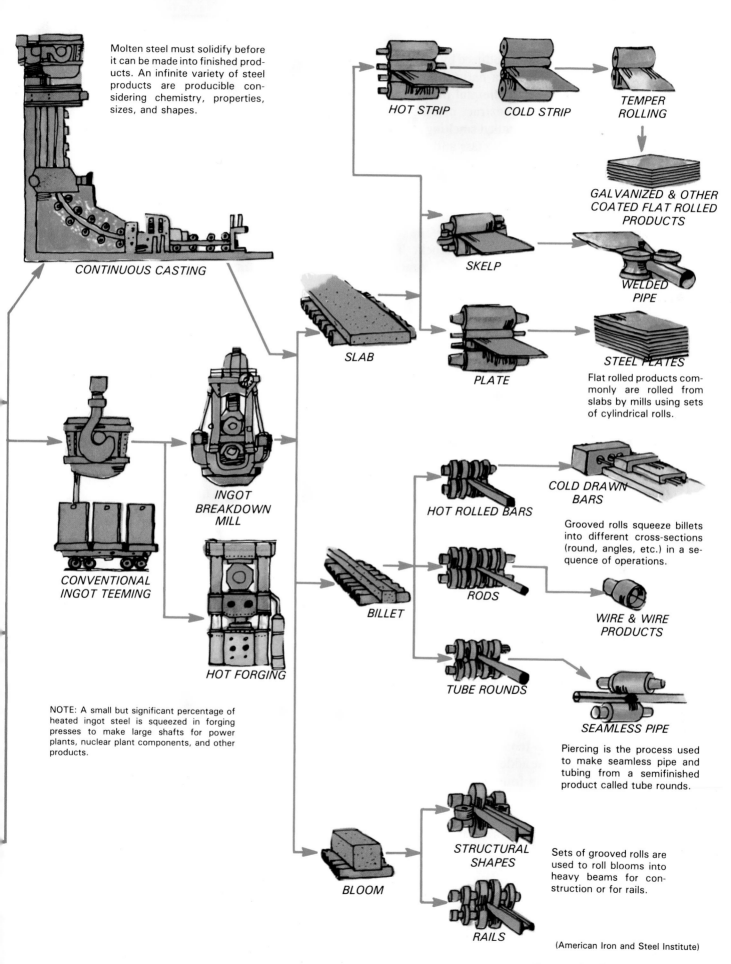

Molten steel must solidify before it can be made into finished products. An infinite variety of steel products are producible considering chemistry, properties, sizes, and shapes.

HOT STRIP

COLD STRIP

TEMPER ROLLING

GALVANIZED & OTHER COATED FLAT ROLLED PRODUCTS

CONTINUOUS CASTING

SLAB

SKELP

WELDED PIPE

PLATE

STEEL PLATES

Flat rolled products commonly are rolled from slabs by mills using sets of cylindrical rolls.

INGOT BREAKDOWN MILL

CONVENTIONAL INGOT TEEMING

HOT ROLLED BARS

COLD DRAWN BARS

Grooved rolls squeeze billets into different cross-sections (round, angles, etc.) in a sequence of operations.

BILLET

RODS

WIRE & WIRE PRODUCTS

TUBE ROUNDS

SEAMLESS PIPE

HOT FORGING

NOTE: A small but significant percentage of heated ingot steel is squeezed in forging presses to make large shafts for power plants, nuclear plant components, and other products.

Piercing is the process used to make seamless pipe and tubing from a semifinished product called tube rounds.

STRUCTURAL SHAPES

Sets of grooved rolls are used to roll blooms into heavy beams for construction or for rails.

BLOOM

RAILS

(American Iron and Steel Institute)

THERMAL PROCESSES

Many industrial materials are produced by processes that use heat to melt and reform a natural resource. This type of processing is called **thermal processing**. This technique is widely used to extract metals from their ores. This process is often called **smelting**. Other thermal processes are used to make glass and cement. For this discussion we will look at steel and glass making. Both of these use a combination of thermal and chemical processes. The thermal energy melts the materials. During the melting process, chemical reactions take place to produce a new material. The new material is then shaped into standard stock. The shaping process uses mechanical techniques. The material is cast, drawn, rolled, or squeezed into new sizes and shapes.

Fig. 15-10. This photo shows a steel furnace being tapped. The furnace is tilted and the steel pours out. (American Iron and Steel Institute)

Producing Steel

Steel is an alloy, a *mixture* of iron and carbon. Adding other elements in small amounts gives the steel specific qualities. These elements include manganese, silicon, nickel, chromium, tungsten, and molybdenum. Steel with nickel and chromium is called stainless steel. Adding molybdenum increases the hardness of the steel. Molybdenum steels are widely used for tools. Adding tungsten makes the steel more heat resistant.

The making of steel requires three steps, as shown in Fig. 15-9:
- Iron smelting.
- Steelmaking.
- Steel finishing.

Iron smelting produces pig iron, the basic input for steel making. **Pig iron** results from thermal and chemical actions that take place in a **blast furnace**. The blast furnace uses a *continuous process*. At no time is the blast furnace empty during the smelting process. A blast furnace operates 24 hours a day, 365 days a year. Every so often raw materials are added to the top and molten pig iron is removed from the bottom.

The operation of a blast furnace is simple. Alternating layers of iron ore, coke (coal with the impurities burnt out), and limestone are added to the furnace. The blast furnace is charged with four parts iron ore, two parts coke, and one part limestone.

Very hot air is blown into the bottom of the furnace. The coke burns, causing the iron ore to melt. The oxygen in the iron ore combines with the carbon to make carbon monoxide gas. During the melting, limestone joins with impurities to form slag. The slag floats on the molten iron and can be drawn off. This leaves molten iron with carbon dissolved in it. This material is called pig iron, which is iron with 3 to 4.5 percent carbon. Pig iron also has 1 to 2 percent other elements, including manganese, silicon, sulfur, and phosphorus.

The steelmaking process starts with the pig iron produced in the iron smelting step. Steelmaking actually removes some carbon from the iron. Heat and oxygen are used to take some of the carbon out of molten pig iron.

The most common steelmaking process uses the basic-oxygen furnace. There are three steps to making steel in a basic oxygen furnace. The first is **charging**. The furnace tilts to one side to receive pig iron, scrap steel, and flux, a material that combines with impurities to form slag. This charge provides the basic ingredients for steel.

The second step is **refining**. The furnace moves into an upright position and the charge is melted. Then, a water-cooled oxygen lance is placed above the molten material. Pure oxygen is forced out of the lance into the iron at supersonic speeds. The oxygen causes the part of the carbon to burn away, producing steel and slag.

The final step is **tapping**, Fig. 15-10. The floating slag is skimmed off the melt. The entire furnace tips to one side and the steel is poured out. It is now ready to enter the steel finishing cycle.

Steel finishing changes molten steel into sheets, plates, rods, beams, and bars. The first step involves pouring the steel into ingots or into the head end of a continuous caster. A continuous caster solidifies the molten steel into shapes called slabs, billets, and blooms. A **slab** is a wide, flat piece of steel, Fig. 15-11. Sheets, plates and skelp are produced from them. Sheets are wide, thin strips of steel, while plates are thicker. Skelps are strips of steel that are used to form pipe.

Billets are square, long pieces of steel. Bars and rods are produced from them. A **bloom** is a short, rectangular piece used to produce structural shapes and rails.

Steel is prone to rust if it is left exposed to the atmosphere. Therefore, many steel shapes are *finished*. They may receive a zinc or tin coating, Fig. 15-12. Zinc coated steel is called **galvanized steel**. It is used for automobile parts and containers, such as fenders, buckets and trash cans. Tin coated steel is called **tin plate**. Tin plate is widely used to make food cans.

Fig. 15-11. These hot slabs are waiting to be rolled into sheets or plates. (Inland Steel Company)

Producing Glass

Glass is another material that is produced using thermal processes. **Glass** is made by solidifying molten silica in an amorphous state. An amorphous material has no internal structure like the regular, uniform lattice structures of metals.

Glass is made from sand (silica), soda ash (sodium carbonate), and lime (from limestone). These ingredients are weighed and mixed to form a batch. For sheet glass the mixture contains about 70 percent silica, 13 percent lime, and 12 percent soda.

The batch is moved into a melting furnace. A typical furnace for flat glass could be 30 ft. (9 m) wide, 165 ft. (50 m) long and 4 ft. (1.2 m) deep. It holds about 1200 to 1500 tons (1088 to 1360 metric tons) of glass at one time. The melting end of the furnace reaches temperatures of 2880°F (1540°C). This heat causes the material to melt and flow together.

From the furnace the glass may go to a secondary manufacturing process. Products like jars, bottles, dishes, glasses, and cookware are manufactured using casting and forming techniques, Fig. 15-13.

Primary manufacturing lines may change the glass into sheets for windows and similar products, Fig. 15-14. Most of this glass is called **float glass**. Float glass is formed by floating the molten glass on a bed of molten tin. The glass flows out of the furnace onto the tin. The glass forms a ribbon as it cools and moves toward the end of the float tank. A typical float tank is 150 ft. (46 m) long and can form glass 160 in. (4064 mm) wide.

The formed glass moves from the float tank to an annealing oven, called a lehr. The temperatures in the

Fig. 15-12. This massive machine applies a tin coating to steel to make tin plate. (American Iron and Steel Institute)

Fig. 15-13. Large quantities of glass are processed into food containers like these jars. (Owens-Brockway)

Separation does exactly what its name says. Separation breaks petroleum into major hydrocarbon groups. The process used is called **fractional distillation**. The petroleum is pumped through a series of tubes in a furnace. There it is heated to about 725°F (385°C). The petroleum becomes a series of hot liquids and vapors. They pass into the bottom of a **fractionating tower**, Fig. 15-15.

This tower can be a tall as 100 ft. (30 m). Within the tower are a series of pans or trays that can hold several inches of liquid. Each pan and its fluid are maintained at a specific temperature. The higher the level in the tower, the lower the temperature.

The hot vapors coming from the furnace rise in the tower. As they rise they are forced to bubble through the liquid in the pans. This action cools the vapors to the temperature of that pan. Hydrocarbons with a boiling point at or below the pan's temperature will condense and stay in the pan. Other vapors will continue to rise. The condensed liquids are continuously drained from the trays.

lehr start at 1200°F (650°C) and falls to 400°F (200°C). This gradual temperature drop relieves internal stresses in the glass. From the lehr the ribbon of glass is cut and packed for shipment.

Other glass products are **fiberglass** and fiber-optic cables. These are strands of glass that are formed, cooled, and annealed. Fiberglass is used as the matrix for composite materials and for insulation. Fiber-optic cables are used to transmit voice, television, and computer data at high speeds.

Refining Petroleum

Petroleum is another resource that is processed using thermal actions. As you learned earlier, **petroleum** is not a uniform material. Instead, it is a mixture of a large number of different hydrocarbons. The process used to isolate these hydrocarbons is called **refining**.

Most petroleum refineries use three processes: separation, conversion, and treating.

Fig. 15-15. These fractionating towers are in Saudi Arabia. (American Petroleum Institute)

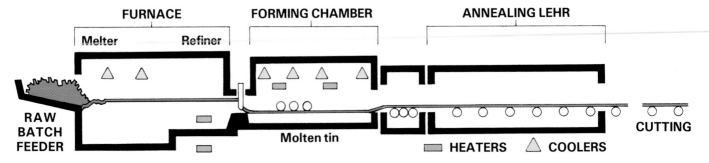

Fig. 15-14. This diagram shows a float glass production line. (PPG)

Technology: How it Works

NUCLEAR ENERGY: Using controlled reactions to release the energy that holds atoms together.

Humans have always marveled at the sun. It is a giant ball of hydrogen gas. The sun's intense heat warms our planet and creates its weather patterns. This heat is created by reactions that turn millions of tons of hydrogen into helium every second. This process is called *nuclear fusion*.

Fusion can combine the nuclei (the centers of atoms) of two atoms with a low atomic number. Fusion produces heavier nuclei, releases neutrons, and great amounts of energy.

The most important fusion process unites atoms of hydrogen, Fig. 1. This element is the lightest of all. Hydrogen atoms contain one proton in the nucleus and one orbiting electron. However, other forms of hydrogen exist. These forms are called *isotopes*. The two isotopes that are used in fusion are deuterium and tritium. Deuterium has *one* neutron in its nucleus while tritium has *two* neutrons. Fusion of these two isotopes creates one helium nucleus. Helium, in its normal condition, has two protons and two neutrons. One neutron is left over. Also, over 17 million electron volts of energy are released.

Creating a controlled fusion reaction is very difficult. The atoms are very small and the nuclei carry positive charges. Since like charges repel, the nuclei resist coming close together. Therefore, they must be accelerated to high speeds to overcome the repelling forces. Also, the extra neutron causes a serious problem. These particles can break free and damage the vessel in which the reaction is taking place.

Fusion is not the only way to use the energy of the atom. A second method is called *fission*, Fig. 2. This process breaks down radioactive materials into smaller molecules. Fission releases some of the energy that holds the atom together.

The most common material used in fission is uranium. Uranium has three main isotopes, U-238, U-235, and U-234. U-235 is the isotope used in nuclear reactors. When a U-235 molecule is bombarded with neutrons, the nucleus breaks apart. It forms two new molecules. The extreme heat that is released is used to create steam to turn turbines in electric generating plants. Fission reactors also provide power on submarines and ships.

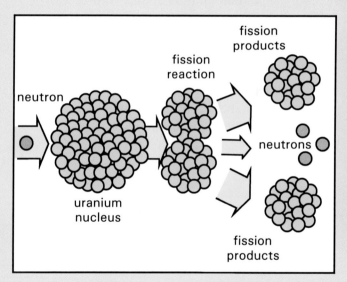

Fig. 2. Nuclear fission breaks apart atoms of radioactive materials. However, the waste products from fission reactors are very harmful to living things.

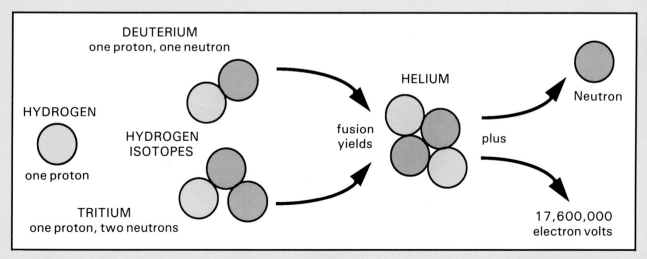

Fig. 1. Nuclear fusion forces the nuclei of atoms together. The extra neutron can be a problem to contain.

Typically, the lighter fractions (products) are taken off the top. These are gasoline and gases such as propane and butane. Other fractions in descending order of temperature are jet fuel, kerosene, diesel, fuel oil, and asphalt. Fig. 15-16 shows a diagram of two distillation towers.

The process of **conversion** changes hydrocarbon molecules into different sizes, both smaller and larger. For example, heavier hydrocarbons may be broken into smaller ones. This process is called **cracking**. Thermal cracking heats heavier oils in a pressurized chamber, Fig. 15-17. The heat and pressure causes the hydrocarbon molecules to break into smaller ones. Catalytic cracking does the same by using a chemical called a catalyst. A *catalyst* is a chemical that helps a reaction to take place. The catalyst is not used up during the reaction. Both catalytic and thermal cracking are used to increase the amount and the quality of products produced from a barrel of petroleum.

A second conversion process is **polymerization**. It is the opposite of cracking. Polymerization causes small hydrocarbon molecules to join together. Refinery gases are subjected to high pressures and temperatures in the presence of a catalyst. They unite (polymerize) to form hydrocarbon liquids. This increases the yield of petroleum products.

Fig. 15-17. This cracking facility in Venezuela processes over 50,000 barrels of heavy oil a day. (American Petroleum Institute)

Treating is the third petroleum refining processes. Treating adds or removes chemicals to change the properties of petroleum products. Sulfur may be removed so kerosene will burn cleaner and smell better. Additives improve the lubrication properties of oils. Other additives help fuels burn quickly and cleanly.

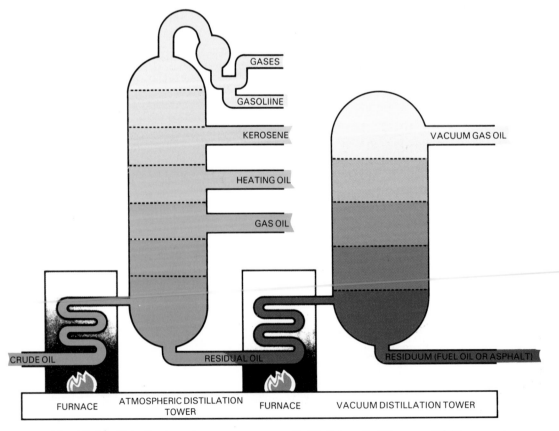

Fig. 15-16. This diagram shows a two-stage distillation unit. (Chevron-USA)

CHEMICAL AND ELECTROCHEMICAL PROCESSES

Some primary processes use **chemical** or **electrochemical** actions. These are used to produce synthetic fibers, pharmaceuticals, plastics, and other valuable products. These processes break down or build up materials *by changing their chemical composition.*

Producing Aluminum

The most common chemical and electrochemical process in use is the one to make aluminum. A diagram is shown in Fig. 15-18. This process is carried out in two stages.

The first stage chemically changes aluminum ore, known as **bauxite,** into aluminum oxide or alumina. The bauxite is crushed and mixed with a caustic soda, sodium hydroxide. The soda dissolves the aluminum oxide forming a sodium aluminate solution that is called *green liquor.* This process leaves behind a residue containing iron, silicon, and titanium. This sludge is called red mud.

The sodium aluminate liquid is pumped into a digester. Some aluminum trihydrate, an aluminum oxide/water compound, is added to start or seed the process. The mixture is agitated with compressed air and cooled. During the process, a chemical action takes place. Aluminum trihydrate and caustic soda are formed. The aluminum compound settles out of the solution. It is removed and dried in a kiln. The drying drives the water out of the aluminum trihydrate leaving pure aluminum oxide. This material is called **alumina** and is the input to the second phase of making aluminum.

In the second stage, alumina from stage one is converted into pure aluminum. This process takes place in large electrolytic cells called pots. The cell has a carbon lining which is the cathode. The anode is also made of carbon.

The pot is filled partially with an *electrolyte*, a material that will conduct electricity. This material is a mixture of molten aluminum fluoride (a salt) and molten cryolite. The electrolyte is kept at 1650°F (900°C). The alumina is dissolved in the electrolyte. Large quantities of electrical current are passed between the anode and cathode. The electrical energy causes the alumina to break into pure aluminum and oxygen. The aluminum settles to the bottom of the pot where it is gathered, purified, and cast into ingots.

The ingots are then formed into sheets, bars, rods, or other shapes. These become industrial materials used for numerous secondary manufacturing processes.

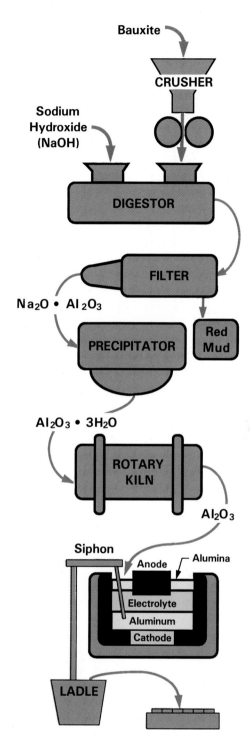

Fig. 15-18. This diagram shows the stages of aluminum production from bauxite to pure aluminum ingots.

SUMMARY

Primary manufacturing processes change natural resources into industrial materials. They accomplish this through mechanical, thermal, chemical, and electrochemical means. From the processes come industrial materials which are often called standard stock. They are the materials inputs that are used to manufacture products and construct structures.

WORDS TO KNOW

All of the following words have been used in this chapter. Do you know their meanings?

Alumina
Bauxite
Billet
Blast furnace
Bloom
Charging
Chemical processing
Core
Cracking
Crossbands
Edger saw
Electrochemical processing
Faces
Fiberglass
Float glass
Fractional distillation
Fractionating tower
Galvanized steel
Glass
Green chain
Hardwood lumber
Head rig
Lumber
Lumber core plywood
Mechanical processing
Particleboard core plywood
Petroleum
Pig iron
Polymerization
Primary processing
Refining
Resaw
Separation
Slab
Smelting
Softwood lumber
Steel
Tapping
Thermal processing
Tin plate
Treating
Trim saw
Veneer
Veneer core plywood

TEST YOUR KNOWLEDGE

Write your answers on a separate piece of paper. Please do not write in this book.

1. List the three types of primary processes.

2. True or false. The center layer of a sheet plywood is called the crossband.

3. Match the process on the left with the type of primary processing it uses from the right.

   _____ Steelmaking.   A. Mechanical.
   _____ Lumber   B. Thermal.
           manufacture.   C. Chemical or
   _____ Petroleum refining.   electrochemical.
   _____ Aluminum smelting.
   _____ Plywood production.
   _____ Glass production.

4. List the three types of cores that are used in plywood manufacture.

5. True or false. Needle-bearing trees are called conifers.

6. Pig iron is produced in a _____ furnace.

7. Bauxite is the raw material for producing _____.

8. True or false. Glass produced on a molten bed of tin is called tin-plate glass.

9. True or false. Steel is a hydrocarbon material that is a mixture of iron ore and carbon.

APPLYING YOUR KNOWLEDGE

1. Build a scale model of a plant to produce an industrial material from a natural resource.

2. Select a material that was not discussed in this chapter. Examples are paper, hardboard, copper, gold, nylon, or polyethylene. Research the processes used to produce the material you choose. Report your results in a chart similar to the one shown below.

Material:
Primary natural resource used to make the material:
Primary production process (steps):
Uses for the material:

3. Develop a list of materials that you see used in the room you are in. List the type of process used to produce each one.

MANUFACTURING PRODUCTS

After studying this chapter you will be able to:
* *Define manufacturing.*
* *List and describe the six types of secondary manufacturing processes.*
* *Describe the basic principles or concepts of casting and molding.*
* *Describe the basic principles or concepts of forming actions.*
* *Describe the basic principles or concepts of separating actions.*
* *List and describe the types of separating machines.*
* *Tell the difference between cutting and feed motions.*
* *Tell the difference between casting and molding, forming, and separating as techniques to give size and shape to materials.*
* *Describe conditioning actions.*
* *Describe finishing.*
* *Categorize the common types of finishes.*
* *List and describe the ways to apply finishes.*
* *Describe assembly actions.*
* *List and describe the major types of fasteners.*

In the last chapter you learned that primary processing produces standard stock. These materials have little worth to the average person. What can you do, by yourself, with a sheet of steel, a two by four stud, a pound of polypropylene pellets, or an ingot of pure aluminum? These materials must be changed into products before they are useful to you. The actions used to change standard stock into products are called **secondary manufacturing processes**, Fig. 16-1.

TYPES OF MANUFACTURING PROCESSES

There are thousands of manufacturing processes. They are used to change the size and shape of materials, to fasten materials together, to give materials desired properties, or to coat the surfaces of products. It would be very difficult to study and understand all of the individual ways to process materials. However, all secondary manufacturing processes can be classified into six groups, Fig. 16-2:
* Casting and molding.
* Forming.
* Separating.
* Conditioning.
* Assembling.
* Finishing.

Each process has actions or concepts that are common to all the other processes in its group. Within these groups are specific processes or techniques. Each one differs in some way from the other processes in the group. In other words, these groups are like the members of a family. They do many things alike and look alike, but each member is unique.

CASTING AND MOLDING PROCESSES

Three groups of processes give size and shape to pieces of material, Fig. 16-3. The first of these is casting and molding. **Casting and molding** give materials shape by introducing a liquid material into a mold. The mold has a cavity of the size and shape that is wanted. The liquid material is poured or forced into the mold. The material is allowed to solidify before being removed.

In this discussion, we will refer to molten and fluid materials as *liquids*. Molten refers to materials that are

Fig. 16-1. Secondary manufacturing processes turn industrial materials into useful products. In these photos the wire on the left is combined with other materials to make the tires shown on the right.

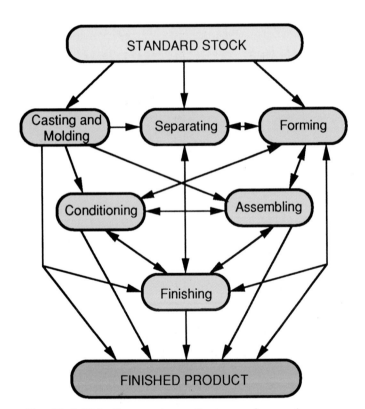

Fig. 16-2. This diagram shows the types of secondary manufacturing processes and how they relate to each other.

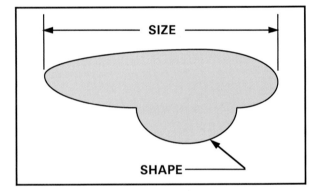

Fig. 16-3. Size is the dimensions of the object, while shape is its form and outline.

- Allowing the material to harden.
- Removing the completed part from the mold.

Molds

All casting and molding processes require a mold to hold the liquid material until it becomes solid. Inside the mold is a cavity of the proper size and shape. Channels, called gates, guide the liquid into the cavity.

There are two major types of molds that are used in casting and molding processes. These are expendable molds and permanent molds.

Expendable Molds. Most cast products are made in molds that are used once. The mold is destroyed to remove the cast item. These molds are called **expendable molds**. Molds can have two or more parts. They are generally made in two steps. First, a pattern of the same shape but slightly larger than the finished product is made. The extra size allows for shrinkage. Most materials *shrink* when they change from a liquid to a solid.

heated to a fluid state. These materials are normally solid at room temperature. Fluid materials, such as water and casting plastics, are liquid at room temperature.

All casting processes involve five basic steps, which are shown in Fig. 16-4. These steps are:
- Producing a mold of the proper size and shape.
- Preparing the material.
- Introducing the material into the mold.

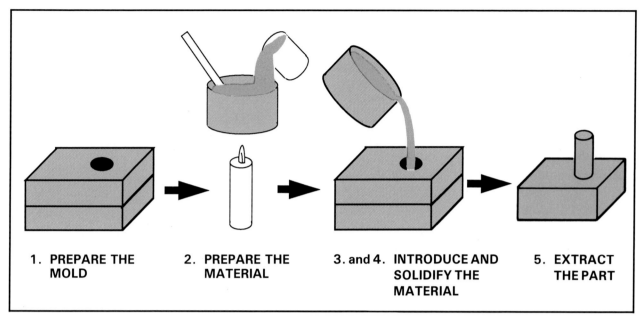

1. PREPARE THE MOLD   **2. PREPARE THE MATERIAL**   **3. and 4. INTRODUCE AND SOLIDIFY THE MATERIAL**   **5. EXTRACT THE PART**

Fig. 16-4. There are five stages in a casting and molding process.

An exception is water, which *expands* as it changes from a liquid to a solid.

The pattern is the foundation for making an expendable mold. The pattern is surrounded with an inexpensive substance such as sand or plaster. Sand is tamped into place, and plaster is allowed to dry around the pattern. When the mold is completed, the mold parts are separated and the pattern is removed. This leaves a cavity of the correct shape to be filled by the liquid material. The mold parts are put back together to make a ready-to-use mold, Fig. 16-5.

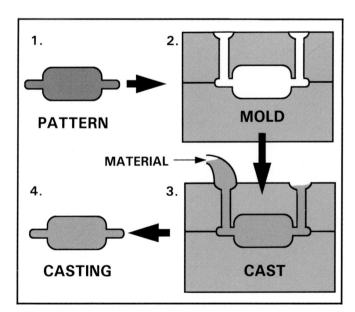

Fig. 16-5. There are four stages in making an expendable mold casting: 1. Making the pattern. 2. The pattern is surrounded with sand or plaster. 3. The liquid material is introduced and allowed to solidify. 4. The mold is broken up and the casting is removed.

Safety in Casting Occupations

People using casting processes observe some basic safety rules.

1. Wear protective clothing and goggles when pouring a casting.
2. Make sure everything to be used in the casting process is free of moisture.
3. Place hot castings where they will not start a fire or burn someone.
4. Keep the casting area orderly.
5. Do not talk to anyone while pouring.
6. Do not stand over the mold when pouring molten metal.

The key to safety in the technology education laboratory is a *SAFE* attitude.

Types of Expendable Molds

Common casting processes that use expendable molds are green-sand casting, shell molding, and investment casting. **Green sand casting** uses sand held together with a binder. Oil and water are used as binders. Oil-bound sand is called *oil sand*. Water bound sand is called *green sand*. The sand is rammed around the pattern to form the mold cavity.

Shell molding uses a sand and resin mixture. This mixture is poured over a heated metal pattern. The heat melts the resin which bonds the sand into a thin shell. The shell halves are cured and assembled to make the mold.

Investment casting uses molds made from plaster. The pattern is normally made from wax. The plaster is poured around the pattern and allowed to dry. The pattern and the plaster shell are placed in an oven. The wax melts and flows out of the plaster mold. This leaves a precisely-shaped mold cavity.

Permanent Molds. A second type of mold is called a permanent mold. **Permanent molds** withstand repeated use. They are often made from steel, aluminum, or plaster, Fig. 16-6. The mold must withstand temperatures above the melting point of the material being cast.

Most metal molds are produced by machining out the cavity. Plaster molds are often produced by casting the mold material around a pattern.

Processes that use permanent molds are die casting, injection molding, and slip casting. Die casting is used to produce aluminum and zinc parts. Injection molding can produce a wide range of plastic parts in many different resins. Slip casting is used to produce clay products. These processes will be discussed later in this chapter.

Types of Liquid Materials

All casting and molding processes require a material in a liquid state. A liquid can flow freely like water. Three types of liquids are used in casting: solutions, suspensions, and molten materials.

A *solution* is a uniform mixture of two substances. Casting plastics and frozen treats are examples of solutions. They are mixtures that will give desired properties when solid. The act of mixing the parts of a fluid for casting is called **compounding**.

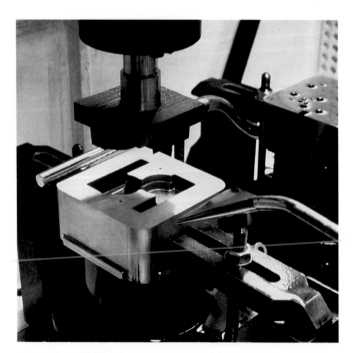

Fig. 16-6. This permanent mold has been machined using the electrical discharge machining (EDM) process. (Agie Losome, Switzerland)

Suspensions are mixtures where the particles will settle out. To keep the components mixed the suspension must be shaken. When the shaking stops, the particles begin to settle out. Slip, which is used in casting ceramics, is a suspension of clay and water.

Some materials, like metals, must be heated to a *molten* state in order to be cast. At room temperature, steel is a solid, and cannot be poured into a mold. Plastics that are injection molded are heated to a liquid state so they can be forced into a mold.

Introducing the Material

The liquid material can now be introduced into the mold. Most expendable molding processes use gravity to fill the mold. The material is poured into the mold and allowed to fill the cavity.

Many permanent molding processes use force to introduce the material into the mold. An example of this technique is shown in Fig. 16-7. This figure depicts the action of an injection molding machine used with plastic resins. The plastic is heated to a liquid state. Then, a ram forces the resin into the mold cavity. After the resin cools and solidifies, then the mold opens. The

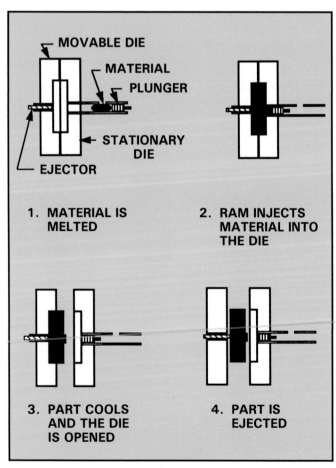

Fig. 16-7. This diagram shows how an injection molding machine works.

finished part is ejected. The mold then closes and the process repeats. Die casting with metals uses the same procedure in a machine similar to an injection molder.

Solidifying the Material

Once the liquid material is introduced into the mold cavity, it must become a solid. There are three ways to do this:
- Cooling.
- Drying.
- Chemical action.

In cooling, the heat energy that caused the solid material to melt leaves. This makes the material return to its solid state. The mold can be cooled by letting the heat radiate into the air. Water can be pumped through passages in the mold to carry the heat away.

Water-based products are also solidified through cooling. Many food products, such as ice pops and ice cream bars, are given their shape in a mold by cooling. This action we call freezing.

Suspended materials are solidified by allowing the solvent to be absorbed into the mold. This action is caused by **drying**. An example of this technique is slip casting, Fig. 16-8. Slip, a clay and water mixture, is poured into the mold. The plaster mold absorbs water from the slip. The action causes a layer of solid clay to build up on the mold walls. The longer the slip stays in the mold, the more moisture it loses, and the thicker the walls become. When the proper wall thickness is reached, the remaining slip is poured out. The product is allowed to dry partially in the mold. Then, the mold is opened and the product is removed. It is placed in a humidity-controlled cabinet for further drying.

The third method of making a material solid is by **chemical action**. For example, casting plastics have hardeners added. These chemicals cause the material to become solid, or set-up. The short molecules in the plastic resin become longer, more complex molecules. This action, called polymerization, changes the liquid plastic into a solid.

Extracting the Product

The solidified product must be removed from the mold, Fig. 16-9. Expendable molds are destroyed to remove the part. Sand molds are shaken apart. Shell and plaster molds are fractured.

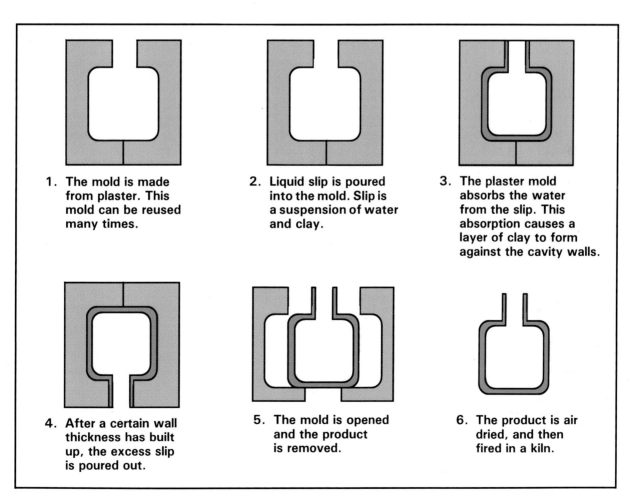

1. The mold is made from plaster. This mold can be reused many times.

2. Liquid slip is poured into the mold. Slip is a suspension of water and clay.

3. The plaster mold absorbs the water from the slip. This absorption causes a layer of clay to form against the cavity walls.

4. After a certain wall thickness has built up, the excess slip is poured out.

5. The mold is opened and the product is removed.

6. The product is air dried, and then fired in a kiln.

Fig. 16-8. These are the steps in making a slip cast product.

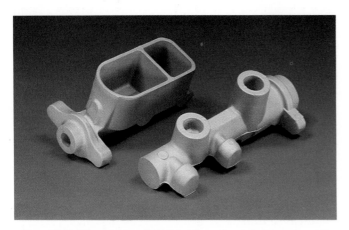

Fig. 16-9. These completed castings have been removed from the mold. They can be machined and assembled with other parts to make useful products. (Stahl)

Permanent molds are designed for easy opening. Often the molding machine automatically introduces the material, causes it to solidify, and then, opens and ejects the finished part.

FORMING PROCESSES

Forming is the second family of processes that give materials size and shape. All **forming** processes apply force through a forming device to cause the material to change shape. This force must be in a specific range, above the material's yield point and below its fracture point.

All materials react to outside forces, Fig. 16-10. Small forces cause a material to flex (bend). When the force is removed the material returns to its original shape. If the force increases, there is a point where the material will not return to its original shape. This point is called the **yield point**. The range between rest and the yield point is called the material's **elastic range.**

Above the yield point the material will be permanently deformed. The greater the force the more the material will be stretched, compressed, or bent. This range is called its **plastic range.** Finally, the material cannot withstand any more force and it breaks. This point is called the **fracture point.**

All forming processes operate in the plastic range of the material. These processes have three things in common:
- A forming device is present.
- A force is applied.
- Material temperature is considered.

Forming Devices

Forming processes are used to produce specific shapes in a material. They must have a way to ensure that the shape is correct and consistent. This is done by using one of two devices:
- Dies.
- Rolls.

Dies. Dies are forming tools made of hardened steel. Dies can be used to form any material that is softer than they are. The three types of dies, as shown in Fig. 16-11, are open dies, mated dies, and shaped dies.

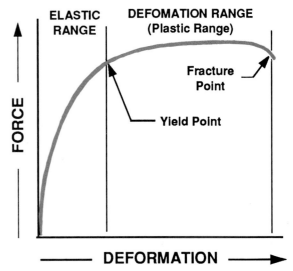

Fig. 16-10. This stress-strain chart shows the elastic and plastic range for a material. Note the yield and fracture points.

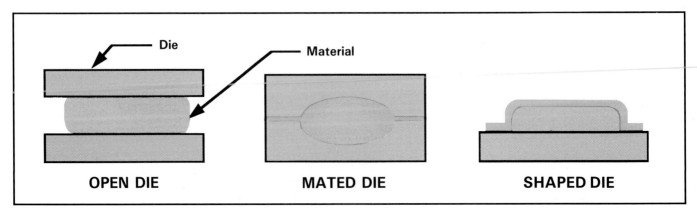

OPEN DIE   **MATED DIE**   **SHAPED DIE**

Fig. 16-11. There are three types of dies.

Open dies are the simplest of all dies. **Open dies** consist of two flat die halves. The material is placed between them and the dies are closed. This action presses the material into a new shape. The most common process using open dies is called **smith forging**, Fig. 16-12. This process is like a blacksmith using a hammer and an anvil to form heated metal. The hammer would be the movable die and the anvil the stationary die.

Mated dies have the desired shape machined into one or both halves of the die set. The material is placed between the die halves. The die set is closed and the material is caused to take on the shape of the die cavities. Drop forging is an example of this type of process.

Fig. 16-12. This large piece of metal is being smith forged. (American Iron and Steel Institute)

In this process the lower die is stationery and the upper die is suspended above it. The material is placed between the two die halves. The upper die is allowed to drop down onto the material. The quick impact caused by the falling die causes the metal to be reshaped. Press forging is very similar, except the upper die is moved by a hydraulic cylinder. This motion produces a squeezing action which shapes the material.

In some cases, mated dies are used to both *form* and *cut* the material. These dies are widely used to produce sheet metal parts. When both forming and cutting is done the processes is called **stamping**. Often there are a number of stations in one die. The material is moved to the next position each time the press cycles. The part slowly takes shape as it moves from one position to the next in the die. This type of die is called a progressive die, Fig. 16-13.

Another process using mated dies is **blow molding**. This process is used to make bottles and jars from plastic and glass. A heated tube of material, called a parison, is formed and lowered into the die. The die halves close, sealing off the bottom of the parison. Air is blown into the center of the tube. The air pressure causes the parison to expand and fill the mold cavity. When it cools the mold is opened and the container is removed.

One-piece **shaped dies** are widely used to form plastic objects. A common process using this type of mold is vacuum forming. A sheet of heated plastic material is placed over the mold. The air between the mold and the sheet is sucked out. This allows atmospheric pressure to force the material tightly over the mold. When it cools, the product is removed, Fig. 16-14.

Fig. 16-13. The photo on the left shows stamping presses in operation. Some presses use progressive stamping dies, as shown on the right. (Minster Machine Co., Alton Tool)

Fig. 16-14. The ice cube tray in this photo was molded using the mold shown at the top. (Brush-Wellman)

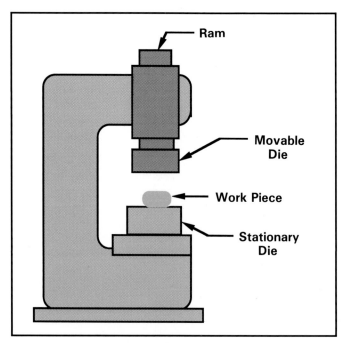

Fig. 16-15. This diagram shows the parts of a forging hammer.

Rolls. Some processes use rolls to form the material. There are two types of rolls: smooth and shaped. Smooth rolls are used to make curved shapes from sheets and shapes of metal. Shaped rolls have shapes machined into their surface. When material passes between shaped rolls the material is squeezed into the shape of the roll. Shaped rolls are used to make pipe, tubing, and corrugated metal.

Forming Force

The force needed to complete a forming action can be delivered in a number of ways. The four most common are:
- Presses.
- Hammers.
- Rolling machines.
- Draw benches.

Presses and hammers are very much alike, Fig. 16-15. They have a bed onto which a stationary die can be attached. The other die part is attached to a ram. A power unit lifts the ram and the material to be formed is placed between the die halves. **Presses** slowly close the die halves by lowering the ram to produce a squeezing action. **Hammers** drop or drive the ram down with a quick action. This motion causes a sharp impact which creates the forming force.

Rolling machines use two rolls that rotate in opposing directions to form the material. The rolls draw the material between them and squeeze or bend it into a new shape. Forming rolls may be smooth, or they may have a pattern machined into them. Smooth rolls are used to

produce metal sheets and foils and to shape sheets and strips into curved products like tanks and pipes.

Shaped rolls produce patterns or bend materials. Typical product produced with shaped rolls are corrugated roofing, bent metal siding, and metal trim.

Drawing machines pull or push materials through die openings. They make wire from rods and extrude other shapes. Wire drawing pulls the rods through a series of dies which have progressively smaller openings. At each die the material is reduced in diameter and increased in length.

Extrusion works much like squeezing toothpaste from a tube. A shaped die is placed over the opening of the machine. The material is held in a closed cavity behind the die. A ram opposite the die forces the material out of the cavity, Fig. 16-16. As the material passes

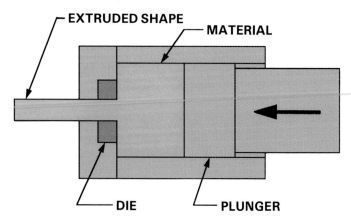

Fig. 16-16. This diagram shows how extrusion works. The material takes on the shape of the die.

through the die it takes on the shape of the opening. Extrusion is used to produce complex shapes from plastic, ceramic, and metallic materials.

Forming forces can be generated by air pressure, electrical fields, or explosive charges.

Material Temperature

Something to consider about forming processes is the temperature of the material while it is being formed. Materials can be formed either *hot* or *cold*. Metals can be formed hot or cold. Plastics and glass are formed hot, while ceramics are formed cold.

All materials have a temperature at which their properties change. The internal structure changes in all materials as they are heated. We must remember that the terms "hot" and "cold" are relative. A "cold" slab of steel at a steel mill may be hot enough to burn you.

When a metal is cold formed, internal stresses are built up. These stresses can cause the material to become brittle. This is called *work hardening*. Work hardening is relieved by heat treating the metal. Heat treating involves heating the metal and allowing it to cool slowly.

When a metal is hot formed, it is important that the metal be heated above its recrystallization point. Hot forming takes place above this point and cold forming takes place below this point. Hot forming prevents work hardening of the metal. As the material cools it forms a normal structure. Therefore, the minimum temperature for hot forming is different for each material. Also, material shaped by hot forming will be stress free, while cold forming builds internal stress in the material.

SEPARATING PROCESSES

Separating processes remove excess material to make an object of the correct size and shape. Casting and forming processes change the shape and size of materials without any removal.

Separating processes remove material by either machining or shearing, Fig. 16-17. **Machining** is based on the motion of a tool against a workpiece to remove material. **Shearing** uses opposing edges of blades, knives or dies to fracture the unwanted material away from the work.

Machining

Machining removes excess material in small pieces. Machining cuts material away using three methods:
- Chip removal: using a tool to cut away the excess material in the form of chips.

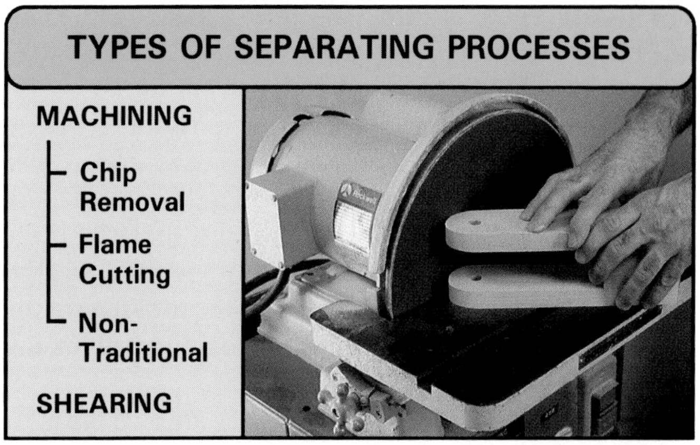

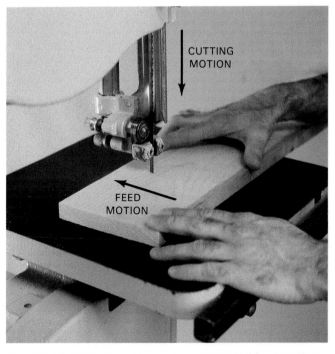

TYPES OF SEPARATING PROCESSES

MACHINING
- Chip Removal
- Flame Cutting
- Non-Traditional

SHEARING

Fig. 16-17. Separating processes include machining and shearing. The photo shows a sanding machine, which uses chip removal to take off unwanted material.

- Flame cutting: using the heat from burning gases to shape and size the material.
- Nontraditional: using electric sparks, chemical action, sound waves, or light waves to separate material.

Machining is the most common separating process. All machining processes have three things in common:
- A tool or other cutting element is always present.
- There is motion between the tool and work.
- The work and the tool are given support, Fig. 16-18.

Tools or Cutting Elements. Machining removes excess material through cutting actions. The most common cutting device is a chip removing tool. **Tools** are based on the fact that a harder material will

CUTTING MOTION

FEED MOTION

Fig. 16-18. This photo shows the cutting and feed motions of a band saw.

cut a softer material. For example, a plastic knife can cut butter, aluminum tools can cut many plastics, steel can cut aluminum, and diamonds can cut steel.

To be efficient, a cutting tool must be properly shaped. The tool has a point, a cutting edge, and a relief angles. The point allows the tool to enter the work. The cutting edge produces the chip as excess material is removed. Relief angles slope away from the point and the cutting edge. They keep the body of the tool from rubbing against the work and cause the chip to roll off the tool.

As you learned in Chapter 7, there are two types of tools: single-point and multiple-point. The single-point tool has one shaped cutting surface, while multiple-point tools have many points. Multiple-point tools can have the points arranged in two ways:
- Evenly spaced, as on a saw blade.
- Randomly spaced, as on abrasive paper and grinding wheels.

Machine Motions

Two types of motion are needed to make a cut. These are the cutting motion and the feed motion.

Cutting motion moves a cutting tool through a material to make chips. **Feed motion** brings new material in contact with the cutting tool. Both motions can occur by either moving the tool or the workpiece.

To understand these actions, think about a band saw, Fig. 16-18. In your mind, start the machine and move a board against the blade. The first tooth moves down through the work and produces a chip. This action is the cutting motion. However, the next tooth would travel in the same path. No additional material would be cut. For cutting to continue, the board has to be moved into the blade. This movement brings new material into contact with the blade. Now, the next tooth has material to cut. Moving the material into the blade is the feed motion of the band saw.

The motion necessary for machining can be made by moving either the tool or the workpiece. There are three types of cutting and feed motion. The first is **linear motion**, where the tool or the work moves in a straight line. The second is **rotary motion**, where the work or the tool rotates. Finally, the work or the tool can move back and forth. This type of movement is called **reciprocating motion**.

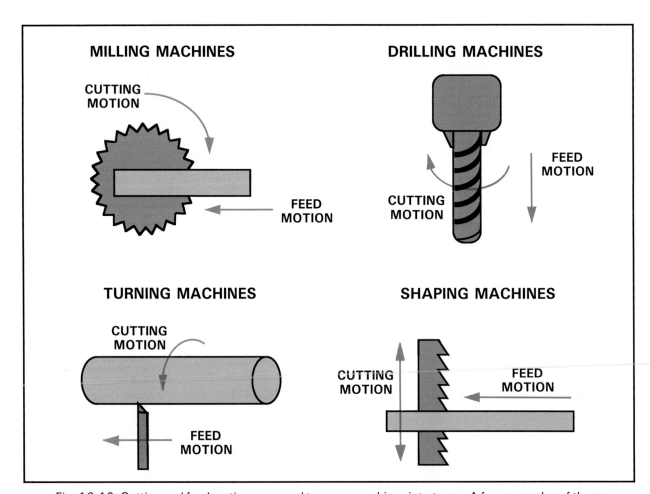

Fig. 16-19. Cutting and feed motions are used to group machines into types. A few examples of these types are shown here.

Technology: How it Works

ROBOT: A programmable part-handling or work-performing device often used to replace human labor in industrial settings.

Robots are a fairly new industrial device used to increase the productivity of operations. Robots also remove humans from dangerous and undesirable working conditions.

The word robot was first coined by the Czech author Karel Capek in the play *Rossum's Universal Robots*. He derived the word, robot, from the Czech word *robota,* which means ''work.''

Some of the first robots were designed in the 1940s to handle radioactive material. The first industrial robot was developed in 1962. The robot's functions were limited to picking up an object and setting it down in a new location. This simple type of robot is called a pick-and-place robot. Today, there are many different types of robots, Fig. 1. They are used in a wide variety of applications including parts handling, welding, and painting.

A robot contains three important units: a mechanical unit, for performing a task (the manipulator); a power unit, to move the robot arm (the power supply); and a control unit, to direct the robotic movement (the controller).

Hydraulic and electric power supplies are the most common units used to raise, lower, and pivot robot arms into various positions. Hydraulic units are usually used to handle heavier objects, and they are generally considered faster. Electric units, though, take up less floor space and run more quietly.

Most robots operate in multiple planes, Fig. 2. The simplest robots rotate around their axis. More complex units can produce motion in two directions–left and right, up and down. Even more complex units add an in and out motion. These three basic motions, or *degrees of freedom*, can be combined with an additional three degrees of freedom (six total) in the end effector (the device at the end of the arm). The six degrees of freedom can place the robot's effector anywhere in its work area.

Fig. 1. Pictured is a common industrial robot.

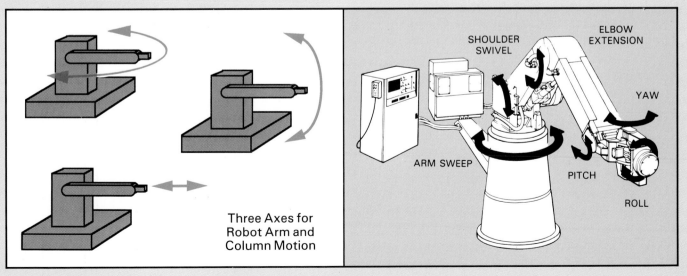

Three Axes for Robot Arm and Column Motion

SHOULDER SWIVEL

ELBOW EXTENSION

YAW

ARM SWEEP

PITCH

ROLL

Fig. 2. These are typical motions of an industrial robot. (Cincinnati Milacron, Inc.)

Safety with Machine Tools

People who work with machine tools use various machines. Some basic rules should be observed during their operation.

1. Wear eye protection.
2. Avoid loose fitting clothing and keep long sleeves rolled up.
3. Wear hearing protection when loud or high pitched noises are present.
4. Keep the laboratory clean. Wipe up any spills immediately.
6. Wait until the machine has stopped to make adjustments, remove the workpiece, or to clean up any scrap. Use a brush to clean chips off of a machine.
7. DO NOT talk to anyone while running a machine.
8. Most of all, if you do not understand how to run a machine, ASK FIRST!

The key to safety in the technology education laboratory is a *SAFE* attitude.

Types of Machines

Machines can be grouped by the cutting and feed motions they use. These groupings include the following, Fig. 16-19:

- **Milling machines** use a rotating cutter for the cutting motion. The feed motion is linear. The material is pushed into the cutter in a straight line. Machines that use the milling principle are: horizontal and vertical milling machines; table saw; wood shaper, planer, and jointer; router; disc sander; and pedestal or bench grinder. The radial saw and the cut-off saw are variations of this group. Rotating multiple-point tools, called blades, are rotated. However, the blade moves to produce the feed and cutting motions at the same time.
- **Drilling machines** rotate a cutter for the cutting motion. The cutter is fed in a linear manner for the feed motion. Machines that use this principle are the drill press and the portable electric drill.
- **Turning machines** rotate the work against a single-point tool to produce the cutting motion. The tool is fed along or into the work to produce the feed motion. Wood and metal working lathes are turning machines.
- **Shaping machines** use a single-point tool that moves back and forth across the work to produce the cutting motion. The cut is usually made on the forward stroke. The tool is lifted sightly for the backstroke. The work is fed linearly under the tool. Metal working shapers are the primary machines using this cutting action. Small saws like hacksaws and scroll saws use a reciprocating motion also.
- **Planing machines** move the work under the tool to make the cutting motion. The feed motion is created by moving the tool across the work in small steps. The metal planer is the machine that uses this technique. Both metal shapers and planers are used to make large machinery.

Flame Cutting

Burning gases can be used to remove unwanted material from a workpiece, Fig. 16-20. A fuel gas and oxygen are mixed in a torch, ignited, and allowed to heat the surface of the metal. When the metal is hot enough, a steam of pure oxygen is forced out of the torch. The stream of oxygen causes the metal to burn. The burning separates the workpiece from the scrap. Cutting material to size and shape using burning gases is called **flame cutting**.

Acetylene gas and natural gas are used as fuel gases for cutting metals. The fuel gas and oxygen are fed through separate regulators to control the pressure. Each gas flows through a hose to the torch. The torch controls the volume and mixture of the gases. The mixture then moves out of the torch. The blast of pure oxygen is controlled separately from the fuel gas and oxygen mixture.

Nontraditional Machining

Chip removing and flame cutting have been used for a long period of time. Since the 1940s, a series of new

Fig. 16-20. This computer controlled flame cutting machine is cutting out parts from steel plate. (Federal Industries)

separating processes were developed. These processes use electrical, sound, chemical, and light energy to size and shape materials. These processes are often called **nontraditional machining**. This term is not very accurate today because many of these processes are widely used. They are rapidly becoming a traditional way to machine materials.

Three common nontraditional processes include electrical discharge machining, laser machining, and chemical machining.

Electrical discharge machining uses electric sparks to make a cavity in a piece of metal. This process requires an electrode, a power source, a tank, and a coolant, Fig. 16-21. The workpiece is connected to one side of the power supply and placed in the tank. The electrode, which is made in the shape of the cavity that is desired, is connected to the other side of the power supply. The tank is filled with coolant. The coolant is a *dielectric* material. A dieletric resists the flow of electricity. The electrode is lowered until a spark jumps between the electrode and the work. When the spark jumps, the dielectric quality of the coolant has been overcome. The spark dislodges small particles of material that are carried away by the coolant. A cavity the same shape as the electrode is created. The electrode is lowered as the cavity is produced until the proper depth is achieved.

Laser machining uses the intense light generated by a laser to cut material, Fig. 16-22. A laser converts electrical energy into monochromatic (meaning single

Fig. 16-22. Lines that are one ten-thousandth of an inch (.0000254 mm) apart are scribed by this laser machining center. (GM-Hughes)

color) light. The light is focused and reflected onto the work. The laser light melts the material where the beam strikes it. The beam of light is guided along a path to produce the desired shape.

Chemical machining uses chemical reactions to remove material from a workpiece. The workpiece is coated with a resist. The resist will not react to the chemicals that are used as the cutting agent. The resist is

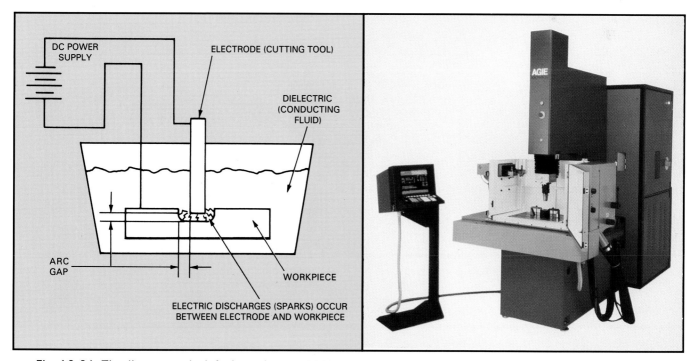

Fig. 16-21. The diagram at the left shows how an EDM works, while the photo on the right shows an EDM machine. (Caterpillar Tractor Co., Agie Losome)

removed from the areas that are to be machined. Then the workpiece is placed in a tank and covered with the cutting chemical. Areas coated with the resist will be untouched. Where the workpiece is exposed to the chemical, material will be removed.

There are two types of chemical machining: chemical milling and chemical blanking. The process is the same for both. The only variable is in how long the workpiece is etched. In chemical blanking, the etching continues until only the part covered by the resist remains. Chemical blanking is used for very thin, delicate parts or for short runs (small numbers of parts). Chemical milling uses the same process, but the part is not totally etched away. Chemical milling is used to remove excess metal in aircraft parts to save weight.

Shearing

The second group of separating processes is shearing. Shearing cuts material to create the desired size and shape. Shearing can be used to cut material to length, produce an external shape, or generate an internal feature, Fig. 16-23. Cutting to length is generally accomplished with opposing blades. The upper blade moves downward to deform and fracture the material where it contacts the lower blade.

Internal and external shapes are often made with a punch and a die. The die has a cavity of the desired shape. The punch fits into this cavity. The material is placed on the die. When the punch moves downward, the material is *sheared* into the shape of the die cavity. Punches and dies are used to produce holes, slots, and notches. Look back to Fig. 16-13. You can observe a number of shearing operations performed with a progressive die set.

CONDITIONING PROCESSES

Casting, molding, and separating operations change the external features of a workpiece. The material is given a new size and shape. In some cases this is not enough. The internal structure of the material needs to be changed. The material may need to be harder, softer, stronger, or more easily worked. To change internal properties, **conditioning processes** are used. The three types of conditioning processes are thermal, mechanical, and chemical conditioning.

Mechanical conditioning uses mechanical forces to change the internal structure of the material. Most metals become harder as they are squeezed, stretched, pounded, or bent. This action is called work hardening. In most cases work hardening is not desired but is unavoidable. If you have ever bent a wire back and forth to cause it to break, you have used work hardening. The repeated bending makes the wire harder and more brittle.

Chemical conditioning uses chemical actions to change the properties of a material. The lenses of many safety glasses are chemically treated to make them more shatterproof. Likewise, the resins used in fiberglass lay-ups undergo chemical action as they cure and become hard.

The most common conditioning processes use heat. These are called **thermal conditioning** and include heat treating, drying, and firing.

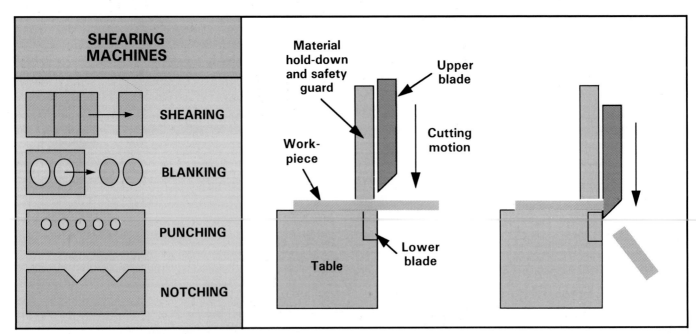

Fig. 16-23. This diagram shows common shearing operations and how shearing works.

Heat treating is a term used to describe the thermal conditioning processes used on metals. These processes include:

- **Hardening** is used to increase the hardness of a material, Fig. 16-24. Hardening steels requires that the part to be heated to a specific temperature. The part is allowed to "soak". This ensures that the entire part is at a uniform temperature. The part is then rapidly cooled in a tank of oil or water.
- **Annealing** is used to soften and remove internal stress in a part. The part is heated to a specific temperature and allowed to soak. Then the part is removed from the oven and allowed to cool slowly to room temperature.
- **Tempering** is used to relieve internal stress in a part. Hardening often creates internal stress that will cause a part to crack under use. This stress is removed by heating the part to a specific temperature and allowing it to slowly cool. The tempering temperature is much lower than the annealing temperature. Tempering is used for metal parts and most glass products.

Firing is a thermal conditioning process used for ceramic products. Most ceramics are made from clay materials that are plastic when wet. After drying, the clay can be heated to a high temperature. The water is driven out of the clay particles and the grains bond together to make a solid structure.

Likewise, certain coatings are fired. Porcelain enamels are fired to give a glass-like finish. The enamel is applied to the part by spraying or dipping. The product is heated and the coating fuses with the part.

Drying is a common thermal conditioning process. Drying removes excess moisture from materials. Ceramic materials and wood products must be dried before they are useful. Drying can happen naturally or be helped by adding heat. For example, lumber is air dried or it is kiln dried. Air dried lumber is stacked outdoors to dry after cutting. Kiln dried lumber is carefully heated in special ovens called dry kilns.

ASSEMBLY PROCESSES

Look around you. How many products with one part do you see? There may be paper clips or straight pins. However, most of the things you see are made from more than one part. These products are *assembled* from two or more parts, Fig. 16-25. The word assemble means "to bring together." A simple product such as a lead pencil is made from five parts. The barrel is two pieces of wood glued around the graphite "lead." The eraser is held on the barrel with a metal band. In fact, a lead pencil uses the two methods by which products are assembled: bonding and mechanical fastening.

Bonding

Bonding holds plastic, metal, ceramic, and composite parts to each other. Many different methods of bonding are used to attach parts together. **Bonding** uses cohesive or adhesive forces to hold parts together. *Cohesive* forces hold the molecules of one material together. *Adhesive* forces occur between different kinds of molecules. Adhesive forces cause some materials to be "sticky."

Fig. 16-24. These hot parts are leaving a heat treating furnace. They will be quenched to harden them. (Bethlehem Steel Co.)

Fig. 16-25. Products are assembled using both mechanical and bonding techniques. These washing machines have mechanisms bolted into place and sound insulation bonded inside.

Bonding Processes. How parts are bonded together is affected by:

- The type of joint.
- The material.
- The bonding process to be used.

Five bonding techniques are used for assembly. The first is **fusion bonding**. This technique uses heat or solvents to melt the edges of the joint. The surfaces are allowed to flow together to create a bond. In some cases, more material is added to the joint to increase the strength. Oxyacetylene, arc, inert-gas, and plastic welding are examples of fusion bonding methods.

A second method is flow bonding. **Flow bonding** uses a metal alloy as a bonding agent. The base metal is cleaned and then heated. The alloy is applied where the two parts meet. The alloy melts, flows between the parts, cools, and creates a bond. Soldering and brazing are examples of flow bonding. The base metal is not melted.

Pressure bonding applies heat and pressure to the bond area. This method is used on plastics and metals. Resistance (spot) welding is an example of pressure bonding, Fig. 16-26. Spot welding has four stages in its cycle.

1. The parts are held together between two electrodes.
2. Electrical current is passed between the electrodes. The current melts the metal between the electrodes.
3. The current is stopped and the molten metal solidifies.
4. The electrodes are released and the welded parts are removed.

Plastics can be pressure bonded using a similar process. No electric current passes between the parts. The parts are held by heated jaws, where they bond together. The jaws release and the parts are removed.

Cold bonding uses extreme pressure to squeeze the two parts to create a bond. This is a little used process. It is only used for small parts made of soft metals like copper and aluminum.

Adhesive bonding uses substances with high *adhesive forces* to hold parts together. These techniques are gaining wide use beyond their original use in woodworking. The advent of synthetic adhesives allows a wide range of plastics, metals, and ceramics to be joined together.

Bonding Agents. All bonding techniques use a **bonding agent**. A bonding agent can be one of three kinds:

- The same material: Parts are held together by a material that is the same as the base material. For example, two metal parts may be welded together, or a steel welding rod may be used to strengthen the bond between two steel parts, Fig. 16-27.

Fig. 16-26. This automobile assembly is being put together using resistance welding. Resistance welding is a type of pressure bonding. (Chrysler Corp.)

Fig. 16-27. This welding operation uses filler rods that are made from the same metal as the parts being welded. (Miller Electric Co.)

- Similar materials: Parts are held in position by the same type of material. For example, a metal (solder) may be used to bond metallic parts (copper wires).
- Different material: Parts are held together by a different type of material. For example, wood may be held together by white glue (a plastic) or ceramic parts may be held together by an epoxy (a plastic).

Joints. Joints are where parts meet. Pieces of material are fastened together at joints. Joints can occur on parts at the ends, sides, and faces. Joints are used to add length, width, or thickness and to make a corner.

The type of joint chosen will affect the type of bonding process that is used. A simple joint can be modified to increase its strength or improve its appearance, Fig. 16-28.

MECHANICAL FASTENING

Mechanical fastening uses mechanical forces to hold parts together. Friction between the parts can be used. For example, a part may be pressed or driven into a hole that is slightly smaller than it is. The friction between the parts will cause the parts to remain together. This type of fit is called a **press fit**. Press fits can be used to hold bearings in place on a shaft.

Fig. 16-29. These workers are assembling automotive driveline parts using mechanical fasteners. (Daimler-Benz)

In other cases the parts may be bent and interlocked to hold the parts together. This type of joint is called a seam. Many sheet metal parts are held together using seams.

The most widely used method to hold parts together are mechanical **fasteners**, Fig. 16-29. Examples of fasteners are staples, rivets, screws, nails, pins, bolts, and nuts.

Mechanical fasteners can be described as permanent, semi-permanent, or temporary. Permanent fasteners are not intended to be removed. Their removal will damage the fastener and/or the parts. Nails, rivets, and wood screws are good examples of permanent fasteners.

Semipermanent fasteners can be removed without causing damage to the parts. They are used when the product must be taken apart for maintenance and repair. For example, a furnace is held together with semipermanent fasteners so it can be serviced. Machine screws, bolts, and nuts are good examples of semipermanent fasteners.

Temporary fasteners are used when frequent adjustments or disassembly are required. Temporary fasteners hold parts in position, but the parts can be taken apart quickly. Wing nuts are a good example of a temporary fastener.

FINISHING PROCESSES

The last secondary process most products go through is **finishing**. These techniques protect the product and enhance its appearance. Finishing processes can be broken into two types. One group changes the surface of the product. The other group applies a coating.

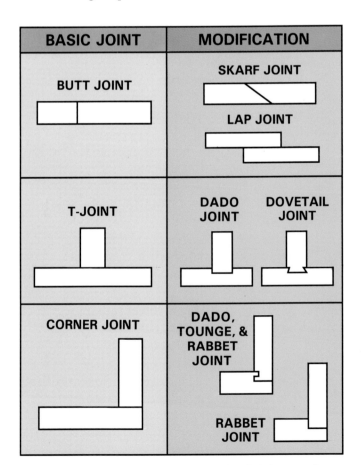

BASIC JOINT	MODIFICATION
BUTT JOINT	SKARF JOINT LAP JOINT
T-JOINT	DADO JOINT / DOVETAIL JOINT
CORNER JOINT	DADO, TOUNGE, & RABBET JOINT RABBET JOINT

Fig. 16-28. These drawings show simple joints can be changed for higher strength and better appearance.

Most metals will begin to corrode if they are not protected in some way. Products made of metals need to be protected so they will last. Metals are easy to protect by changing the surface chemically. For example, anodizing converts the surface of aluminum products to aluminum oxide. This type of finish is called a **converted surface finish**.

The other type of finishing applies a coating to the product. A film of finishing material is applied to the product or base material. These **coatings** protect the surface and can add color. **Finishing processes** involve cleaning the surface, selecting the finish, and applying the finish.

Safety with Finishing Materials and Equipment

People who work with finishing processes observe basic safety rules.

1. Wear eye and face protection.
2. Apply finishes in a well ventilated area. Wear a respirator and protective clothing if solvents are toxic.
3. Never apply a finish near open flame.
4. Keep the finishing area orderly and clean up any spills immediately.
5. Store all rags and chemicals in approved containers.
6. Use a spray booth to remove toxic fumes.
7. Dispose of any waste solvents and finishes properly. Wash your hands to remove any finishes you have handled.
8. Most of all, if you do not understand how to apply a finish, ASK FIRST!

The key to safety in the technology education laboratory is a *SAFE* attitude.

Cleaning the Surface

Finishing materials are applied to the surface of a product. This surface must be free of dirt, oil, and other foreign matter. These unwanted materials may be removed by chemical cleaning or mechanical means. Chemical cleaning is often called pickling. **Pickling** involves dipping the material in a solvent which will remove the unwanted materials. Then the clean part is rinsed to remove the solvent.

Mechanical cleaning includes: abrasive cleaning (sand blasting or sanding), buffing, or wire brushing.

Finishing Materials

There are hundreds of finishing materials. However, they are either organic or inorganic materials. The inorganic materials include metal and ceramic coatings.

Fig. 16-30. This roll of steel has been coated with tin to make tin plate. Tin plate is used to make cans for food. (American Iron and Steel Institute)

For example steel is coated with zinc to produce galvanized steel, or with tin to make tin plate, Fig. 16-30. Clay products, like floor and wall tile, are coated with a glaze. Electric range tops are coated with a ceramic material called porcelain enamel.

All of us are familiar with the organic finishes. These materials include:

- **Paints**: Any coating that dries through polymerization (hardening).
- **Varnish**: a clear finish made from a mixture of oil, resin, solvent, and a drying agent.
- **Enamel**: A varnish that has color pigment added.
- **Lacquer**: a solvent based, synthetic coating that dries through solvent evaporation.

Applying Finishing Material

Finishing materials may be applied in many number of ways. Metallic coatings are often applied through dipping or plating. **Dipping** involves running the stock through a vat of molten metal. This technique is widely used to produced galvanized steel.

Plating is an electrolytic processes. The parts are hung on racks and lowered into a cell full of electrolyte, Fig. 16-31. The plating metal is the anode and the part is the cathode. Electrical current is run through the cell. The electrical current causes the metal to move from the anode into the electrolyte. From there it moves across to the cathode and is deposited as a uniform coating of the product.

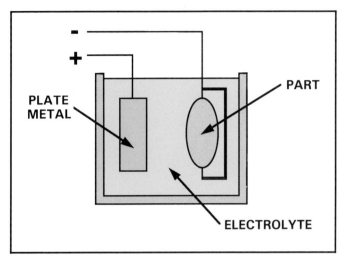

Fig. 16-31. This is a schematic drawing of an electrolytic cell. It is used to apply metal plating to parts.

Organic materials can be applied through brushing, rolling, spraying, flow coating, and dip coating.

Most of us are familiar with brushing and rolling. The material is gathered in the brush or on the roller. It is then applied by wiping the brush or rolling the roller across the surface of the material.

Spraying uses air to carry fine particles of finishing materials to the surface of the product. This can be a manual process where a worker uses the spray gun. Automatic spraying systems, such as the one shown in Fig. 16-32, are becoming widely used.

Flow coating passes the product under a flowing stream of finishing material. The material flows over the surface and the excess runs into a catch basin to be used again. Dip coating is the same process as is used for

galvanizing. In both cases excess coating is allowed to drain off the product for re-use.

After parts have a finished applied, they are allowed to dry in air, or they can be run through a heated drying oven.

SUMMARY

Secondary manufacturing processes convert industrial materials into finished products. They add worth to material by making them useful to consumers.

The secondary processes can be grouped as casting and molding, forming, separating, conditioning, assembling, and finishing. The first three change the size and shape of the materials. Conditioning change the properties of materials by altering their internal structure. **Assembling** processes attach parts together to make assemblies and products. Finishing produces coatings that protect and improve a product's appearance.

The output of secondary processes may be delivered directly to the consumer, Fig. 16-33, or become an input to construction activities. The use of manufactured products in construction will be discussed in the next chapter of this book.

Fig. 16-32. This automobile body is being sprayed with a coating. Automatic systems such as this one give great results and do not put workers at risk from paint fumes. (Chrysler Corp.)

Fig. 16-33. This product of secondary manufacturing will be distributed directly to consumers. The product is now usable to the consumer. (Reynolds Metals Co.)

WORDS TO KNOW

All of the following words have been used in this chapter. Do you know their meanings?

Adhesive bonding
Annealing
Assembling
Blow molding
Bonding
Casting and molding
Chemical action
Chemical conditioning
Chemical machining
Coatings
Cold bonding
Compounding
Conditioning processes
Converted surface finishes
Cutting motion
Dies
Dipping
Drawing machines
Drilling machine
Drying
Elastic range
Electrical discharge machining
Enamel
Expendable molds
Extrusion
Fastener
Feed motion
Finishing processes
Firing
Flame cutting
Flow bonding
Flow coating
Forming
Fracture point
Fusion bonding
Green sand casting
Hammers
Hardening
Investment casting
Lacquer
Laser machining
Linear motion
Machining
Mated dies
Mechanical conditioning
Mechanical fastening
Milling machine
Non-traditional machining
Open dies

Paint
Permanent molds
Pickling
Planing machine
Plastic range
Plating
Press
Press fit
Pressure bonding
Reciprocating motion
Rolling machines
Rotary motion
Secondary manufacturing processes
Separating
Shaped dies
Shaping machine
Shearing
Shell molding
Smith forging
Spraying
Stamping
Tempering
Thermal conditioning
Tool
Turning machine
Varnish
Yield point

TEST YOUR KNOWLEDGE

Write your answers on a separate piece of paper. Please do not write in this book.

1. List the six types of secondary manufacturing processes.
2. True or false. All separating processes use a tool.
3. Match the cutting and feed motions to the machine. (Note: there will be two answers per machine.)

Cutting Motion	Feed Motion	
_____	_____	Milling machine
_____	_____	Drill press
_____	_____	Wood lathe
_____	_____	Table saw
_____	_____	Scroll saw
_____	_____	Hack saw
_____	_____	Wood jointer

   A. Linear tool.
   B. Linear work.
   C. Rotating tool.
   D. Rotating work.
   E. Reciprocating tool.
   F. Reciprocating work.

4. List five ways organic finishes can be applied.

5. True or false. Casting changes the shape of material by applying a force that is above the yield point and below the fracture point.
6. Solidifying a cast material by removing the solvent is called _____.
7. List three thermal conditioning processes used on metals.
8. True or false. The two types of motion needed to produce a cut are cutting motion and feed motion.

APPLYING YOUR KNOWLEDGE

1. Build a simple product, such as a kite, using secondary manufacturing processes. List each step you would follow and the type of process used.
2. Select a simple product made from more than one part that you see in the room.
   A. List the parts it is made of.
   B. Select one part and list the steps that you think were used to manufacture it.
   C. Complete a form like the following for one of the steps.

Product:
Part name:
Production process:
Step needed to complete the process:

Construction processes are used to build innovative structures like this one. (Temcor)

CONSTRUCTING STRUCTURES

After studying this chapter you will be able to:
- *Define construction.*
- *List and describe the two types of construction.*
- *Describe the major types of buildings that are constructed.*
- *Describe the structures built by heavy engineering.*
- *List and describe the steps involved in constructing a structure.*
- *Describe how a site is prepared for a construction project.*
- *Diagram and explain the types of foundations.*
- *Describe the major types of walls and roofs used in buildings.*
- *Explain the types of utility and mechanical systems, and how to install them.*
- *Describe the types of heavy engineering structures.*
- *Explain how heavy engineering structures differ from buildings.*
- *Describe the types of roads, bridges, and dams.*

Human beings have three basic needs: food, clothing, and shelter. Each of these can be satisfied using technology. Agriculture and bio-related technology help us grow, harvest, and process food. Manufacturing helps us to produce natural and synthetic fibers. These fibers become the inputs to clothing and fabric manufacture. Materials and manufactured goods can be fabricated into dwellings and buildings using construction technology. **Construction** uses technological actions to erect a structure on the site where it will be used.

TYPES OF STRUCTURES

Construction builds two types of structures. These include buildings and heavy engineering structures, Fig. 17-1. **Buildings** are enclosures to protect people, materials, and equipment from the elements. Buildings also provide security for people and their belongings. **Heavy engineering structures** help our economy function effectively.

BUILDINGS

Buildings are grouped into three types: residential, commercial, and industrial, Fig. 17-2. These groupings are based on how the buildings are used.

Residential Buildings

Residential buildings are buildings that people live in. These buildings can be single-family or multiple-unit dwellings. The multiple-unit dwellings include apartments, town houses, and condominiums.

A residential building can be either owner occupied or rented from the owner. The owner of a dwelling is responsible for its upkeep. In some types of dwellings, such as condominiums, the costs of upkeep are shared between the owners. Each owner belongs to and pays fees to an *association*. This group elects officers who manage the maintenance of common areas like entryways, garages, parking areas, and lawns. The association is also responsible for exterior repairs and insurance on the building. The individual owners maintain their own living quarters and insure their personal belongings against fire and theft.

Commercial Buildings

Commercial buildings are used for business and government purposes. These buildings can be publicly

TYPES OF CONSTRUCTION PROJECTS

BUILDINGS

HEAVY ENGINEERING

Fig. 17-1. Construction erects buildings and heavy engineering structures.

TYPES OF BUILDINGS

RESIDENTIAL

COMMERCIAL

INDUSTRIAL

Fig. 17-2. Construction is used to build residential, commercial, and industrial buildings. (Marvin Windows and Doors, Inland Steel Company)

or privately owned. Commercial buildings range in size from small to very large. Retail stores, offices, courthouses, schools, libraries, and warehouses are commercial buildings.

Industrial Buildings

Industrial buildings house the machines that make products. These buildings are used to protect machinery, materials, and workers from the weather. The building supports the machines and supplies the utility needs of the manufacturing process. Many industrial buildings are specially built for one manufacturing process.

HEAVY ENGINEERING STRUCTURES

Heavy engineering structures support the transportation and communication systems that make the world work. These structures include highways, rail lines, canals, pipelines, power transmission and communication towers, hydroelectric and flood control dams, and airports. They provide the paths for the movement of water, people, goods, information, or electric power.

CONSTRUCTING BUILDINGS

Most construction projects follow the same basic steps. These steps, as shown in Fig. 17-3, include:
- Preparing the site.
- Setting foundations.
- Building the framework.
- Enclosing the structure.
- Installing utilities.

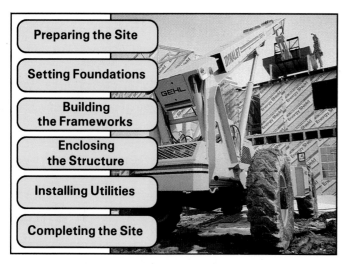

- Preparing the Site
- Setting Foundations
- Building the Frameworks
- Enclosing the Structure
- Installing Utilities
- Completing the Site

Fig. 17-3. These are six of the steps in a construction project. (Gehl Co.)

- Finishing the interior and exterior.
- Completing the site.

Each type of structure needs to have specific actions taken during each step. This helps complete the structure on time. We will look at the steps used to construct a small single-family home. Later in the chapter, other construction activities will be discussed.

A common type of building is a single-family home. It is designed to meet a number of needs of the owners. These needs, as shown in Fig. 17-4, include protection from the weather, security, and personal comfort. To meets these needs a home must be properly designed and constructed. The construction process starts with locating, buying, and preparing a site.

PREPARING THE SITE

The location for a home needs to be carefully selected. It should meet the needs of the people who will live there. For example, a family with children may think about the schools serving the area. The parents will consider the distance to work, shopping, recreation, and cultural facilities. The condition of other homes in the neighborhood, building codes, and covenants are other factors to think about.

Once the site is chosen, it is purchased from the original owner. This may require working with a real estate agent and obtaining a bank loan or other financing. The financing will probably include the money to erect the home. This is important because most banks will not loan money to build a house on land that is mortgaged.

Next, the site is cleared to make room for the structure. The location of the new building is marked out. Then, that area is cleared of obstacles. When it is possible, the building should be located to save existing trees and other plant life. The site may require grading, Fig. 17-5, to level the site. Grading prepares areas for

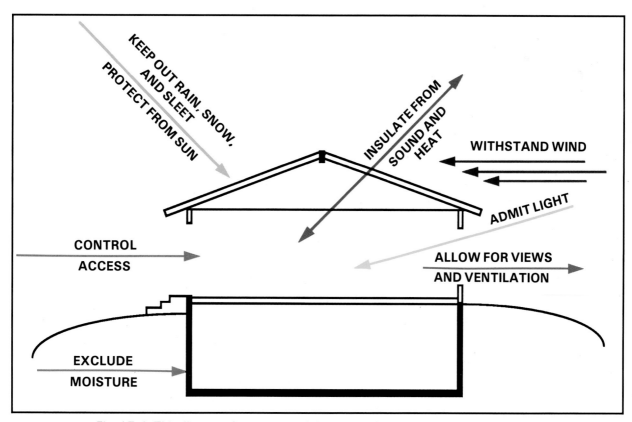

Fig. 17-4. This diagram shows some of the needs a home must meet for its owner.

Fig. 17-5. Before a building can be built the site must be cleared and graded.

sidewalks, landscaping, and helps water to drain from the site. These preparations are needed for the next step, setting foundations.

FOUNDATION TYPES

The foundation is the most important part of any building project. The foundation serves as the "feet" of the building. Try to stand on just your heels. You will be unstable and wobble. Likewise, a building without a proper foundation will settle unevenly into the ground. It will lean, become unstable, and may fall to the ground. The Leaning Tower of Pisa in Italy is an example of a building that has a poor foundation. Over time, the tower has settled and is leaning several feet to one side.

A complete foundation has two parts: the footing, and the foundation wall, Fig. 17-6. The footing spreads the load over the bearing surface. The bearing surface is the ground on which the foundation and building will rest. This can be rock, sand, gravel, or a marsh. Each type of soil offers unique challenges for the construction project.

The type of foundation to use is selected to match the soil of the site. Three types shown in Fig. 17-7 are:

- **Spread foundations**: This type of foundation is used on rock and in hard soils such as clay. The foundation walls sit on a low flat pad called a footing. On wide buildings posts support the upper floor between the foundation walls. These posts also rest on pads of concrete called footings.
- **Slab foundations**: These types of foundation are used for buildings that are built on soft soils. They are sometimes called floating slabs. The foundation becomes the floor of the building. Such foundations allow the weight of the building to be spread over a wide area. This type of foundation is used in earthquake areas because it can withstand vibration.
- **Pile foundations**: These foundations are used on wet, marshy, or sandy soils. Piles are driven into the ground until they encounter solid soil or rock. Piles are large poles made of steel, wood, or concrete. They are widely used for high-rise buildings, marine docks, and homes in areas that flood easily.

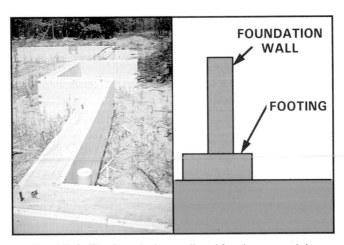

Fig. 17-6. The foundation wall and footing spread the building's weight onto the bearing surface. The concrete foundation shown in the photo has been insulated to reduce heat loss.

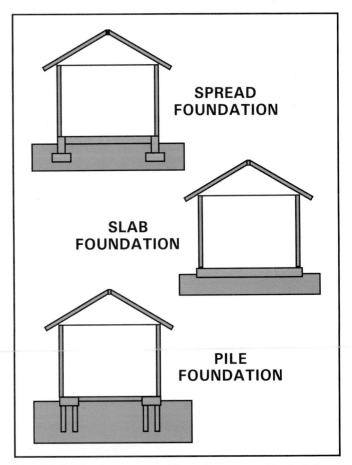

Fig. 17-7. The three types of foundations used for buildings are spread, slab, and pile.

SETTING THE FOUNDATION

Each type of foundation is built in a unique way. Let's consider a spread foundation. The site is surveyed to locate the foundation, Fig. 17-8. Then the site is excavated in preparation for the footings and the walls. If the building is to have no foundation, excavation does not go as deep. Buildings with basements require deeper excavations. Footing forms are next set up. Forms are a lumber frame to hold the wet concrete until it cures

Fig. 17-9. This worker is excavating a hole for a pool.

Fig. 17-8. This worker is surveying a site for a new building. The survey will locate where the foundation will be placed. (Inland Steel Co.)

(hardens). Forms give the footings or slabs height and shape. Concrete is poured and leveled off. When the concrete is cured, the forms are removed. Walls of poured concrete or concrete block are built atop footings. Slabs are ready for above-ground superstructures. Fig. 17-9 shows an excavation for a pool.

Wooden foundations use no concrete for either footings or walls.

LUMBER

STEEL

REINFORCED CONCRETE

Fig. 17-10. The materials used for frameworks are lumber, steel, and reinforced concrete.

BUILDING THE FRAMEWORK

The foundation becomes the base for the next part of the building, the framework. Erecting the framework gives the building its size and shape. The framework includes the floors, interior and exterior walls, ceilings, and roof. Also, the location of doors and windows are set up at this time.

The framework can be built out of three different materials, Fig. 17-10. Small and low cost buildings have frameworks made from **lumber**. Industrial and commercial buildings have either **steel** or **reinforced concrete** frameworks.

Building the framework involves three steps. First, the floor is constructed, Fig. 17-11. Homes with slab foundations use the surface of the slab as the floor. Homes with basements or crawl spaces use lumber floors.

Lumber floors start with a wood **sill** that is bolted to the foundation. Floor joists are then placed on the sill. They extend across the structure. **Floor joists** carry the weight of the floor. The size and spacing of the joists will be determined by the span (distance between outside walls) and the load on the floor. On top of the joists a **subfloor** is installed, usually made from plywood or particleboard. After the building is enclosed flooring material will be installed on top of the subfloor.

The wall frames are placed on top of the floor. These frames support both exterior and interior walls. Wall framing is often made of 2 x 4 or 2 x 6 construction grade lumber, Fig. 17-12. A framed wall has a strip at the bottom called the **sole plate**. Nailed to the sole plate

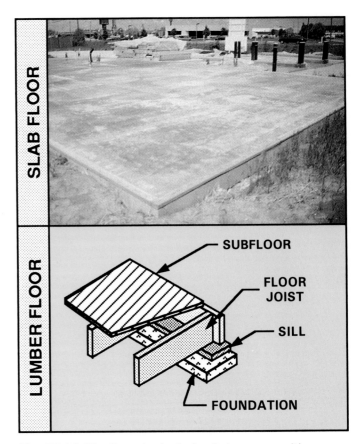

Fig. 17-11. The floors in single-family homes are either concrete slabs or lumber.

are uprights called **studs**. The length of the studs is set by how high the ceilings will be. At the top of the wall the studs are nailed to double ribbons of 2 x 4s called a top plate or wall plate. Door and window openings require headers above them. **Headers** carry the weight

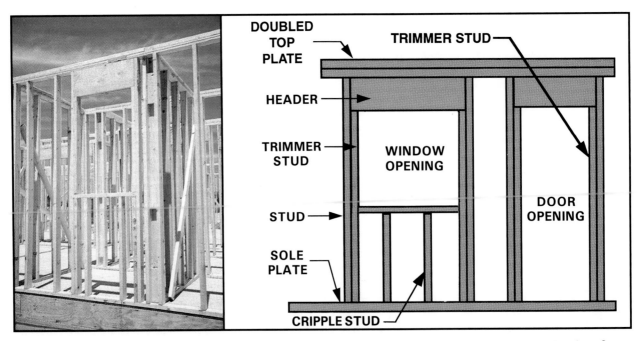

Fig. 17-12. These are the parts of a wood framed wall. How many of these parts can you find in the photo?

from the roof and ceiling across the door and window openings. Headers are held up by shorter studs called trimmer studs.

The walls support the ceiling and roof, Fig. 17-13. The **ceiling** is the inside surface at the top of a room. The roof is the top of the structure that protects the house from the weather.

The ceiling is supported by **ceiling joists**. These joists rest on the outside walls and some interior walls. Interior walls that help support the weight of the ceiling and roof are called *load-bearing*, or *bearing walls*.

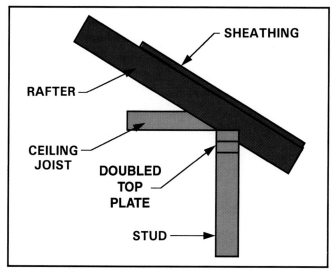

Fig. 17-13. This illustration shows how the roof meets the wall frame.

The roof forms the top of the building. There are many types of roofs including flat, gable, hip, gambrel, and shed, Fig. 17-14. The type of roof is chosen for its appearance and how it withstands the weather. For example, flat roofs are poor choices in areas with heavy snow. This type of roof cannot easily support the weight of deep snow. Likewise, a hip roof would look out of place on Spanish-type homes. This kind of roof would not give the "Spanish-style" look.

Roof construction involves two steps. First, the roof frame is built with rafters. **Rafters** are angled boards that rest on the top plate of the exterior walls. Often a special structure called a truss is used. A **truss** is a triangle-shaped structure that includes both the rafter and ceiling joist in one unit. Trusses are manufactured in a factory and then shipped to the building site.

The rafters or trusses are covered with plywood or waferboard sheathing. This step completes the erection of the frame.

ENCLOSING THE STRUCTURE

After the framework is complete the structure needs to be enclosed. The roof and wall surfaces need to be covered. This process has two steps: enclosing the walls, and installing the roof.

Enclosing the Walls

All homes have both interior and exterior wall coverings. These coverings improve the looks of the building and keep out the elements (rain, snow, wind, and sun).

The first step is enclosing the exterior walls, Fig. 17-15. This involves **sheathing** (covering) all the exterior surfaces. Plywood, fiberboard, or rigid foam sheets are used to sheath the walls. Most foam sheets have a reflective backing to improve the insulation value of the sheet. Most homes constructed today have a layer of plastic over the sheathing to prevent air from leaking in.

Installing the Roof

Normally the roof is put in place before the utilities are installed, Fig. 17-16. The actual roof surface has two parts. Sheathing is applied over the rafters. This sheathing may be plywood or waferboard. Now the roofing material is installed. Builder's felt is often applied over the roof sheathing. Wood or fiberglass shingles, clay tile, or metal roofing is then installed over

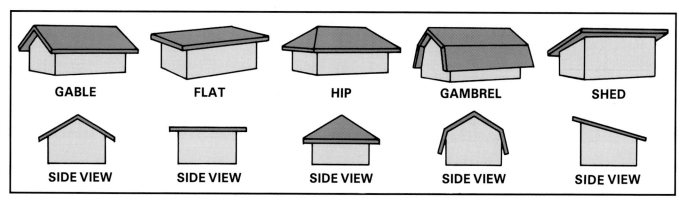

Fig. 17-14. These are some popular types of roofs used on homes.

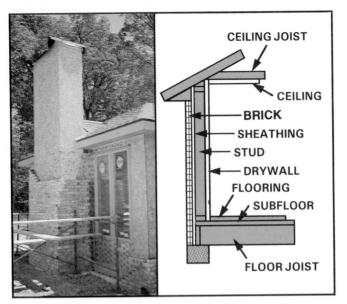

Fig. 17-15. This is a cross-section diagram of a finished wall. The photo shows brick being applied as a siding material.

CEILING JOIST
CEILING
BRICK
SHEATHING
STUD
DRYWALL
FLOORING
SUBFLOOR
FLOOR JOIST

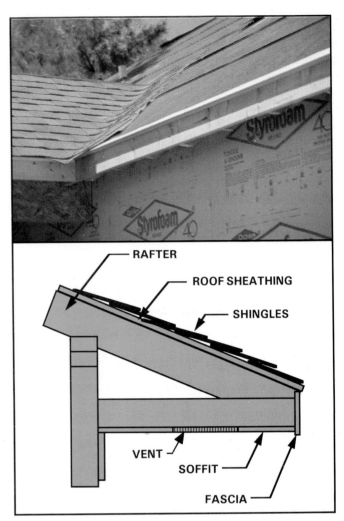

Fig. 17-16. These are the parts of a finished roof. The photo shows asphalt shingles being installed on a new roof.

RAFTER
ROOF SHEATHING
SHINGLES
VENT
SOFFIT
FASCIA

the sheathing and felt. Flat and shed roofs often use a built-up roof. A built-up roof starts with laying down sheets of insulation. Roofing felt is laid down, followed by a coat of tar, which is covered with gravel.

On many structures the overhang of the roof is also finished. A **fascia** board is used to finish the ends of the rafters and the overhang. The **soffit** is installed to enclose the underside of the overhang. The soffit can be made of aluminum, vinyl, or plywood. Soffits must have ventilation holes or vents to prevent moisture and heat build up in the attic.

Once the sheathing and roof are installed, the openings for doors and windows are cut out. Then the doors and windows are set in place. Now the house is secure and weather tight.

INSTALLING UTILITIES

Normally the utilities are installed after the building has been enclosed. This prevents theft and damage from the weather. Some parts of the utilities are installed earlier, such as large plumbing lines. The utility system includes four major systems:

- Electrical.
- Plumbing.
- Climate control.
- Communications.

Electrical

The electrical system delivers electrical power to the different rooms of the home. The power is brought into the house through wires to a meter and distribution

panel. This panel splits the power into 110 volt and 220 volt circuits. Each circuit has a circuit breaker to protect against current overloads.

Appliances such as clothes dryers, electric ranges, water heaters, and air conditioners require 220 volt power. Circuits for smaller appliances use 110 volts. Outlets may have power fed to them at all times. Outlets can also be controlled by switches. Fig. 17-17 shows a 110 volt circuit with wall (duplex) outlets and a ceiling light. You will note that the outlets will always have power. However, the circuit to the light has a switch.

Most 110 volt circuits are limited to 15 or 20 amps. Therefore, a number of different circuits are required to supply various parts of the home. A kitchen might have one or two circuits because of how many appliances are used there. One circuit might feed two bedrooms since there are few appliances in these rooms.

Plumbing

The plumbing system has two separate parts. One part supplies potable water, Fig. 17-18. **Potable water** is

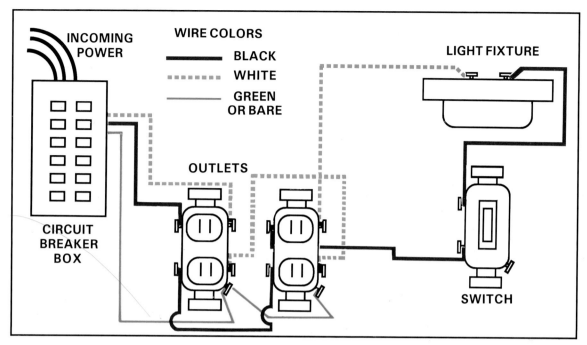

Fig. 17-17. This diagram shows a 110 volt electric circuit. Note how the light is controlled by the switch but the outlets are not.

safe for drinking. The other part of the system carries away waste water. Plumbing fixtures and systems are designed to prevent mixing of potable and waste water or sewer gas from leaking into the dwelling.

The potable water system starts with a city water supply or a well for the house. The water enters the house through a shutoff valve. The water may pass through a water conditioner. This device removes impurities such as iron and calcium.

The water line is split into two branches. One line feeds the water heater. The other line feeds the cold water system. Separate hot and cold water lines feed fixtures in the kitchen, bathrooms, and utility room. Toilets, however, receive only cold water. Most water lines have shutoff valves before they reach the fixture.

For example, the water lines under a sink should have a shutoff valve. The valve allows repairs to be made without stopping the water flow to the rest of the house.

The second part of the plumbing system is the **waste water** system. This system carries used water away from sinks, showers, tubs, toilets, and washing machines. The waste water is routed to a city sewer line or to a septic system. At each of these fixtures and appliances, a device called a trap is provided. A trap is a U-shaped piece of pipe that remains full of water. The water in the line stops gases from the sewer system from leaking into the home. Waste water systems have a network of vents to prevent the water from being drawn out of the traps. The vents also allow sewer gases to escape above the roof without causing any harm.

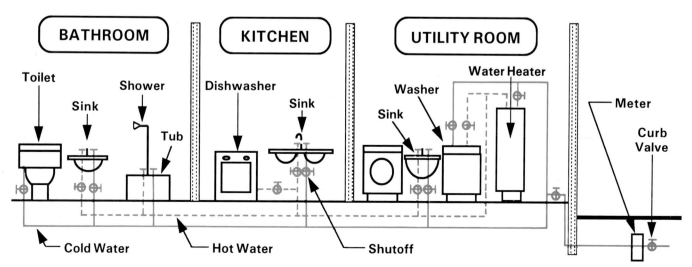

Fig. 17-18. This schematic diagram shows the potable water system for a home. Each fixture has a shutoff valve.

Homes that use natural gas have a third type of plumbing. Gas lines carry natural gas to furnaces, stoves, water heaters, and other appliances. There are shutoff valves at the entrance and at the major appliances.

Climate Control Systems

In many homes the climate control system is used to heat the building in winter and cool it in summer. This may be done with a single unit or with separate heating and cooling units.

Heating Systems. Rooms in a home can be directly or indirectly heated. In a direct heating system, the fuel is used in the room to be heated. Direct heating may use a stove or a fireplace that burns wood or coal. However, burning wood or coal can be expensive and causes considerable air pollution.

Other direct heating methods use electrical power. These systems use resistance heaters that are installed in the walls or along the baseboards. Also, ceiling radiant wires or panels may supply the heat.

Indirect systems heat a conduction medium such as air or water. This medium then carries the heat to the rooms. The heat is then given off to the air in the room, Fig. 17-19. The energy source for these systems are electricity, coal, oil, wood, natural gas, or propane.

Furnaces that heat air as a conduction medium are called **forced-air heating** systems. Forced-air furnaces draw air from the room. This air is heated as it moves through the furnace. A fan delivers the heated air through ducts to various rooms.

Hot water heating uses water to carry the heat. The water is heated in the furnace and pumped to various rooms. The water passes through room heating units which have metal fins that surround the water pipe. The fins dissipate the heat into the room.

Some homes are heated with solar energy. There are two types of solar homes: active and passive. Passive solar homes use no mechanical means to collect and store heat from the sun. Active solar homes use pumps or fans to move a liquid or air to collect solar heat. After the heat is collected the liquid or air is moved to a storage device. Solar heating systems are very effective in areas that have ample sunshine.

Cooling Systems. Many buildings have cooling systems to cool the air during the warm parts of the year. Cooling systems use compressors, evaporators, and condensers much like a refrigerator. The system has a fan that draws the air from the room. The air passes over a cold evaporator. This is similar to a forced air furnace. Instead of the air being heated, it is cooled. The cool air is returned to the room.

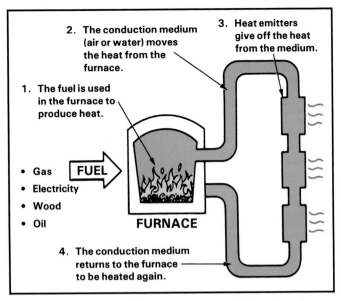

Fig. 17-19. This figure shows how an indirect heating system works.

A new development in climate control is a unit called a **heat pump**. A heat pump works as a cooling and a heating system. A heat pump can be operated in two directions. Operating in one direction the heat pump acts like an air conditioner. The heat pump takes warmth from inside the house and discharges it outside. In the winter the heat pump works in reverse. It takes warmth from the outside and brings it into the house. Heat pumps can use air, water, or the ground as a heat source. Heat pumps that use air work best in areas that do not get very cold, such as the southern and central U.S. ground water heat pumps use well water as the heat source. The water is pumped from the well to the heat pump. The water has heat removed or the water receives excess heat, and then the water returns to the well. Ground coil systems are buried in the soil to take or give off heat. Ground water and ground coil systems can be used in colder climates. Otherwise, heat pumps need a small furnace as a backup or auxiliary heat source.

Communication Systems

Most homes have communication systems such as telephone, radio, and television. These systems require special wiring. Telephone wiring and television cables are normally installed during the construction of the building. Installing them after a building is finished is costly. It takes considerable work to feed the wires through attics, under floors, and inside walls. Some homes have intercom systems that allow two-way communication between rooms. Many intercom systems allow radio programs to be played throughout the house.

FINISHING THE EXTERIOR AND INTERIOR

The final exterior finishing step is installing siding and trim. Siding is the finish covering used on a wood building. Many siding materials are in use. Wood shingles and boards, plywood, hardboard, brick, stone, aluminum, vinyl, and stucco are all used as siding. Look back at Fig. 17-15. You will see bricks being installed over plywood sheathing. Trim is the strips of wood that cover the joints between window and door frames and the siding.

The interior walls are the next walls to be finished. Insulation is placed between the studs and around the windows and doors of all exterior walls. Insulation reduces heat loss on cold days and heat gain on hot days. The most common type of insulation is fiberglass. It is available in blankets or batts. A vapor barrier of polyethylene film is attached to the studs over the insulation. The vapor barrier prevents moisture from building up in the insulation.

Once insulation and utilities are in place, the interior wall surfaces can be covered. The most widely used interior wall covering is gypsum wallboard, commonly known as **drywall**. Drywall has replaced plaster in most applications. Drywall is a sheet material made of gypsum bonded between layers of paper. The sheets of drywall are nailed or screwed onto the studs and ceiling joists. The fastener heads and drywall seams are then covered with a coating called joint compound. The compound is applied in several thin coats. This is done to make smooth surfaces and joints between the sheets of drywall.

The inside and outside of the house is now ready for the finishing touches. Interior wood trim is installed around the doors and windows. Kitchen, bathroom, and utility cabinets are set in place. Floor coverings such as ceramic tile, wood flooring, carpet, or linoleum is installed over the subflooring. Baseboards are installed around the perimeter of all the rooms. The exterior siding and wood trim is painted. Interior trim is painted or stained. The walls are painted or covered with wallpaper or wood paneling. Lighting fixtures, switch and outlet covers, towel racks, and other accessories are installed. The floors and windows are cleaned. Now the home is finished and ready to be lived in.

COMPLETING THE SITE

Completing the building is the major part of the project. However, other work remains to be done. The site must be finished. Earth is moved to fill in areas around the foundation. Sidewalks and driveways are installed.

The yard area needs to be landscaped. **Landscaping** helps to prevent erosion and improves the appearance of the site. Trees, shrubs, and grass are planted. Landscaping can divide the lot into areas for recreation and gardening. Landscaping can be used to screen areas for privacy, direct foot traffic, and shield the home from wind, sun, and storms.

Look at Fig. 17-20. The top view shows dirt being moved onto the site for landscaping activities. The bottom view shows a finished landscaped area. Notice how the trees and lawn improve the appearance. Also, note that a grassy mound is used to guide people onto the sidewalk.

Fig. 17-20. The site is finished by grading the lot (top) and planting landscaping (bottom).

Technology: How it Works

TOWER CRANE: A self-raising crane used to lift materials on high-rise building projects.

Building high-rise structures offers a number of unique challenges. One involves lifting into place the structural steel members that will be part of the building. This is often done with tower cranes like those shown in Fig. 1.

A tower crane has several parts: the tower sections, the main jib (cross arm), the slewing (turning) gear, and the cab. The base unit of the tower crane is lifted into place by a truck-mounted crane. The height of the crane needs to be increased as the building goes up. The crane does this under its own power. As shown in Fig. 2, several steps are involved.

As additional height is needed, a device called a climbing frame is used. The frame is positioned beneath the cab. Hydraulic cylinders lift the jib and cab unit above the tower. Then, as shown in the middle drawing of Fig. 2, another section is lifted by the crane and swung into the opening in the climbing frame. The new section is bolted in place to produce a stable tower. This procedure is repeated as many times as needed. The climbing frame is generally removed when it is no longer needed.

The tower crane uses a trolley, cables, and a hoist (a power-driven drum) to lift loads. The trolley is a frame with pulleys that can be moved along the main jib. A cable extends from the cab out to the trolley and down to the load. The cable is wound around the hoist drum to lift the load. When the load is at the right height, the main jib turns and the trolley moves along the jib to position the load over the structure. Then, the load is lowered into position. When the project is completed, the tower crane is disassembled and moved to the next construction site.

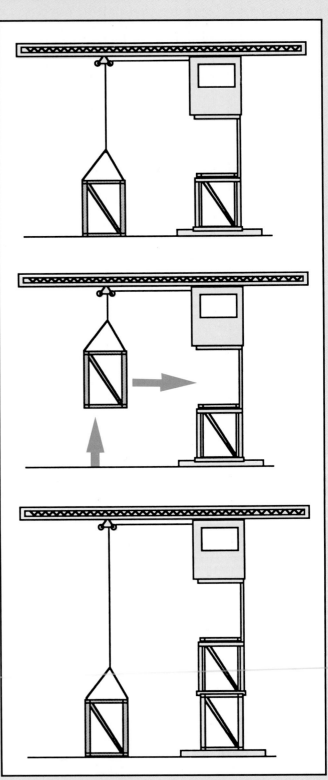

Fig. 1. These tower cranes are ready to lift building materials into place.

Fig. 2. This drawing shows how a tower crane increases its height. The climbing frame lifts the cab and jib, and a new section is put in place.

OTHER TYPES OF BUILDINGS

You see commercial, industrial, and residential buildings all around you. However, if you look around your town or city, you will probably see other types of buildings, Fig. 17-21. These may include:

- Monuments: Structures that pay tribute to the accomplishments or sacrifices of people or groups.
- Cultural buildings: These buildings house theaters, galleries, libraries, performance halls, and museums. These buildings host musical, dramatic, and dance performances, literary activities, and art exhibits.
- Government buildings: These buildings house government functions. Examples include city halls, post offices, police stations, fire houses, state capitols, courthouses, and government office buildings.
- Transportation terminals: These buildings are used to aid in the loading and unloading of passengers and cargo from transportation vehicles. Examples are airports, train and bus stations, freight terminals, and sea ports.
- Sports arenas and exhibition centers: These facilities are used for sporting events, concerts, trade shows, and conventions.
- Agricultural buildings: These include barns and storage buildings used to house livestock, shelter machinery, and protect farm products (grain, hay, etc.).

These special buildings are built using the same construction steps used for a single-family home. First, the site is prepared. Next, foundations are laid and the framework is built. The structure is enclosed, utilities are installed, and the interior and exterior is finished. Finally, the site is completed.

A special type of building is the **manufactured home**. As you will remember, manufacturing produces products in a factory. The completed product is transported to their place of use. This is exactly how manufactured homes are produced, Fig. 17-22. Most of the structure is built in a factory. This type of home is usually built in two halves. The floors, walls, and roof are erected; then the plumbing and electrical systems are installed. The interior and exterior of the structure is enclosed and finished. This includes installing flooring, painting walls, setting cabinets and plumbing fixtures, and installing appliances and electrical fixtures.

The two halves of the structure are transported to the site. The foundation is already in place. Each half is lifted from its transporter and placed on the foundation. The two halves are finally bolted together. The final trim that connects the halves is installed. The utilities are hooked up and the home is ready for the homeowner.

Similar techniques are used to produce temporary classrooms, construction offices, and modular units that can be assembled into motels or nursing homes.

UNITED STATES CAPITOL

STATUE OF LIBERTY

HOOSIER DOME

SYDNEY OPERA HOUSE

GRAIN ELEVATORS

Fig. 17-21. Can you identify what kind of buildings these are?

BUILD IN FACTORY	MOVE TO SITE	LIFT FROM TRUCK

SET ON FOUNDATION	FINISH	READY TO MOVE IN

Fig. 17-22. This series of photos shows the steps in building a manufactured home.

HEAVY ENGINEERING STRUCTURES

Construction activities do not always produce buildings. We need and use many other types of constructed structures. These structures are sometimes called civil structures, or heavy engineering structures. The dams, bridges, rail lines, roads, power lines, pipelines that allow modern commerce and society to function are examples of nonbuilding structures. Each of these structures is the result of **heavy engineering construction.**

These projects can be grouped in a number of ways. For this discussion, we will group them into transportation, communication, and production structures.

TRANSPORTATION STRUCTURES

Transportation systems depend on constructed structures. These structures include railroad lines, highways and streets, waterways, and airport runways. Other constructed works help vehicles cross uneven terrain and rivers. These include bridges and tunnels. Pipelines are land transportation structures that are used to move liquids or gases over long distances.

Let's look at some examples of these constructed works. We will discuss roadways and bridges.

Roadways

Roads are almost as old as civilization. People first used trails and paths to travel. Later, more extensive road systems were developed. The first engineered roads were built by the Romans over 2000 years ago. Their influence remained until the 18th century, when modern road building started. Today's roads have their roots in the work of the Scottish engineer John McAdam. He developed the crushed stone road. His roads were built of three layers of crushed rock that was laid in a ribbon about 10 inches (25 cm) thick. Later, this roadbed was covered with an asphalt-gravel mix. Asphalt roads are very common today. A more recent development is the concrete roadway.

Building a road starts with selecting and surveying the route. Next, the route is cleared of obstacles such as trees, rocks, and brush. The roadway is graded so that it will drain. Drainage is important to prevent road damage from freezing and thawing. Also, a dry road bed withstands heavy traffic better than a wet, marshy one. Another reason for grading is to keep the road's slope gentle. Elevation changes are described using the term *grade*. Grades are expressed in percentages. A road with a 5 percent grade would gain or lose 5 feet of height for every 100 feet of distance. Most grades are kept below 7 percent.

Once the road bed is established, the layers of the road are built, Fig. 17-23. The graded dirt is compacted and a layer of coarse gravel is laid. This is followed with finer gravel that is leveled and compacted. Next, the concrete or asphalt top layer is applied. Concrete roads are laid in one layer. Asphalt is generally applied in two layers: a coarse undercoat and a finer top coat. Finally, the shoulders or edges of the road are prepared. The shoulders can be gravel or asphalt.

Bridges

Another constructed structure vital for transportation is the bridge. Bridges provide a path for vehicles to move over obstacles. This may be a marshy area, a ravine, another road, or a body of water. Bridges can carry a number of transportation systems. These include highways, railroads, canals, pipelines, and foot paths.

Generally there are two types of bridges: fixed and movable. A fixed bridge does not move. Once the bridge is set in place it stays there. Movable bridges can change their position to accommodate traffic below it.

This type of bridge is used to span ship channels and rivers. The bridge is drawn up or swung out of the way so that ships can pass.

Bridges have two parts, Fig. 17-24. The substructure spreads the load of the bridge to the soil. The abutments and the piers are parts of the substructure. The superstructure carries the loads of the deck to the substructure. The deck is the part used for the movement of vehicles and people across the bridge.

The kind of superstructure a bridge has indicates the type of bridge. The types of bridges are beam, truss, arch, cantilever, and suspension, Fig. 17-25.

Beam bridges use concrete or steel beams to support the deck. This type of bridge is widely used when one road crosses another one. Beam bridges are very common on the interstate highway system.

A **truss bridge** uses small parts arranged in triangles to support the deck. These bridges can carry heavier loads over longer spans than beam bridges. Many railroad bridges are truss bridges.

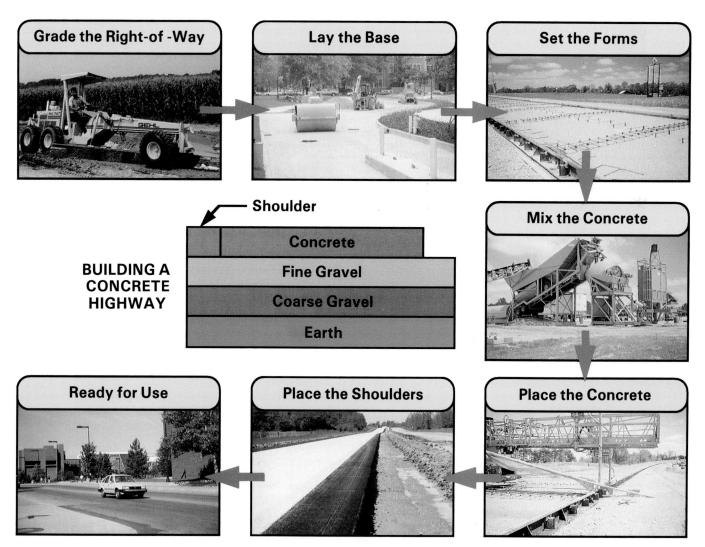

Fig. 17-23. These are the steps in building a road.

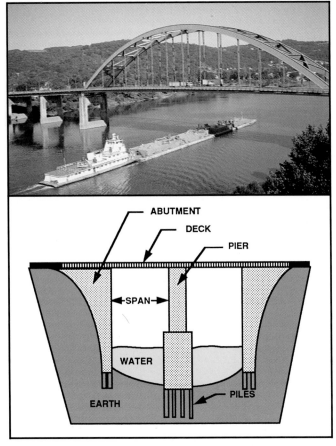

Fig. 17-24. This diagram shows the parts of a bridge. An arch bridge is shown in the photo. (American Electric Power Co.)

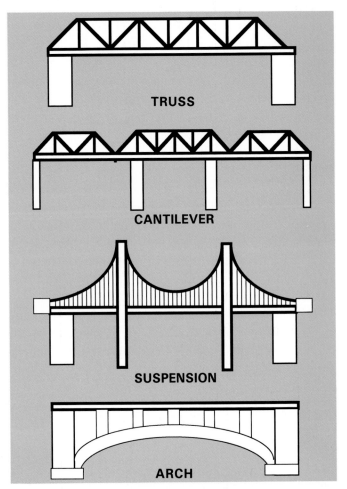

Fig. 17-25. These are four types of bridges.

Arch bridges use curved members to support the deck. The arch may be above or below the deck. Arch bridges are used for longer spans. One of the longest arch bridges spans over 1650 feet (502 m).

A **cantilever bridge** uses trusses that extend out like arms. The ends of the arms can carry small span or hook up to each other. The load is transmitted by the arms to the center. None of the load is carried by the ends of the arms.

Suspension bridges use cables to carry the loads. A large cable is suspended from towers. From the large cable, smaller cables drop down to support the deck. Suspension bridges can span distances as great as 4000 feet (1220 meters) and higher.

COMMUNICATION STRUCTURES

Most telecommunication technology relies on constructed towers to support antennas. These towers are usually placed on a concrete foundation. A steel tower is built on top of the foundation. Once the tower is complete the signal wiring can be installed. Similar techniques are used to construct towers for power transmission lines, Fig. 17-26.

PRODUCTION STRUCTURES

There are types of structures that are used for production activities that are not buildings. For example, petroleum refineries are a mix of machinery and pipelines. Irrigation systems are constructed to bring water to farms in dry areas. Evaporation basins are built to recover salt and other minerals from sea water.

An important production structure is the dam. Dams are used for flood control, water supply, making recreational lakes, or generating electricity. There are several types of dams. One type is called a **gravity dam**. Its lake side is vertical while the other side slops outward. The sheer weight of the concrete the dam is made from holds the water back. The dam on the left of Fig. 17-27 is a gravity dam.

Two more types of dams are the rock dam and the earth dam, shown in Fig. 17-27. A rock dam looks like two gravity dams placed back-to-back. Both sides slope outward. Rock and earth dams must be covered with a waterproof material to prevent seepage. Clay is often used for this covering.

A **buttress dam** uses its structure to hold back the water. This type of dam is not solid. It uses walls of concrete to support a concrete slab or arches against the water.

Tall dams that hold back large quantities of water are called arched dams. The arched shape increases the strength of the dam. The arched shape also spreads the pressure onto the walls of the canyon where the dam is built.

Fig. 17-26. This helicopter is helping to construct a tower for an electricity transmission line. (American Electric Power Co.)

SUMMARY

Construction is a vital production activity. Construction provides us with homes, offices, factories, highways, railroads, pipelines, bridges, dams, and other structures. Construction can be divided into projects that produce buildings and heavy engineering structures.

Most structures are built using the same steps. The site is cleared and prepared for construction. The foundation for the structure is produced. Then, the framework or superstructure is erected. Utilities are installed and the structure is enclosed. The building is finished and site is landscaped.

WORDS TO KNOW

All of the following words have been used in this chapter. Do you know their meanings?

Arch bridge
Beam bridge
Building
Buttress dam
Cantilever bridge
Ceiling
Ceiling joist
Commercial buildings
Construction
Drywall
Fascia
Floor joist
Forced-air heating
Gravity dam
Headers
Heat pump
Heavy engineering construction
Heavy engineering structure

Fig. 17-27. These drawings show two common types of dams. (American Electric Power Co.)

Hot water heating
Industrial buildings
Landscaping
Lumber
Manufactured home
Pile foundation
Potable water
Raft foundation
Rafter
Top plate
Reinforced concrete
Residential buildings
Sheathing
Sill
Soffit
Sole plate
Spread foundation
Steel
Studs
Subfloor
Suspension bridge
Truss
Truss bridge
Waste water

TEST YOUR KNOWLEDGE

Write your answers on a separate piece of paper. Please do not write in this book.

1. List the two kinds of constructed works.
2. True or false. A condominium is a residential structure that is generally owned by the people living in it.
3. Match the part of a building's frame with its correct description:

    _____ The board that rests on the foundation.
    _____ The boards that span the building and support the floor.
    _____ The vertical members in the wall.

    A. Floor joists.
    B. Rafter.
    C. Header.
    D. Sill.
    E. Ceiling.
    F. Studs.
    G. Rafter plate.

    _____ The boards that form the slopes or shape of the roof.
    _____ The board at the top of a window opening.
    _____ The "top" of the room.

4. The material attached to the outside of rafters and studs is called _____.
5. True or false. Spread, boat, and pier are the most common types of foundations.
6. The two types of water systems that are part of a plumbing system are called _____ and _____ systems.
7. List the six major steps in constructing a structure.
8. Name four types of bridges.
9. True or false. Dams are constructed out of earth, rock, or concrete.
10. A home built in a factory is called a _____.

APPLYING YOUR KNOWLEDGE

1. Use a chart like the one below to list and describe a few of the constructed structures you see as you travel from your home to school.

STRUCTURE	TYPE Building/Heavy engineering	DESCRIPTION- USE

2. Select one structure that you saw in completing the previous assignment. Make a drawing or model of the structure and label the major parts.

18
CHAPTER

USING AND SERVICING PRODUCTS AND STRUCTURES

After studying this chapter you will be able to:
- *List the types of activities involved in properly using manufactured and constructed items.*
- *Describe the factors to consider when selecting a product or structure.*
- *List the three major service activities needed to maintain the usefulness and value of a product or structure.*
- *Describe the ways to dispose of obsolete and worn out products and structures.*
- *Explain the importance of recycling.*
- *Describe the activities involved in altering products and structures.*
- *Explain the importance of owner's and service manuals.*

People use manufactured products and constructed structures every day. They choose the items that meet their needs. People make their choices to live better. However, sometimes people misuse some of their belongings. People can also choose the wrong item, use it improperly, or maintain it poorly.

If technology is to help us live better, we must choose the correct product or structure for our needs or for the task at hand. This involves six steps: selecting, installing, maintaining, repairing, altering, and disposing of technological products, Fig. 18-1.

SELECTING TECHNOLOGICAL PRODUCTS

The next time you go to a store, stop and look around you, Fig. 18-2. There will be thousands of products on display. Each of these products is developed for a specific use. Some products help to fulfill basic needs or perform essential tasks. Stoves cook our food. Automobiles and buses transport us to work and school. Other products are designed to make life more pleasant. Games, video tapes, and recorded music help to entertain us. Colorful pictures and art works brighten our homes and workplaces.

Choosing the best product to fill a need or want is quite a challenge. *Needs* come first, as they are the things that are necessary for living. *Wants* are requirements that are not necessary for living or to complete a task, but nice to have. It is not always easy to determine what is a need and what is merely a want. Once you have separated needs and wants, you must decide how much money can be spent on meeting the

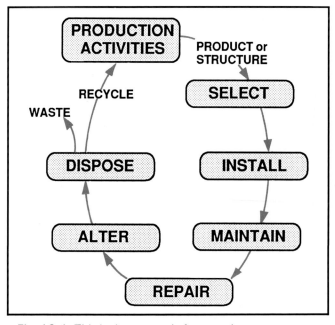

Fig. 18-1. This is the use cycle for a product or structure.

needs. You probably cannot afford to pay for all of your wants at one time. Therefore, you will have to rank them by their importance to you. Essentials, like food and clothing, will rank high on your list. Decorative items may rank low on your list.

Knowing what you need leads to the next step. This is selecting the best of the many products that fit your needs. Some people buy products on impulse. A better way is to analyze the products. This allows you to think about the products you want to buy. You should consider three important factors:

- **Function**: How well will the product meet your needs? How well does it work? Is the product durable and easy to maintain? How well will it do the job you have in mind?
- **Value**: Does the performance of the product match its price? Do other products meet your needs as well but cost less? Is it worth the selling price?
- **Appearance**: Is the design and color of the product pleasing to you? Is it something you would be proud to own?

INSTALLING TECHNOLOGICAL PRODUCTS

Some products are ready to use when you buy them. Other products must be installed. For example: Items such as cookware, clothing, and tools are sold *ready to use*. No set up is required. Other products require installation, such as the dishwasher shown in Fig. 18-3. The purchaser or a service person must complete several tasks to make a product ready to use.

First, products must be *unpacked*. Products are shipped from the factory in protective crates and boxes.

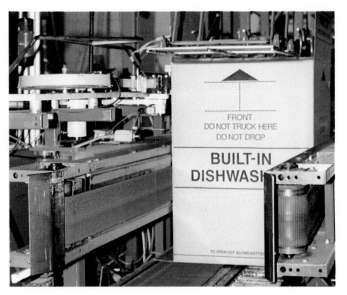
Fig. 18-3. This dishwasher must be installed before it can be used. (White Consolidated Industries)

Boxes, crates, and rigid foam components protect the product from damage during shipment. The boxes and packing materials must be removed and properly discarded or recycled.

Second, some products require *utilities* in order to work. The products are attached to electrical, water, and waste water lines. Natural gas or compressed air may also be needed.

Third, the product may need to be *positioned* and secured to the floor or inside a cabinet. It may also need to be leveled to operate properly.

Fourth, the product may require *adjustment*. Electrical meters may need to be set on zero. Clearances between doors may need to be adjusted. The space between moving parts should always be checked.

Fig. 18-2. There are thousands of products designed to meet our needs and wants.

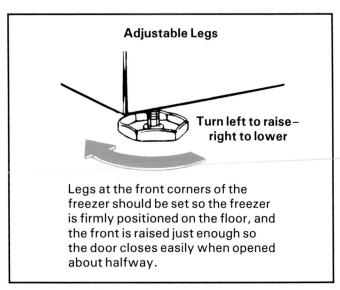

Adjustable Legs

Turn left to raise— right to lower

Legs at the front corners of the freezer should be set so the freezer is firmly positioned on the floor, and the front is raised just enough so the door closes easily when opened about halfway.

Fig. 18-4. This is an example of installation instructions for a product. (General Electric Co.)

Finally, most products must be tested. For example, a sample roll of film may be shot to test a camera's operation and adjustments. A water conditioner may be cycled to see that its controls are operating.

Directions for the installation and use of many products are in the owner's manual, Fig. 18-4. This information may be on one sheet of paper or in a large book. The complexity of the product determines the size of the owner's manual.

MAINTAINING TECHNOLOGICAL PRODUCTS

We want products to work properly when we need them. This often requires a maintenance program. The goal of **maintenance** is to keep products in good condition and in good working order, Fig. 18-5. Clothing and dishes are washed to maintain their usefulness and extend their life. Filters in furnaces and air conditioning units are changed so the units function properly. The oil in automobile engines is changed to maintain the oil's ability to lubricate moving parts. Locomotives and buses are serviced to maintain their performance.

Most maintenance is done on a schedule, Fig. 18-6. The schedule is designed to keep the product working properly. Therefore, it is sometimes called **preventive maintenance**. It is designed to *prevent* breakdowns.

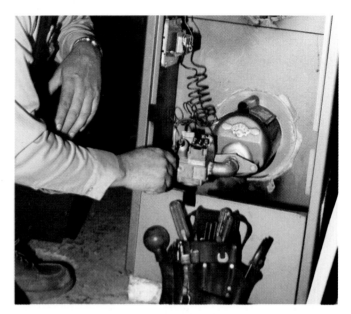

Fig. 18-6. Maintenance of many products should be done periodically. This technician is performing a yearly check of a gas furnace.

Many products come with a maintenance manual. This document lists:
- The types of maintenance needed.
- Methods for performing maintenance.
- A time schedule for each maintenance task.

Buildings and other constructed structures need maintenance just like manufactured products. Buildings must be cleaned and painted. Windows are cleaned. Roofs are sealed to prevent leaks. Bridges and communication towers are painted to prevent rusting. Railway tracks are leveled and switches are lubricated. Streets and driveways receive periodic coatings to seal out water, Fig. 18-7.

Fig. 18-5. This mechanic is performing routine maintenance on an aircraft engine. The maintenance keeps the engine in good working order. (United Parcel Service)

Fig. 18-7. This worker is helping to apply a slurry seal to a street. (Rohm and Haas)

Technology: How it Works

BAR CODE READER: A system used to read numerical codes on packages and tags.

Business and industry strive to be as efficient as possible. This allows companies to produce and sell products at competitive prices. One recently introduced labor-saving device is the bar code reader, Fig. 1. Most people have seen a bar code reader at the supermarket checkout. The reader, Fig. 2, left, is connected to a cash register (called a point of sale terminal) and one or more computers, Fig. 2, right.

Fig. 1. This bar code reader is reading bar codes on boxes before they are shipped to the customer.

The bar code system uses binary numbers. Binary numbers represent numbers and letters with a series of ones and zeros. The bar code uses white spaces to represent zeros and black bars to represent ones. To see an example of a bar code, look on the back cover of this book.

A laser beam reads the bars. The beam hits a spinning mirror that spreads the beam out so it can read the bar codes. The laser beam strikes the package as it is moved past the reader. The black bars absorb the laser light. White spaces reflect back into the reader and onto a detector. The detector receives the reflected light and changes the light into pulses of electricity which are amplified and decoded.

The bar code information is fed to the in-store computer. This computer has in its memory the description and price for each product in the store. Price information is sent to the point-of-sale terminal where it is printed on the receipt and shown to the customer on a digital display. The in-store computer keeps track of the sales information it receives from the scanner. A record of it can be sent to the store's central office. The central office can use this information to track sales and to order new inventory.

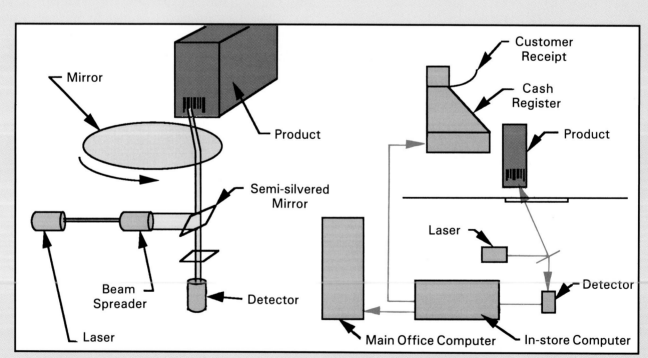

Fig. 2. The diagram on the left shows how a bar code reader works. It sends the laser beam toward the bar codes, and then receives the reflections. The detector turns the reflections into electrical signals. The diagram on the right shows how the signals from the scanner are fed to the in-store computer. The in-store computer can send the information from the scanner to the store's central office.

DIAGNOSING	ADJUSTMENT OR REPLACEMENT	TESTING

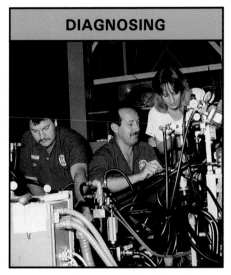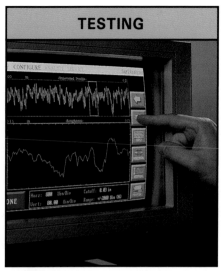

Fig. 18-8. These are the steps in repairing a product. (General Electric, Hewlett-Packard, Federal Products Corp.)

REPAIRING TECHNOLOGICAL PRODUCTS

No product or structure will work all the time or last forever. Some products are used until they stop functioning and are then discarded. For example, few of us try to salvage bent paper clips or bolts with stripped threads.

However, many products are too costly to discard the first time they stop working. Throwing away a bicycle or a car every time a tire goes flat would be very expensive. It costs less money to repair the product. **Repair** is the process of putting a product back into good working order. This requires three steps, as shown in Fig. 18-8:

- **Diagnosis:** The cause of the problem is determined.
- **Replacement or adjustment:** Worn or broken parts are replaced. Misaligned parts are adjusted.
- **Testing:** The repaired product must be tested to ensure that it works properly.

The information needed to repair a product is contained in the product's service manual. This manual provides a parts list, as shown in Fig. 18-9, so that repair parts can be ordered. It also gives directions for completing common repairs, Fig. 18-10.

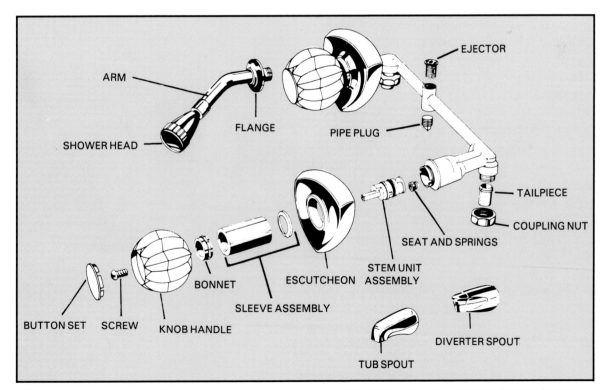

Fig. 18-9. This exploded pictorial view provides a parts list. This helps a service person order repair parts.

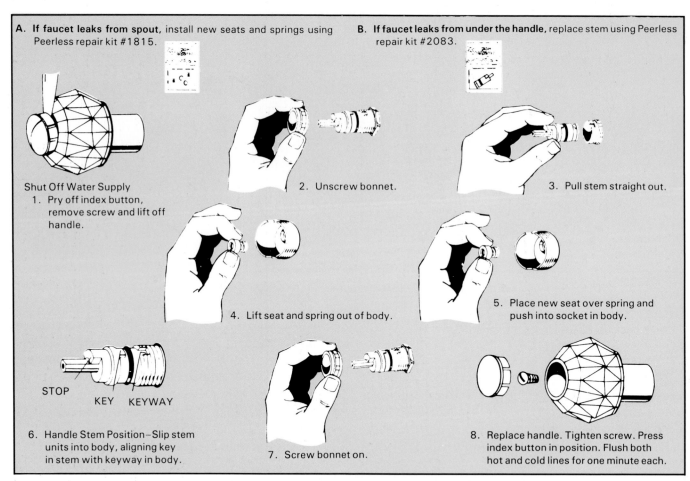

A. **If faucet leaks from spout,** install new seats and springs using Peerless repair kit #1815.

B. **If faucet leaks from under the handle,** replace stem using Peerless repair kit #2083.

Shut Off Water Supply
1. Pry off index button, remove screw and lift off handle.

2. Unscrew bonnet.

3. Pull stem straight out.

4. Lift seat and spring out of body.

5. Place new seat over spring and push into socket in body.

STOP   KEY   KEYWAY

6. Handle Stem Position–Slip stem units into body, aligning key in stem with keyway in body.

7. Screw bonnet on.

8. Replace handle. Tighten screw. Press index button in position. Flush both hot and cold lines for one minute each.

Fig. 18-10. These directions allow a home owner to repair a bathroom faucet.

Constructed structures also need periodic repair. The walls of buildings may crack or be damaged, Fig. 18-11. Roofs begin to leak or windows get broken. Bridges may need rebuilding. Many streets, highways, and parking areas need patching and resurfacing, Fig. 18-12. These problems are repaired so the building or structure will last longer.

ALTERING TECHNOLOGICAL PRODUCTS

Some products become obsolete as time passes. Their useful life can be extended by **altering** the product. For example, a person may put on or lose weight. Their clothes may not fit properly. A tailor can alter the clothes to fit the person better. Buildings are altered also. A building that is outdated can be changed. The needs of the owners may change. The rooms may be too small, or the windows to large. Contractors can alter (remodel) the building to meet current needs, Fig. 18-13. Remodeling can involve restoring or changing the appearance of a building.

Some altering is done to change the performance of a product or structure. Memory chips may be added to

Fig. 18-11. This photograph shows a faulty wall of a building being replaced.

Fig. 18-12. The parking lot shown in this photo is being repaired by applying a cold-mix asphalt (left) and compacting it (right).

Fig. 18-13. The exterior of this historic building has been preserved. Its interior has been altered to provide modern offices for a law firm.

improve a computer's speed. A better set of tires and wheels can be installed on a car to improve its handling. New lighting fixtures and better heating and cooling systems can lower a building's energy use. Scrubbers are added to electrical generating plants to reduce sulfur emissions. These actions help products and structures to last longer and work better.

DISPOSING OF TECHNOLOGICAL PRODUCTS

All products and structures eventually reach the end of their useful life. They become so worn that they no longer can be repaired or altered. At that point, the owner has the responsibility to properly dispose of them, Fig. 18-14.

The first choice for disposal should be **recycling**. This means that the materials in the product or structure are reclaimed. The reclaimed material is used to make new products.

Recycling can reduce the strain on both resources and landfill disposal sites. The average home produces

Fig. 18-14. This employee is placing a can in a container as part of a company sponsored recycling program. (Boise Cascade)

an enormous amount of waste. Typically, household garbage contains the following materials:

- Paper and paperboard–42 percent.
- Food and yard wastes–24 percent.
- Glass–9 percent.
- Metals–9 percent.
- Plastics–7 percent.
- Other–9 percent.

Newsprint is the largest single component in land fills. Newsprint accounts for nearly 15 percent of the volume. The paper can be de-inked (the ink is removed) and used to make cardboard and new newsprint. Magazines and coated book paper are more difficult to recycle. This is because of the coatings used to give the paper a glossy finish.

Some food wastes and most yard wastes can be placed in a compost bin. The waste will decompose and become excellent garden fertilizer. Glass is easily recycled into new glass. It readily melts and can be reformed into containers and sheet glass. Metal is also easily recycled. Nearly half of all aluminum is now coming from recycled beverage cans.

Plastics provide a unique recycling challenge. There is not just one plastic. There are hundreds of types in use today. Household packaging is divided into seven categories for recycling. Each of these are identified by a code number inside a triangular symbol, Fig. 18-15.

Some materials in household waste are very dangerous. These materials are called household hazardous waste. Paints, solvents, engine oil, batteries, and other hazardous chemicals are examples of this type of waste. They should be disposed of through a hazardous waste center.

In addition to consumers doing their part, industry has joined the recycling campaign. Steelmakers use recycled iron and steel as base material in the steel-making process. Lumber mills use scraps from its processes to make paper, particleboard, and hardboard. Lumber mills also burn scrap lumber to provide power for the mills.

It is important that all companies dispose of their hazardous materials through licensed subcontractors. Paving contractors are using material ground from highway surfaces as aggregate for new blacktop. Old concrete becomes excellent flood control fill and is also used as aggregate.

SUMMARY

Using products makes our lives better. However, we must use and dispose of products wisely. We should select products that meet our needs and our ability to pay for them. Instruction manuals should be carefully

 PET or PETE (Polyethylene terephthalate) All two-liter soda bottles and most plastic containers that have replaced glass.

 HDPE (High-density polyethylene) detergent, liquid bleach, milk, and engine oil containers.

 V (Vinyl, polyvinyl chloride) Meat wrappers and many other translucent and transparent wraps and containers.

 LDPE (Low-density polyethylene) shopping and garment bags, bread wrappers, and most shrink wrap packaging.

 PP (Polypropylene) Margarine tubs, straws, plastic bottle caps, twine and rope.

 PS (Polystyrene) Foam cartons, packing peanuts, clear plastic bowls, plated, and utensils.

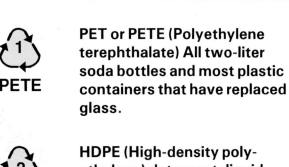 **OTHER (Non recyclable plastics like some squeezable bottles.**

Fig. 18-15. These symbols are used on plastic containers and materials to aid in sorting for recycling.

read and the instructions followed. All products should be used only for the purpose for which they were designed. Each product and structure should receive periodic maintenance and necessary repairs. Finally, each product should be disposed of properly after it has served its purpose. Whenever possible, products should be recycled. Recycling helps to reduce the strain on natural resources and disposal sites.

WORDS TO KNOW

All of the following words have been used in this chapter. Do you know their meanings?

Altering
Appearance
Diagnosis
Function
Maintenance
Preventive maintenance
Recycling
Repair
Value

TEST YOUR KNOWLEDGE

Write your answers on a separate piece of paper. Please do not write in this book.

1. List the five steps in installing a product.
2. True or false. You should consider function, value, and appearance when selecting a product.
3. Changing the oil in an engine is part of a _____ maintenance program.
4. True or false. All products can be recycled.
5. Adding memory to a computer is called _____.
6. True or false. Only companies should recycle materials.

APPLYING YOUR KNOWLEDGE

1. Select one day in your life. List all the items that you throw away. Determine their type and if they can be recycled. Record you results on a chart like the one below.

DAY:	
TYPE OF MATERIAL	**RECYCLABLE** **Yes or No**

2. Select a complex product that you use often. List:
   a. The preventative maintenance it requires.
   b. The repairs that it needs now or may need in the future.

Section Five - Activities

ACTIVITY 5A - DESIGN PROBLEM– MANUFACTURING TECHNOLOGY

Background:

A major task in developing a manufacturing system is to select and properly sequence the production operations.

Situation:

Balsum Manufacturing Company has purchased the rights to the game shown in Fig. 5-A. Golf tees or 1″ long 1/4″ dia. dowels are used as pegs. You will need five light-colored and five dark-colored pegs. This game is a modification of TIC-TAC-TOE that was developed in central Africa. The rules are: Each player uses four pegs. The object of the game is to get three pegs in a horizontal, vertical, or diagonal row. Play begins with each player alternately placing a peg in a hole. If no one has completed a row when all eight pegs are in place, the second phase starts. Then, each player alternately moves a peg into an empty hole adjacent to one of his or her pegs until one player makes a row and wins.

Challenge:

List the operations that you would use to produce the product. If there is time, set up a simple production line to make one product for each member of the class.

SAFETY: Use the safe and appropriate procedures demonstrated by your teacher.

ACTIVITY 5B - FABRICATION PROBLEM– MANUFACTURING TECHNOLOGY

Background:

Manufacturing adds worth to materials by changing their form. This is done by a number of different techniques.

Challenge:

Study the drawing (Fig. 5-B) and operation sheet (Fig. 5-C) for the recipe card holder. The recipe cards hang from the dowel on split key rings. Build a holder individually or use a continuous production line. Be sure to use safe and proper practices demonstrated by your teacher.

Materials:

5/8″ thick x 5″ wide random length pine (base)
5/8″ thick x 2″ wide random length pine (uprights)
1/4″ diameter birch dowel
3″ x 5″ note cards
Split key rings

Procedure:

Use the procedure shown on the operation process chart. Note that the parts are finished before assembly. This makes it easier to produce a high-quality finish.

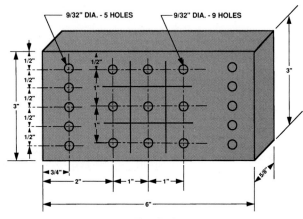

Fig. 5-A.

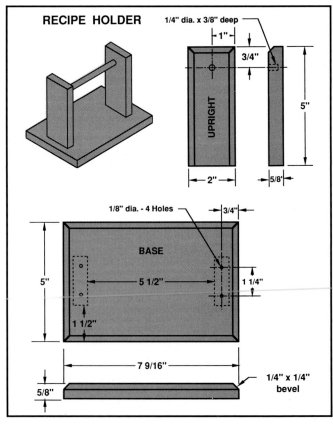

Fig. 5-B.

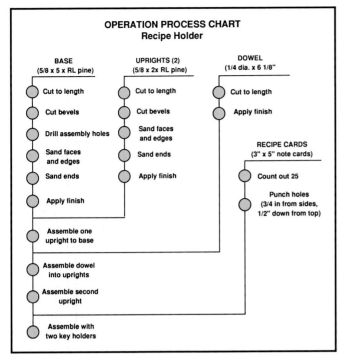

OPERATION PROCESS CHART
Recipe Holder

BASE
(5/8 x 5 x RL pine)

- Cut to length
- Cut bevels
- Drill assembly holes
- Sand faces and edges
- Sand ends
- Apply finish

UPRIGHTS (2)
(5/8 x 2x RL pine)

- Cut to length
- Cut bevels
- Sand faces and edges
- Sand ends
- Apply finish

DOWEL
(1/4 dia. x 6 1/8")

- Cut to length
- Apply finish

RECIPE CARDS
(3" x 5" note cards)

- Count out 25
- Punch holes (3/4 in from sides, 1/2" down from top)

- Assemble one upright to base
- Assemble dowel into uprights
- Assemble second upright
- Assemble with two key holders

Fig. 5-C.

ACTIVITY 5C - DESIGN PROBLEM– CONSTRUCTION TECHNOLOGY

Background:

During harvest time, farmers haul their grain to local and regional elevators. Often, more grain is brought in than can be processed. This extra grain is stored on the ground next to the elevator until it can be processed.

Situation:

The managers of Southwest Grain Elevators expect to receive 20 percent more grain than they will be able to process during the month of August. In previous years they stored the excess grain on the ground. Storing the grain in this manner results in a 10 percent reduction in quality.

Challenge:

Design and construct a model of an inexpensive shelter for 8000 cubic feet (equal to a pile 25 feet wide, 40 feet long, and 8 feet high) of grain. Fig. 5-D shows how one grain elevator in Indiana solved the problem.

ACTIVITY 5D - FABRICATION PROBLEM– CONSTRUCTION TECHNOLOGY

Background:

More and more people are living in multiple family dwellings. Sound transmission between dwellings is a major concern. A number of ways have been developed to reduce sound being transmitted from one apartment to another.

Fig. 5-D.

Challenge:

Select one of the high STC (sound transmission class) partition designs shown in Fig. 5-E. Construct a model wall using the design you selected. Use the tools and techniques demonstrated by your teacher.

Materials and Equipment:

Scale 2 x 4 and 2 x 6 stock (scale may be 1/2 or 1/4)
Scale 1/4″ and 1/2″ gypsum board
Scale 2 1/2″ fiberglass insulation
Pencils, rulers
Back saw, brads, hammer

Procedure:

1. Select the type of wall that you will construct.
2. Build the model according to the design.
3. Test the sound transmission by placing a speaker on one side of the wall and a tape recorder and microphone on the other. Create a sound with the speaker. Observe the level of sound on the microphone side by using the level meter on the recorder.

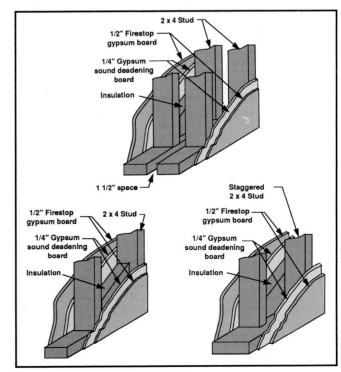

Fig. 5-E.

section 6

Applying Technology: Communicating Information and Ideas

19. Using Technology to Communicate

20. Printed Graphic Communication

21. Photographic Communication

22. Telecommunication

Section 6 Activities

HEWLETT-PACKARD

People use technology to communicate with each other. Electronic media, such as radio and television, can send messages around the world. (Voice of America, Bowling Green State University-Visual Communication Technology Program)

19
CHAPTER

USING TECHNOLOGY TO COMMUNICATE

After studying this chapter you will be able to:
- *Define communication.*
- *Describe communication as a system.*
- *Describe noise in terms of communication.*
- *List and describe the purposes of communication.*
- *List and describe the components of communication systems.*
- *List and describe the four types of communication systems.*
- *Describe the four types of communication technology.*

Fig. 19-1. This is a reproduction of a painting found in an ancient Kiva (a Pueblo Indian structure) in the southwest part of the United States.

Earlier in the text, you learned that there are data, information, and knowledge. You discovered that **data** are unorganized facts; **information** is organized data; and **knowledge** is information applied to a task. How do you obtain information? Where does information come from? These are important questions.

Think back over the day. Did you read a newspaper or listen to a radio newscast? Did you see a traffic signal or sign? Did you look at a speedometer? Did you listen to music or watch television? If you did any of these, you were using **communication technology**. Communication technology is a system that uses technical means to transmit information or data from one place to another or from one person to another.

Communication has always been an essential part of human life. Humans first communicated with gestures and grunts. Later they developed language, increasing their ability to communicate. However, these forms of communication do not involve technology. There is no technical means between the sender of the message and the receiver.

Possibly the first use of communication technology was cave paintings. The "artists" used sticks, grass, or their fingers to apply paint to cave walls, Fig. 19-1. The result was a message that could be stored. At some later date, a person could retrieve the message. Over time, communication technology has expanded in range and complexity.

PURPOSES OF COMMUNICATION

Communication is used for a number of purposes. Most often it is used to convey ideas, exchange information, and express emotions, Fig. 19-2. An **idea** is a mental image of what a person thinks something should be. You probably have ideas about what kind of music is good, how people should behave, or what activity is

EMOTION	INFORMATION	IDEA

Fig. 19-2. Communication technology can be used to convey ideas, information, or emotions. (American Greetings Co., General Electric Co., Ball State University)

fun. You probably also have opinions on how to do various tasks, such as washing an automobile, riding a bicycle, or mowing a lawn. People have ideas about how to protect the environment, whether to allow capital punishment, and other issues. Communication media can be used to share these ideas.

Information is vital to taking an active part in society. It provides a concrete foundation for decision-making and action. Information can be as simple as the serving time for lunch. It can be as complex as the moon's effect on the tides. Communication media can be used to exchange information among people.

Finally, you may want to communicate **emotions**. Ideas and information are important. However, feelings are just as vital to many people. Communication media can convey these feelings. For example, a photograph can communicate the excitement of a sporting event. People can communicate affection for each other through greeting cards. Communication media can make us laugh or cry, make us excited or calm, make us feel good or feel bad.

IMPACTS OF COMMUNICATION

Each communication message is designed to impact someone. The communication can do one or more of three basic tasks:

- **Inform** by providing information about people, events, or relationships. We read books, magazines, and newspapers to obtain information. Radio news programs, television news programs, and documentaries are designed to provide information.
- **Persuade** people to act in a certain way. Examples are, "Say No to Drugs," "Buckle-Up" or "Give a Hoot, Don't Pollute." Print and electronic advertisements, billboards, and signs are typical persuasive communication media.
- **Entertain** people as they participate in or observe events and performances. Television programs, movies, and novels are common entertainment-type communication.

These three goals may be merged. Two new words in our language arise from this merging of goals. The first is *infotainment*, which means providing information in an entertaining way. You may learn as you watch a quiz show on television, or play with computer simulations. Both of these are enjoyable ways of gaining new and useful information.

The second term, *edutainment*, takes communication one step beyond infotainment. Edutainment is more than allowing the information to be available in an entertaining way. It creates a situation in which people want to gain the information. The television program, Sesame Street, is a good example of edutainment.

MODEL OF COMMUNICATION

Communication can be thought of as a simple process, Fig. 19-3. The action starts with the **encoding** of a message. Encoding means that the message is changed into a form that can be transmitted. The encoding may be electronic impulses on a magnetic tape, an arrangement of letters on a printing plate, or exposed chemicals on photographic film.

The message is then **transmitted** to the receiver. Transmission involves a communication **channel** or carrier. The channel might be electromagnetic waves broadcast through the air, electrical signals carried by a wire, pulses of light on fiber-optic cable, or printed text on paper.

The other end of the communication channel is the **receiver.** The receiver gathers and **decodes** the message. Examples of receivers are telephones, which change electrical impulses into sound, and fax machines, which change electrical impulses into printed messages. The human mind decodes the written and graphic messages contained in photographs and printed media.

Communication models can be more complex than the one shown in Fig. 19-3. These models show **interference** in the communication channel. Interference is anything that impairs the accurate communication of a message. Static on a radio is interference. Noise in a movie theater is interference. Smudged type would be interference in printed messages.

It is important to be aware of the difference between *information* and *noise.* **Noise** is unwanted sounds or signals that become mixed in with the desired information. When you listen to the radio, both information and noise are present. In this example, they are both in the form of sound. The information is the sound you want to hear, while the noise is the sound that you do not want to hear. Noise on a television transmission turns into static on your television screen. Noise, then, is a type of interference. Also, noise can involve personal taste. Some people will call the music you listen to "noise." You may feel the same way about the music they prefer. You are both correct, since unwanted sound is noise.

TYPES OF COMMUNICATION

One way to look at communication is in terms of the sender and receiver, Fig 19-4. We are all familiar with people communicating with people. This type of communicating is the most common. It works through our electronic media and our printed products. This type of communication is called **human-to-human communication**. It is used to inform, persuade, and entertain other people.

However, there are other types of communication. Have you ever reacted to a traffic light, a warning light on the dashboard of your car, or the bell that indicates the end of a class period? If so, you have participated in **machine-to-human communication**. This type of communication system is widely used to display machine operating conditions.

Have you keyed material into a computer or set the temperature on a thermostat? If so, you have engaged in **human-to-machine communication**. This type of communication system starts, changes, or ends a machine's operations.

Finally, computer-controlled operations use **machine-to-machine communication**. Modern industry is

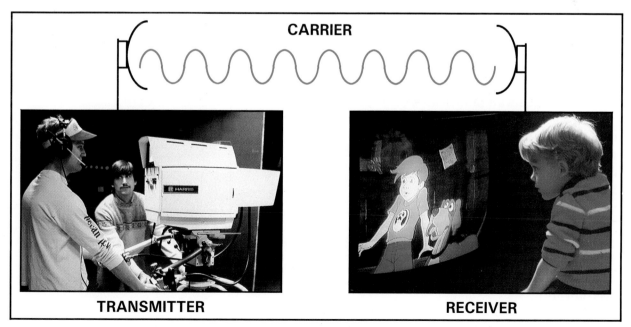

Fig. 19-3. All communication systems have transmitters, channels, and receivers. This shows the television communication system. (Ball State University, Westinghouse Electric Co.)

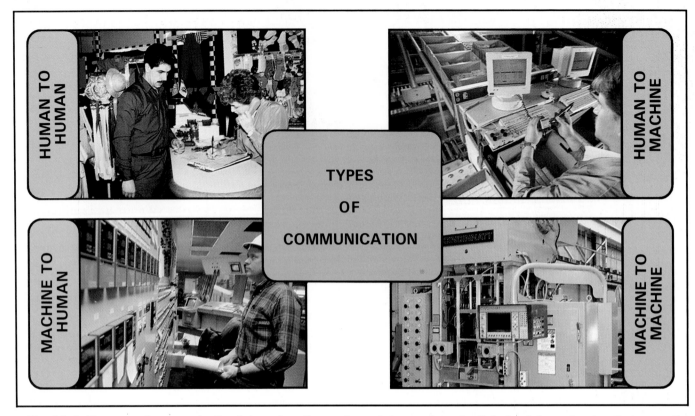

Fig. 19-4. The type of communication is based on the sender and the receiver. A—Relaying information to others over the telephone. B—Relaying information to a computer. C—Checking a readout from a plant system. D—Controlling a manufacturing system through a computer. (United Parcel Service, Siemens, Inland Steel Co., Cincinnati, Inc.)

becoming more computer-based. Humans enter programs and data into the computer, then the computer directs and controls an apparatus. A computer controlling a printer, or a thermostat controlling a furnace, are examples of this type of communication. Typical examples of computer-controlled operations are computer-aided drawing (CAD), computer-aided manufacturing (CAM), computer-integrated manufacturing (CIM), and robotics.

COMMUNICATION SYSTEMS

We are bombarded with information every day. It comes in many printed and electronic forms. However, the communication technology used to deliver the information can be divided into four main types, Fig. 19-5. These types are:
- Printed graphic communication.
- Photographic communication.
- Telecommunication.
- Technical graphic communication.

PRINTED GRAPHIC COMMUNICATION

Much of communication technology was developed to satisfy the need for mass communication. People wanted to tell their message to large numbers of other people. The first mass communication system developed was printing or **printed graphic communication**, Fig. 19-6. In Chapter 7, you were introduced to movable type, the foundation of printing.

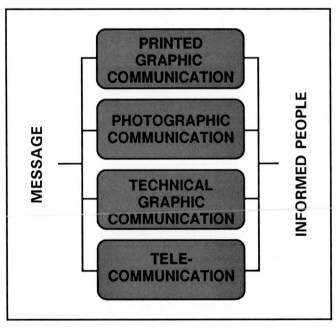

Fig. 19-5. The four basic communication technology systems.

Fig. 19-6. These magazines are examples of printed graphic communications.

The term "printing" originally meant putting an image on paper with inked type. It now includes all of the processes used to reproduce a two-dimensional image on a material. That material may be paper, metal, plastic, cloth, or wood. These processes will be discussed further in Chapter 20.

PHOTOGRAPHIC COMMUNICATION

Some communication technologies use light to convey their message. The most common of these is photography. Photography captures light as an image on a recording medium that we call film. When the film is developed, the image becomes permanent. The film can be used in making prints of the image or in projecting the image onto a screen.

Photography is the process of producing the image. **Photographic communication** is the process of using photographs to communicate a message, Fig. 19-7. These processes will be discussed in Chapter 21.

TELECOMMUNICATION

One type of communication system is playing an increasing role in modern life. This system is **telecommunication**, communication at a distance, Fig. 19-8.

Fig. 19-7. What feelings are communicated by this photograph?

Fig. 19-8. This designer is preparing technical illustrations for a new product. (Ohio Art Co.)

Technology: How it Works

COMPACT DISC (CD) SYSTEM: An information reproduction system that uses a laser to read optically encoded information, changes the optical signals into digital signals, and then recreates the information.

In 1877, Thomas Edison became the first person to record sound. The first records were wax cylinders with grooves that captured the voice and music, Fig. 1. These cylinders gave way to flat disc records, which were cheaper to produce. The first discs rotated at 78 rpm (revolutions per minute). These were overtaken by longer playing 45 rpm and 33 1/3 rpm records.

In all of these earlier systems, the information is encoded by producing wavy grooves in the surface of the record. The code is read by a stylus, or needle, that traces the grooves. As the stylus is vibrated (rapidly moved left and right) by the wiggles in the grooves, it sends continuous, or *analog,* electrical signals to the playback unit. The playback unit amplifies this rising and falling signal and feeds it to the speaker. The electrical energy is converted to the sound you hear in the speaker.

This system has a major drawback. Dirt and groove wear cause the sound to distort. This produces a scratching or popping noise through the speakers.

As a result, analog records are now being largly replaced by compact disc recordings, or CDs. The CD is much smaller, 4 3/4 in. diameter, than the old 7 in. 45 rpm records or the 12 in. 33 1/3 rpm records. Yet, a CD can hold far more music than the larger records.

Compact disc systems are very different from past systems. First, the information on the CD is recorded as a *digital* signal, a signal in discrete steps. Digital signals can be very accurately reproduced. Second, CDs use a beam of light, rather than a stylus, to pick up the coded message from the disc. Thus, there is no physical contact with the CD. This prevents disc wear.

When music is recorded on a CD, its frequency is sampled 44,100 times a second. This reading is converted to a *binary* number. A binary code uses a series of zeros and ones to represent any number. The ones and zeros can be represented by electricity flowing (on), or not flowing (off). A compact disc contains billions of tiny pits in its surface. When the laser detects a change from the flat surface to a pit, or vice versa, it signals a one to the system. Flat stretches, either over the surface or over the pits in the surface, signal a zero to the system.

This code is read by a laser. When the laser hits the disc, the beam is reflected through a prism to a light-sensitive diode, Fig. 2. This diode allows a voltage flow relative to the amount of light that strikes it. Little light is reflected from pitted areas. Nearly all the light is reflected from the flat surface. The diode translates the light signals into electrical signals. These signals are converted from digital to analog and then fed to the speaker.

Compact disc technology is not limited to simply reproducing sound. The CD is becoming a common storage medium for large amounts of read-only memory (ROM) for computers. Currently, entire encyclopedias, as well as the contents of every phone book in America, can be purchased on discs. With this potential, CD systems are taking on more duties than just being quality replacements for Edison's wax cylinders.

Fig. 1. An early Edison phonograph.

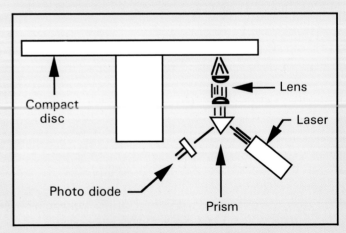

Fig. 2. Structure of a typical CD player.

Fig. 19-9. These people are using telecommunication in a corporate training session. (Hewlett Packard)

Telecommunication includes a number of specific types of communication. Probably the most widely used are radio, television, and the telephone. Telecommunication is the subject of Chapter 22.

TECHNICAL GRAPHIC COMMUNICATION

People and companies often want to communicate specific information about a product or its parts. This information may convey the size and shape of a part. It may suggest how the parts are assembled to make a product or structure. It may tell how to install, adjust, operate, or maintain a device. This type of information is often communicated through engineering drawings or technical illustrations, Fig. 19-9. The methods that prepare and reproduce this media are called **technical graphic communication** systems. Technical graphic communication is discussed at length in Chapter 12.

SUMMARY

We are rapidly moving from the industrial age into the information age. This new age is characterized by the wide-scale availability of communication media. People receive and send a constant barrage of data and information every day.

This transfer is done in terms of human-to-human, human-to-machine, machine-to-human, and machine-to-machine communication. Each of these systems has a transmitter that encodes and sends the message. The message then travels over a carrier to a receiver. The receiver collects and decodes the message.

The most common forms of communication systems are printed graphic, photographic, telecommunication, and technical graphic. Our lives are affected daily through these systems.

WORDS TO KNOW

All of these words have been used in this chapter. Do you know their meanings?

Channel
Communication technology
Data
Decode
Edutainment
Emotions
Encode
Entertain
Human-to-human communication
Human-to-machine communication
Ideas
Inform
Information
Infotainment
Interference
Knowledge
Machine-to-human communication
Machine-to-machine communication
Noise
Persuade

Photographic communication
Printed graphic communication
Receiver
Technical graphic communication
Telecommunication
Transmitter

TEST YOUR KNOWLEDGE

Write your answers on a seperate sheet of paper. Please do not write in this book.

1. Define communication technology.
2. True or false. Knowledge is organized data.
3. Match the product on the left with the communication system it represents.

   _____ Newspaper.   A. Printed graphic.
   _____ Radio.   B. Photographic.
   _____ Motion   C. Telecommunication.
           picture.   D. Technical graphic.
   _____ Catalog.
   _____ Blueprint of
           a house.
   _____ Vacation photo.
   _____ Parts drawing in
           an owner's manual.

4. What three main things can be communicated?
5. True or false. The simplest communication system includes a transmitter, channel, and a receiver.
6. An unwanted sound or signal is called _____.
7. List the four basic types of communication.
8. True or false. Television is an example of machine-to-human communication.

APPLYING YOUR KNOWLEDGE

1. Select a piece of information or an emotion you want to communicate. List four ways you could communicate it and the communication system you would use. Record your work on a chart like the one below.

Information or emotion to be communicated:	
Technique #1	____ Printed graphic ____ Photographic ____ Technical graphic ____ Telecommunication
Technique #2	____ Printed graphic ____ Photographic ____ Technical graphic ____ Telecommunication
Technique #3	____ Printed graphic ____ Photographic ____ Technical graphic ____ Telecommunication
Technique #4	____ Printed graphic ____ Photographic ____ Technical graphic ____ Telecommunication

2. Design a communication message that will persuade a person to your point of view on an issue.

CHAPTER

PRINTED GRAPHIC COMMUNICATION

After studying this chapter you will be able to:
- Describe printed graphic communication.
- Describe the many products produced by printed graphic communication processes.
- List and describe the six major types of printing processes.
- Describe the steps involved in designing a printed graphic message.
- Explain the differences between copy and illustrations.
- Define an image carrier.
- Explain how printed messages are prepared for production.
- Explain how printed graphic messages are produced.
- Describe desktop and electronic publishing.

Communication means many things to many people. We talk about communication messages, communication media, communication networks, and communication systems. Central to all of these is the word *communication*. Simply stated, communication is the passing of data and information from one location to another or from one person to another. When the spoken word is used to communicate, we call it language or verbal communication. When a technical means is used to convey information, it is called communication technology.

As you learned in the last chapter, there are a number of communication systems that use technology. The oldest of these systems is **printed graphic communication**, or **printing**.

Most printed communication use an alphabet to convey the message. An alphabet is a series of symbols developed to represent sounds. The first alphabet was developed in Syria around 1200 B.C. Today, there are a number of different alphabets including the 26-letter one used for the English language. These letters can be arranged to convey a message through writing.

The process of printing takes written words and places their image on a material. The printing is applied to a material called the **substrate**. Originally, almost all printing was done on paper. Today, printing is done on a variety of substrates including: paper, glass, plastic, cloth, ceramics, metal, and wood. The result of this mixture is a broad range of printed products. These products include: newspapers, magazines, books, brochures, pamphlets, labels, stickers, clothing designs, and signs. Each of these is carefully designed and produced by a specific printing process.

PRINTING METHODS

The first printing was done by carving an image into a block of some material. Next, the block was covered with ink. The inked block was then pressed on paper to

Fig. 20-1. This museum exhibit shows early block printing.

produce the "printed word." This technique, shown in Fig. 20-1, was the forerunner for all printing processes. Today, there are six major printing processes, Fig. 20-2. These processes are:

- Relief.
- Lithography.
- Gravure.
- Screen.
- Electrostatic.
- Ink jet.

RELIEF PRINTING

Relief printing was introduced in Chapter 7 during the discussion of information processing equipment. Relief printing, the oldest of all printing processes, uses an image that is on a raised surface, Fig. 20-3. Ink is applied to this raised surface and it is pressed against the substrate. The pressure forces the ink to adhere to the substrate, producing the printed message.

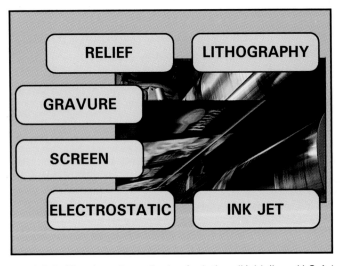

Fig. 20-2. The six main methods of printing. (Heidelberg U.S.A.)

Relief printing requires that the image be reversed, or "wrong reading," on the printing block, or **image carrier**. Once the inked block has been pressed against the substrate, the message will be forward, or "right reading." Look at a rubber stamp and you will see an example of a reversed or "wrong reading" relief image.

The two main types of relief printing are letterpress and flexography. **Letterpress** uses metal plates or metal type as the image carrier. Until the last 30 to 40 years, letterpress was the most common printing processes. It now accounts for less than 5 percent of all printing.

Flexography is an adaption of letterpress. It uses a rubber or plastic image carrier. Flexography is widely used for printing packaging materials. Today, this process accounts for about 20 percent of all printing.

LITHOGRAPHIC PRINTING

Lithographic printing, or **offset lithography**, is the most widely used method of printing today. This process prints from a flat surface, Fig. 20-4. Offset lithography is based on the principle that oil and water do not mix. This printing method was discovered by Alois Senefelder. Senefelder used a limestone slab as his image carrier. He first wrote a reverse, "wrong reading," image on the stone with a grease pencil. Next, he wet the stone, which dampened the stone surface except for the areas with grease marks. Then, he rolled ink over the stone surface. The ink adhered to the grease-marked areas but not to the wet stone. An inked image remained which could be transferred to paper. Finally, paper was placed over the stone, and pressure was applied. This caused a direct transfer of the image from the carrier to the substrate. Today, some artists still use Senefelder's basic process to produce prints of their work.

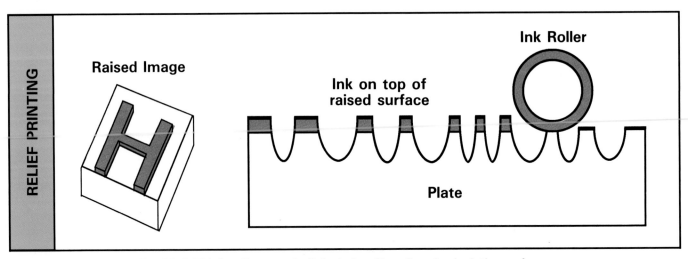

Fig. 20-3. This is a diagram of relief printing. Note the raised printing surface.

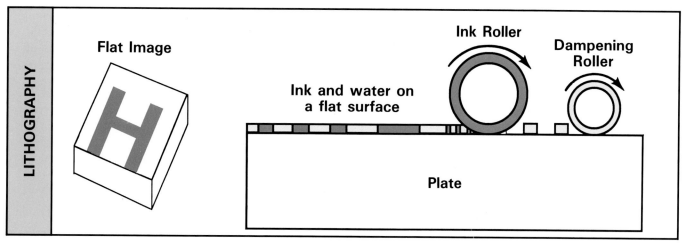

Fig. 20-4. This is a diagram of lithographic printing. Note the flat printing surface of this method.

The basic lithographic process has been refined. A "right reading" image is produced on a image carrier called an offset plate. The inked image on the plate is first transferred to a special rubber roller called an offset blanket. This reverses the image. The image becomes "right reading" again when it is transferred to paper as it is fed through the press. Because the original image is first *offset* to a blanket, the process is called *offset* lithography. Modern offset presses can produce more than 500,000 copies from a single offset printing plate.

GRAVURE PRINTING

Finely detailed items, like paper money and postage stamps, are usually printed using the **gravure**, or intaglio, process, Fig. 20-5. This process is the opposite of relief printing. In gravure printing, the message is chemically etched or scribed into the surface of the image carrier. The carrier is then coated with ink. Next, the surface ink is scraped off with a doctor blade. This leaves ink only in the recessed areas of the carrier. When paper is pressed very tightly against the carrier, it picks up the ink in the cavities.

Gravure is an expensive process. However, it is economical for very long production runs because one gravure plate can withstand several million impressions. In addition to money and stamps, some magazines are printed with this process.

SCREEN PRINTING

Screen printing is a very old printing process, dating back more than 1000 years. Screen printing uses a stencil with openings that are the shape of the message. The stencil is mounted on a synthetic fabric screen. Paper is then placed beneath the screen, and ink is applied to the screen's upper surface. Finally, a squeegee is pulled across the stencil. This forces ink through the openings in the stencil to produce the printed product, Fig. 20-6.

Screen printing is used to print on fabrics, T-shirts, drinking glasses, PC (printed circuit) boards, and many other products. It is also used to print small quantities of very large products, such as posters or billboards.

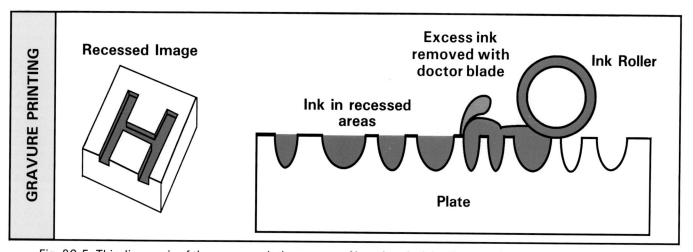

Fig. 20-5. This diagram is of the gravure printing process. Note that the inked image is inset into the image carrier.

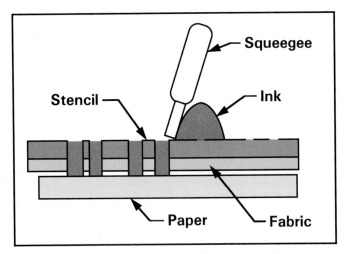

Fig. 20-6. This is a diagram of screen printing, or silkscreening.

Screen printing is often called **silkscreening**. Originally, the stencils were mounted on silk fabric. Currently, however, stronger and more durable fabrics have almost completely replaced silk as the screen fabric used in the process.

ELECTROSTATIC PRINTING

A commonly used process today is **electrostatic printing**. Electrostatic printing is often called "xerox copying." This is not the proper term, because Xerox® is the trade name for one brand of electrostatic printers.

The electrostatic printing process uses a machine with a special drum. The image to be copied is placed on this machine. Then a strong light is reflected off the image and onto the drum. The reflected light creates an electrically charged likeness of the image on the drum's surface. This charge is transferred to a sheet of paper that is then passed over fine particles called toner. The toner is attracted to the charged image on the paper. Finally, the paper passes through a heating unit that fuses the toner onto it.

The early office copiers were fairly slow. Today, high speed electrostatic copiers can produce several thousand copies an hour. Some of the newest electrostatic copiers can be directly interfaced with (joined to) microcomputers. The computer sends a chosen image directly to the copier. The paper original is eliminated, and even the most complex page layouts are easily and quickly reproduced.

INK JET PRINTING

The newest printing process is **ink jet printing**. In this process the printed message is generated by a computer. The computer then directs a special printer that sprays very fine drops of ink onto the paper. Since the printing head never touches the substrate, a wide range of materials can be printed by this process.

Ink jet printing is used for coding packages and producing mailing labels. The process can use multiple printing heads allowing the printer to produce several thousand characters per second.

PRODUCING PRINTED GRAPHIC MESSAGES

Printed messages and products are the result of a series of planning and production activities. These activities, Fig. 20-7, can be grouped into three main steps:
- Designing the message.
- Preparing to produce the message.
- Producing the message.

DESIGNING THE MESSAGE

Communication is successful only when the intended audience receives the given data or information. There are a number of ways to communicate an idea or concept. Think of the concept, church, Fig. 20-8. We can say the word and cause people to form a mental image of a church. However, this process requires us to be near the audience. A larger audience over a greater distance can be reached if we use printed graphic communication. Printed graphic communication can include additional words which we call text, or **copy**. We can also illustrate the message. **Illustrations** are pictures and symbols, which add interest and clarity to the printed communication.

The design of each communication follows a set procedure. This process starts with an **audience assessment**. For an audience assessment, the designer must determine who the audience is, what they like and dislike, what their interests are, plus a number of other factors, Fig. 20-9. In short, graphic designers must *get to know* their audience. This background information

Fig. 20-7. The three main steps in printing.

Fig. 20-8. Here are several ways to communicate the idea "church."

will help the designer decide *what information to tell* and *how to tell it*. The designer, generally, has a great deal of information available. The designer's challenge is to select the useful information to include in the message. For example, suppose you are developing a promotional flyer for recycling. You would, first, have to decide who your target audience is. Will it be private citizens, business, industry, or government? Each of these groups have their own set of interests and behaviors. Your message will have to appeal to their individual interests in order to help change their personal behav-

ior. It is difficult to write one message for all people. Most communication messages are targeted to a specific audience.

After getting to know your audience, you must, next, get them to receive the message. This prompts the "how" part of the design. To promote your recycling program, should you use a humorous appeal or an environmental concerns appeal? A message needs an attention-getting aspect, or a "hook," to have a strong chance of making an impact.

Designing the message involves two main steps, **design** and **layout**. In the design stage the message is developed. In the layout stage the message is put together.

Design

Graphic design deals with the appearance or "look" of the page. The design should attract readers and hold their attention. An important message that is never read is useless.

To develop a message, you must consider a number of design principles, Fig. 20-10. These principles include proportion, balance, contrast, rhythm, variety, and harmony.

Proportion. This principle deals with the relative size of the parts of the design. It is concerned with the height-width relationships of the parts within the design. Good proportion will have an eye pleasing relationship between large and small elements within the message.

Balance. People seem to enjoy a visual balance, so strong media messages will be designed with this principle in mind. Balance is accomplished through having the information on both sides of a center line appear equal in visual weight. There are two ways to achieve balance, Fig. 20-11. The first way is *formal* balance.

Fig. 20-9. An audience assessment allows this designer to tailor his message to the audience. (AC-Rochester)

DESIGN PRINCIPLES
PROPORTION
BALANCE
CONTRAST
RHYTHM
VARIETY
HARMONY

Fig. 20-10. When designing graphic communication, check for these six principles.

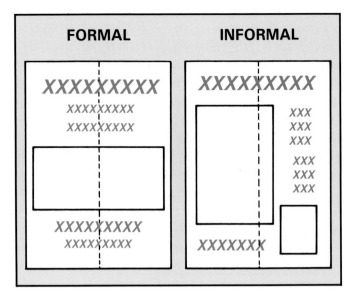

Fig. 20-11. These are representations of formal and informal balance. Notice how there is a similar visual weight on both sides of the center line.

With formal balance both sides of the center line are close to mirror images of each other. The second way is *informal* balance, in which there are equal amounts of copy and illustrations on each side of the layout, yet both sides of the center line do not mirror each other. This layout still produces a feel of equal weight or balance when viewed.

Contrast. This is used to emphasize portions of the message. This can be done by changing the color of the important elements, or by printing the important elements in **bold**, *italic*, or enlarged type.

Rhythm. The principle of rhythm deals with the flow of the communication. Rhythm in an advertisement might repeat certain elements in the design. It produces a sense of motion and guides the eye to an important feature.

Variety. People receive thousands of messages each day. Variety is the technique that makes the message unique and interesting. For example, in a full color publication, a black-and-white message stands out. Unique layouts will also catch the reader's eye.

Harmony. We enjoy hearing harmony in music. Harmony refers to the notes blending together to form a pleasing sound. Likewise, harmony in graphic design is achieved by blending the parts of the design to create a pleasing message. The parts of a harmonized message will fit and flow together.

Layout

Layout is the physical act of designing the message. It involves positioning the copy and illustrations to form a message that communicates effectively. This task starts with sketching. First, a series of **thumbnail sketches** are prepared, Fig. 20-12. These allow the graphic designer to experiment with various arrangements of copy and illustrations. These thumbnails are then used in preparing **rough sketches** which integrate

and refine the ideas generated in the thumbnails. Rough sketches are drawn in the final proportion.

Finally, a **comprehensive layout** is made of the best rough sketch, Fig. 20-13. The comprehensive, or "comp," is the layout for the final design. The comp deals with several elements. One is typography. This is the typeface, size, and style of type that will be used in the layout. Also, illustrations must be sized and cropped to fit into the layout. Cropping is selecting and separating the important information in a photograph. Finally, color for the various parts must be specified.

PREPARING TO PRODUCE THE MESSAGE

After the design has been developed, it must be made ready for printing. This involves three basic steps: composition, pasteup, and image carrier preparation.

Fig. 20-12. These are typical thumbnail sketches used in designing a graphic communication. (Ohio Art Co.)

Fig. 20-13. Shown here are comprehensive layouts for a communication product. (Ohio Art Co.)

Composition

Composition consists of the activities that change written words and illustrations into forms that can be used in various printing processes. In most cases, this involves typesetting and illustration preparation.

Typesetting produces the words of the message. This includes generating, justifying, and producing the type. A typeface is specified in the design. It is selected from hundreds of different available styles, Fig. 20-14. Today, the written words are usually developed through electronic typesetting machines. Whether to use letters with serifs or letters without serifs (sans serif) is one option in choosing a typeface. Serifs are the small flares at the ends of printed letters. The body of this test is printed in a typeface with serifs, while the headings are printed without serifs. Additional groupings include traditional, contemporary, decorative, and script.

Justification (alignment of the copy) on the left, right, or both edges of the paper, is another part of typesetting. Letters and informal messages are gener-ally left-justified, leaving an uneven or "ragged" right margin. Books, magazines, and newspapers justify both edges of the copy. Artistic messages are sometimes right-justified to attract attention.

Finally, the type font must be selected. This may be plain, bold, italics, condensed, expanded, or one of several other styles, Fig. 20-15.

The type is produced in one of two ways. The first way is **hot-metal composition**. Hot-metal composition involves casting molten metal into type to produce the message. This process is seldom used today. The most common techniques use **cold composition**. Cold composition includes all other means used. These techniques extend from transfer letters, strike-on (typewriter) type, and laser printer output, to sophisticated computer-driven phototypesetters. This book was composed using phototypesetting equipment.

Illustration preparation is often required for useful communication. This preparation may include sizing and converting the line art or photographs. They might need to be enlarged or reduced to fit the space in the layout. Other than sizing, line art, often, needs no preparation to be placed in a layout. Photographs, however, must be changed into different forms. They cannot be directly used in printing processes. All photographs are first changed into halftones.

The large brown dog ran fast.

The large brown dog ran fast.

The large brown dog ran fast.

The large brown dog ran fast.

The large brown dog ran fast.

THE LARGE BROWN DOG

Fig. 20-14. Some common type faces. Where would you use each one?

TYPE STYLE	
SERIF	**SANS SERIF**
Johannes Gutenberg began the printing revolution more than 500 years ago. Gutenberg's method of printing is called letterpress.	Johannes Gutenberg began the printing revolution more than 500 years ago. Gutenberg's method of printing is called letterpress.
LEFT JUSTIFICATION	**FULL JUSTIFICATION**
Johannes Gutenberg began the printing revolution more than 500 years ago. Gutenberg's method of printing is called letterpress.	Johannes Gutenberg began the printing revolution more than 500 years ago. Gutenberg's method of printing is called letterpress.
PLAIN	**BOLD**
Johannes Gutenberg began the printing revolution more than 500 years ago. Gutenberg's method of printing is called letterpress.	**Johannes Gutenberg began the printing revolution more than 500 years ago. Gutenberg's method of printing is called letterpress.**

Fig. 20-15. Notice the contrasts in these three typesetting options.

To produce a halftone in a black-and-white photograph, a negative of the picture is created with the photograph being exposed through a halftone screen. When the photograph is printed, the various tones in the photograph are changed into a series of dots. The light areas are made up of small dots while dark areas are made up of large dots. Look at a photograph printed in a newspaper under a magnifying glass. You will see that it is actually a series of dots.

Color photographs are separated into four printers, or masters. These, as shown in Fig. 20-16, are yellow, magenta (red), cyan (blue) and black. During the separation, the four are run through a halftone screen. Later, each color will be printed independently to recreate the color photograph.

Line drawings for multi-color screen prints are produced in several layers. One layer is necessary for each different color that will be printed, Fig. 20-17.

Pasteup

The second step in preparing the message produces a **pasteup**, or mechanical. In this step, all the type and illustrations are assembled onto a sheet that looks exactly like the finished message, Fig. 20-18. This assembled layout is called a pasteup.

Black-and-white (or any single-color) printing uses one pasteup for each image carrier (printing plate or screen). Color printing requires one pasteup for each color. The pasteup is the master copy used to prepare the image carriers, which will, in turn, produce the final printed product.

New computer-driven page composition systems generate the layouts and "pasteup" in a single step. The computer operator edits the text, manipulates the copy

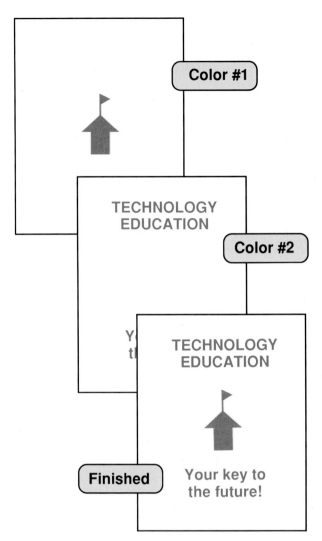

Fig. 20-17. Each color on a line drawing is printed separately.

and locates the spaces for illustrations right on the computer screen. The layout is sent electronically to the printer, which then produces a finished page layout. With these computer systems, the need for manual locating and pasting down of type and illustrations is eliminated. Also, the layouts are stored in the computer's memory, so changes are quick and easy. The use of these systems, often called electronic publishing, will be further discussed later in this chapter.

Preparing Image Carriers

Most printing processes use an image carrier. The image carrier is the feature that makes each process unique. Let's look at how relief, offset lithography, gravure, and screen image carriers are prepared.

Preparing Relief Image Carriers. A relief image carrier has a raised printing surface. This surface can be produced by setting type, as described in Chapter 7. However, the letterpress (metal type) printing process is becoming less common each year.

Fig. 20-16. A color photograph (such as this flower) is separated into four single color printers.

Fig. 20-18. A graphic designer is working on this pasteup.

As you learned earlier in the chapter, flexography is an emerging relief printing process. Flexography uses a flexible plastic or rubber printing plate. This image carrier is produced in one of two basic ways, casting or etching. Casting uses a mold that is formed and cured around metal type. The mold is then removed from the type and filled with a plastic material. The plastic cures (hardens) into a sheet that is removed from the mold. This becomes the printing plate.

Etching is a photochemical process. A negative of the desired image is placed over a light-sensitive material. The material and negative are held together tightly and placed under a strong light. The light-sensitive material is exposed through the open areas on the negative. A chemical developer hardens the exposed material. A detergent is then applied to wash away the unexposed material. A raised image of the message is left.

Preparing Lithographic Image Carriers. Offset lithography uses a flat printing plate, produced through a photochemical process, Fig. 20-19. The pasteup is converted into a negative and mounted on a sheet of paper. The paper over the type and illustration area is then cut

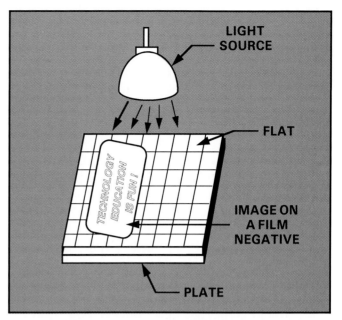

Fig. 20-19. Offset plates are produced by exposing the plate through a photographic negative, and then developing the resulting image.

away. The resulting form is called a flat. The flat is placed on top of a sheet of metal (offset plate) that has a light-sensitive coating. The flat-and-plate assembly is placed in a vacuum table that holds them together tightly. This assembly is exposed to a very bright light. The image area on the negative allows light to pass through and expose sections of the plate. The plate is then chemically developed, producing the image carrier.

Preparing Gravure Image Carriers. The gravure image carrier, or printing cylinder, has the image engraved into it. This is generally done with a electro-mechanical scanner, Fig. 20-20. This computer-controlled machine reflects light off the copy. The light intensity is then picked up by a sensor. Black areas reflect little or no light, while white areas reflect a high

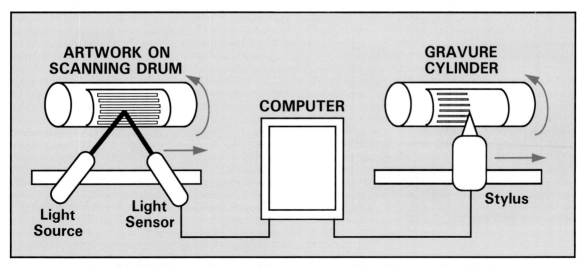

Fig. 20-20. This is a diagram of the process used in making a gravure plate.

percentage of light. The gray areas reflect varying amounts of light in proportion to their shade. The amount of light reflected is communicated to the computer through the sensor. The computer processes the information and then uses the information to drive a stylus touching the gravure cylinder. The stylus carves the desired pattern into the cylinder producing the image carrier.

Preparing Screen Image Carriers. Most screen carriers are produced using a photochemical process. First, a positive transparency of the image is produced. This transparency is placed on top of photosensitive film with a clear plastic backing. Light is passed through the positive, exposing the film. When the film is developed, the portion that was not exposed to light will harden. The remaining emulsion is them rinsed away. The wet film is placed under the screen fabric. The fabric is pressed into the film, and it is allowed to dry. The clear plastic film backing is then removed, leaving a negative of the image on the screen.

PRODUCING THE MESSAGE

The third activity in printed graphic communication is producing the message. Production involves printing the message (message transfer) and finishing the product (product conversion). Two basic forms of substrate are used in the production of printed graphic messages, Fig. 20-21. The first form is sheet. This form includes all the flat, rectangular pieces of substrate. They can be sheets of paper, glass, plastic, metal, etc. The other form is web. The web form uses a roll of substrate that is continuously fed into the printing process.

Both of these forms require specific printing press equipment. A typical press has a feeder unit, which moves the substrate into the press; a registration unit, which positions the material for correct placement (register) of the message; the printing unit, which transfers the message from the carrier onto the substrate, and finally, the delivery unit, which stacks the printed product.

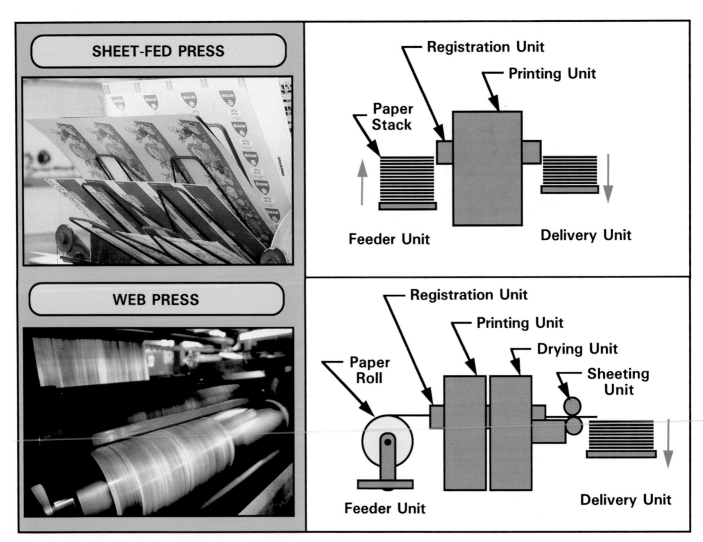

Fig. 20-21. Shown are two printing press systems. The top system is printing on a sheet substrate. The bottom system is printing on a web (roll) substrate. (Ohio Art Co., Tribune Co.)

Technology: How it Works

FAX MACHINE: A device that sends copies of documents over telephone lines or radio waves using digital signals.

A communication system that is new to the general public is the facsimile machine, more commonly known as a *fax*. A fax machine is like a copier, except it has two parts. The two parts, the scanner and the printing unit, work independently. The scanner on one fax sends the image of a page to the printing unit of another fax machine. The two fax machines can be many miles apart.

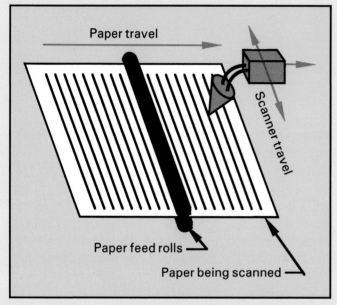

Fig. 1. This is a schematic diagram of how a fax scanner works. Some fax machines use moving scan heads, others move the paper over a row of sensors.

A fax machine changes the black and white images on a page into *digital* information. Digital information has two states, represented by ones and zeros. The two states can be "on" and "off", or a higher voltage and lower voltage. A fax uses an optical scanner that looks at the image on a page, Fig. 1. The scanner breaks the page into small lines. Each line is broken into small dots, called pixels. The scanner examines each pixel to see if it is black or white. A digital signal is generated by the black and white pixels. A white pixel might be represented by a "1" and a black pixel by a "0" (or vice-versa).

A modem is used to convert the digital signals into audible tones that can be sent over telephone lines or using radio waves. Newspapers and police departments have sent "wirephotos" over the phone lines for years. Amateur radio hobbyists use fax machines to send documents over the airwaves. Weather maps are also sent via radio. Instead of using telephone lines, the tones from the modem are fed into a transmitter and broadcast. Another person with a receiver feeds the tones back into a fax machine for printing. Today, anyone with a fax machine can send letters, pictures, and other printed information across town or around the world.

The receiving fax machine has a modem that converts the tones back into digital signals. The digital signals are changed back into black and white pixels by a printing system. The pixels are grouped back into lines, and the lines are printed onto a page. The printing system in a fax machine can be a thermal printer, or may use a photosensitive drum and toner like an office copier, Fig. 2.

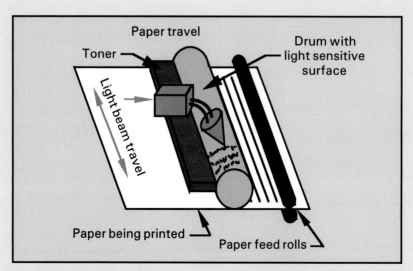

Fig. 2. The printing unit of a fax turns the digital signals from the modem into black and white pixels on paper. This diagram shows a photosensitive drum printing unit.

Web presses have two added features, drying units and sheeting units. Drying units set the ink so that it does not smear, and then the sheeting, or cutting, units cut the continuous ribbon of material into sheets. Some printing equipment also includes automatic folding machines, which fold and assemble the sheets into products such as newspapers or magazines.

Message Transfer

The action of placing the message on the substrate is called message transfer. The transfer process varies with each printing process. We will look at the message transfer process for four printing processes: relief, offset lithography, gravure, and screen printing.

Relief Message Transfer. Most relief printing is done on flexographic presses. This press transfers very thin ink from a reservoir to an application, or anilox, roll, Fig. 20-22. This roll distributes a thin coating of ink on the raised surface of the rotating plate. As the plate revolves, it contacts the web of paper that is passing over the impression roll. The pressure between the plate and impression roll transfers the image to the substrate.

A special flexography press, called the Cameron press, is sometimes used to print books. It prints with a flexography "belt," which contains all the pages of the book. The paper and the flexography image carrier are moved through two press units. The first unit prints on one side of the paper while the second unit prints on the reverse side. A dryer is at the end of each printing unit to set the ink.

Each cycle of the image carrier prints one complete book. The paper is then sent through assembly and binding units at the end of the press to complete the

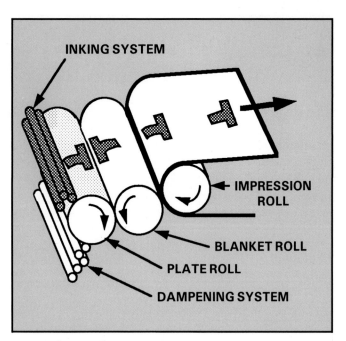

Fig. 20-23. Shown here is the operation of an offset lithography press. Notice how the image is first offset before it is printed.

book. The Cameron press is used to print lower quality publications such as paperback novels.

Offset Lithography Message Transfer. As you learned earlier in the chapter, offset lithography prints from a flat surface, using the principle that oil and water do not mix. The offset press has several systems in its printing unit, Fig. 20-23. A plate cylinder holds the image carrier. The dampening system applies a special water-based solution, called the fountain solution, wetting the plate. Next, the inking system applies ink to the plate's surface. Finally, the impression system accepts the substrate, moves it through the printing unit, and onto the delivery unit. The impression system works in two steps. First, it transfers the inked image from the plate to an offset (blanket) roll. The image is then transferred from the blanket to the substrate.

Two or four offset units may be arranged in a single press. This allows color printing to be done efficiently. A four-color press can print full-color materials in a single pass. A two-color press can print the base color (usually black) and a highlight color in one pass. They can also be used for four-color work, however, the press must print the first two colors (yellow and cyan) on one pass. Then the press must be cleaned and the ink colors changed. The second pass will print the other two colors (magenta and black) needed for a four-color run.

Gravure Message Transfer. The gravure image area is a series of very small ink reservoirs engraved in the surface of the carrier. This carrier is a metal drum which is partly suspended in a tray of ink, Fig. 20-24. As the roll

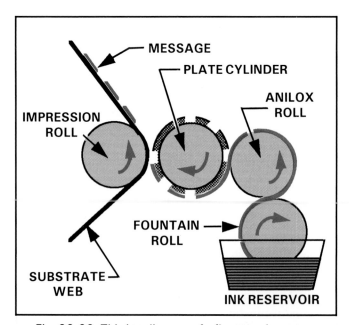

Fig. 20-22. This is a diagram of a flexography press.

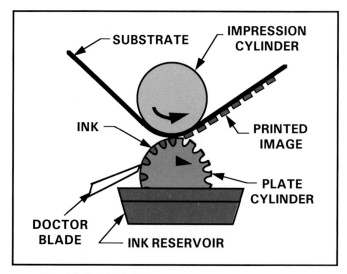

Fig. 20-24. This is how a gravure press operates.

Fig. 20-25. The irregular shape of this package is die cut.

revolves, it picks up a layer of ink. A doctor blade wipes off the excess, leaving ink only in the cavities in the roll. The roll then contacts the substrate under high pressure. This action transfers the ink onto the substrate.

Often several gravure units are assembled into a single press. Each unit is used to print a different color of a multiple-color run.

The image on a gravure roll can be a continuous, repeated pattern. Therefore, this process is often used to print wallpaper, hardboard wall paneling, vinyl flooring, and vinyl shelf covering.

Screen Message Transfer. Many people have done screen printing in art classes or as a hobby. The process, as described earlier in this chapter, is quite simple. Ink is pressed through a stencil onto the substrate. Production screen printing presses use one or more stations. The substrate is automatically fed into position under the screen. A screen is lowered onto the substrate and a squeegee automatically moves across the image. The screen is then lifted and the substrate is removed, or, if more than one color is being used, it is moved to another printing station. Each station lays down a different color until the final image is produced on the product.

Finishing the Product

Most printed graphic communication products are not finished when they leave the printing press. They require additional work. This work may include:

- Cutting: Trimming the substrate into rectangular sheets, or cutting assembled products to their final size.
- Die-cutting: Shearing irregular shapes or openings in the material, Fig. 20-25. This process is often done with cutting dies on a relief press.

- Folding: Creasing and folding the product to form pages. Often each printed sheet will contain a number of pages, Fig. 20-26. Properly folded sheets are called signatures.
- Drilling or punching: Creating holes in the substrate for insertion in binders.
- Assembling: Gathering and placing sheets or signatures into the final order, Fig. 20-27. This may be done manually or with automated machines.
- Binding: Securing sheets and signatures into one unit. This may be done by gluing, stapling, or sewing the units to form a book or magazine.
- Stamping: Transferring an image to a book or album cover with a hot stamping technique. This is often done using heated type and colored foil. The heat melts the foil and transfers the type image onto the product.
- Embossing: Pressing a raised pattern into the substrate. Paper or other material is pressed between mated dies to produce a pattern.

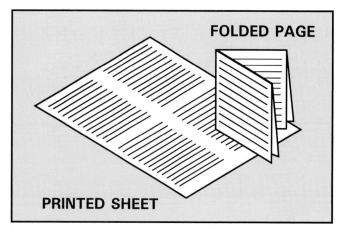

Fig. 20-26. A single sheet containing multiple pages is folded to make a signature.

Fig. 20-27. This worker is using an automatic machine to assemble signatures into a book. Note his ear protection to prevent hearing loss.

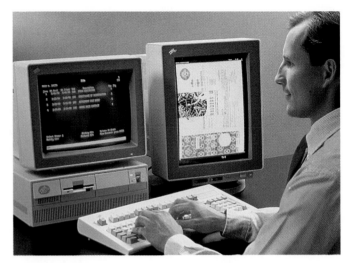

Fig. 20-28. Shown is an electronic publishing system. The system is used to combine text and illustrations into an attractive final product. (IBM)

COMPUTER-BASED PUBLISHING

Computers have given rise to several new graphic communication systems. These systems include desktop publishing, electronic publishing, and pagination systems.

The term **desktop publishing** is most often used to describe a fairly simple computer system that produces type and line illustration layouts for printed messages. The type and line illustrations are generated using two separate software packages.

This system includes a computer, a laser printer, and the desktop publishing software. Simply adding this software to the computer systems found in many businesses, makes desktop publishing available to most companies. They use desktop publishing to create newsletters, announcements, flyers, and other simple communication products.

More complex systems can function as typesetting and layout systems. These systems are called **electronic publishing**. They produce and combine text and illustrations into one layout. Electronic publishing produces a higher quality product than desktop publishing produces.

A common electronic publishing system, as shown in Fig. 20-28, will include:
• Computer keypad to input text copy and to provide commands to the system.
• Processing unit (microcomputer) to store and process data.
• Monitor to display the text and illustrations that are being manipulated.
• Page layout software to direct and control computer and printer actions.

• Mouse to easily send commands to the computer.
• Inkjet printer to run proofs and sample layouts.
• Laser printer to product high quality single page layouts.
• Scanner to digitize and input illustrations into the

The system may also use several additional software programs. Many systems include illustration software to create line illustrations. Also, computer clip art files may be used as a source of additional line illustrations.

The most complex and expensive computer systems are **pagination systems**. These systems allow the operator to very accurately merge text and illustrations, Fig. 20-29. The abilities of a pagination system may include: cropping or size adjustments to photographs, color enhancement, image grade enhancement, wrapping text around images, and a complete layout control. Often the output of the system can be directly converted to film or printing plates.

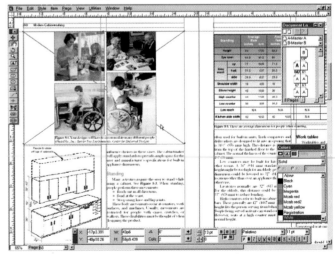

Fig. 20-29. This design was set up and produced using a pagination system.

SUMMARY

Each of us is constantly served and impacted by printed graphic communication products. These products may be books and magazines, flyers and pamphlets, or signs and posters. Each of these products followed a three step production process. The product is designed. The image carriers are developed. The product is printed and finished.

Each printed product is produced using one of six common printing methods. These methods are relief, offset lithography, gravure, screen, electrostatic, and ink jet printing.

With current printing methods we find learning easier, information more available, and entertainment broader.

WORDS TO KNOW

All of the following words have been used in this chapter. Do you know their meanings?

Audience assessment
Balance
Cold composition
Comprehensive layout
Contrast
Copy
Design
Desktop publishing
Electronic publishing
Electrostatic printing
Flexography
Gravure printing
Harmony
Hot-metal composition
Illustration preparation
Illustrations
Image carrier
Ink jet printing
Layout
Letterpress
Lithographic printing
Offset lithography
Pagination
Pasteup
Printed graphic communication
Printing
Proportion
Relief printing
Rhythm
Rough sketch
Screen printing
Silkscreening
Substrate
Thumbnail sketch
Typesetting
Variety

TEST YOUR KNOWLEDGE

Write your answers on a separate piece of paper. Please do not write in this book.

1. What are the two major types of illustrations?
2. True or false. Photographs cannot be directly used in printing processes.
3. Using a computer and a software package to develop a layout for a printed product is called _____.
4. True or false. Printing presses that use paper on a roll are called spool-fed presses.
5. True or false. Using metal type to generate the text is called cold composition.
6. Having equal amounts of copy and illustration on each side of a layout is called _____ balance.
7. List the three types of sketches and layouts used in designing a message.
8. On what principle is offset lithography based?

APPLYING YOUR KNOWLEDGE

1. Choose or invent a school event to promote. Complete a chart of the event similar to the one below. Then prepare rough and refined sketches for a poster to promote the school event.

What is the name of your event?
Who will be your audience?
What is the theme of your event?
What is your design approach (humorous, etc.)?
What type of promotional material (flyer, poster, etc.) will you use?

2. Design a logo or symbol for an organization or a cause. Make a block cut relief image carrier to print your image.

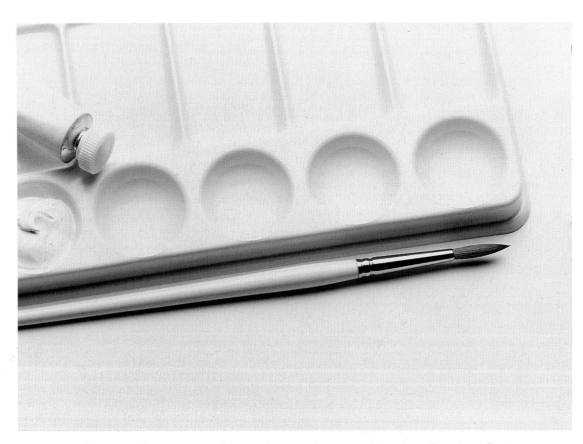

Photographs can convey images that words cannot. (American Greetings Co.)

PHOTOGRAPHIC COMMUNICATION

After studying this chapter you will be able to:
- Explain the value of pictures in communication.
- Describe photography.
- Differentiate between photography and photographic communication.
- Describe three characteristics of visible light.
- List the three steps in developing a photographic communication message.
- Describe how photographic communication messages are designed.
- Describe how photographic images are captured.
- Explain the principles of a camera.
- Describe how black-and-white films and color films work.
- Explain how photographic messages are reproduced.

"A picture is worth a thousand words," is an old saying, Fig. 21-1. This statement suggests that it is often more effective to convey your message visually than to describe it. Pictures are an efficient method of communication. In some cases, a picture alone is used to communicate an idea or feeling. In other cases, the picture is used to supplement the written word.

The word "picture" once was used to refer only to paintings. Now, just about any two-dimensional visual representation is called a picture. A **photograph** is a common type of picture. The act of producing a photograph is called **photography**. Using these photographs to convey an idea or information is called **photographic communication**.

To distinguish photography from photographic communication, you may consider the following. Your family may take snapshots at family events and during vacation travels. The main goal of the snapshots is to capture a moment in time. Later, you can look at these pictures and remember days long passed. This type of photograph is designed to provide a historical record, rather than to communicate information.

However, if you have flown on a commercial jet, you have seen a safety information card. It contains a series of photographs showing how to fasten the seat belt and exit the plane during an emergency. These photographs are designed to communicate a specific procedure. They are used as photographic communication.

LIGHT AND PHOTOGRAPHY

Photography is based on the understanding and appropriate use of light. Light is a wave of energy. We define light as electromagnetic radiation with wavelengths falling in a range including: infrared, visible,

Fig. 21-1. This photograph shows a replica of an early Egyptian hieroglyph.

ultraviolet, and X-ray. Light waves have three central characteristics, Fig. 21-2:

- **Amplitude** is the height of the wave. This indicates its strength or intensity. The height is measured from the center to one peak of the wave.
- **Wavelength** is the distance from the beginning to the end of one wave cycle. The wavelength of the light determines whether the wave falls into the visible range, and if it does, what color it is. Color is more properly referred to as hue. Each hue, or color, has a set wavelength. For example, red waves are 650 nanometers (nm), or 650 billionths of a meter, in length. Closely associated with wavelength is frequency. This is the number of waves that pass a point in one second.
- **Direction** refers to the path that the light wave travels. Light waves travel in straight lines in all directions from their source. The direction may change when the wave encounters another material or force, Fig. 21-3. If the wave passes through the material, the wave is said to be **transmitted**. Smooth, clear glass allows waves to pass directly through it without bending. This is called specular transmission. Etched glass allows the light to pass, but it randomly bends the light's path. This is called diffuse transmission, since the light is diffused, or spread out. Some materials **reflect** the light that strike them. If the angle of reflection is equal to the angle at which the wave struck the object, it is called specular reflection. Good mirrors exhibit specular reflection. Other substances reflect *and* diffuse the light. Rough surfaces exhibit this property, called diffuse reflection.

Finally, many materials **absorb** light waves. For example, black objects absorb all light that strike them. Colored objects absorb only certain wavelengths. The other wavelengths are reflected and received by your eyes. This property allows your eyes to see the object's "color."

TYPES OF ELECTROMAGNETIC WAVES

Light is but one of a number of different types of electromagnetic waves. Electromagnetic waves include gamma rays, X rays, ultraviolet light, visible light, infrared rays, and radio waves. Each of these has a specific frequency range and can be divided into even more specific groups, Fig. 21-4.

When all the wavelengths of visible light are present, we have white light. Sunlight is an example of white light. However, this light can be divided into a group of wavelengths, each a different color. Have you ever seen a rainbow? If so, you have seen white light divided into its six basic color components: violet, blue, green, yel-

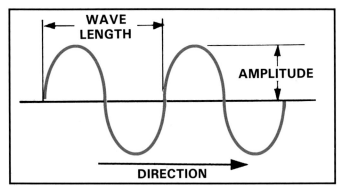

Fig. 21-2. The wavelength and the amplitude are two descriptive characteristics of a light wave.

low, orange, and red. This separation is caused by the raindrops in the air. As light passes through them, they reflect and separate the various wavelengths. The shorter violet rays appear on the bottom of the rainbow, while the longer red waves are on top.

FUNDAMENTALS OF PHOTOGRAPHIC COMMUNICATION

Photographic communication media are produced using a common set of steps. These steps include:
- Designing the message.
- Recording the image.
- Reproducing the message.

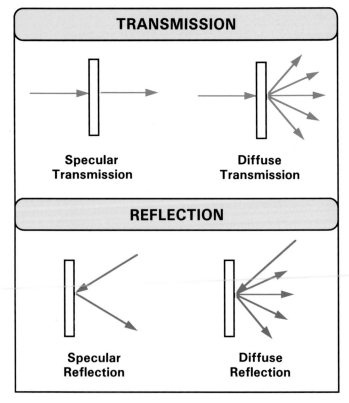

Fig. 21-3. Light waves can be transmitted, reflected, or absorbed when they strike an object.

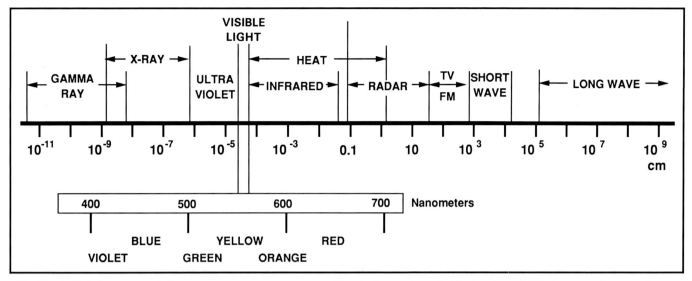

Fig. 21-4. This is a chart of electromagnetic waves grouped by their wavelength. Notice that visible light is only a small set of electromagnetic radiation.

DESIGNING A PHOTOGRAPHIC MESSAGE

Photographic messages must be designed with the same considerations used in other communication media. This approach includes answering such questions as:

- Who is the intended audience?
- What is the message to be communicated?
- What type of photograph will best communicate the information or idea?

A photograph shows a record of light. It captures the light reflected from a series of objects the instant before the film is exposed. Some people see this as a window to reality, while others see it as a flat representation of a scene. The goal in designing a photographic communication message is to make the message a realistic window, rather than the flat representation.

To be a photographic communication, the photograph must describe something. This can be a process (way of doing something), an idea (a mental image), or an event.

Several techniques are used in helping photographs communicate. First, photographs are **composed**, or designed, with a **point of interest**. The point of interest is the place where your eye is drawn. This is where the central message is communicated. Look at the two photographs in Fig. 21-5. Where is your eye drawn in each of them? In one photograph, the sky is the central focus. The other uses the sky simply as a background to the central focus, the horseback riders. What message does each point of interest communicate to you?

Another technique uses **distance** and **scope** (panorama) to add interest. Look at the photographs in Fig.

Fig. 21-5. What is the point of interest for each of these photographs.

21-6. They both show the same mountain from approximately the same vantage spot. The mountain is the point of interest for both pictures. However, in one the mountain is the dominant feature, while the other picture shows it as part of a larger world. What diverse ideas do these pictures inspire?

Fig. 21-6. A close-up and distant view of the same mountain. How are the photographs similar? How are they different?

A third technique uses color to help communicate. Look at the photographs in Fig. 21-7. Both are pictures of a "concept car." Which one communicates the idea more clearly to you? Why?

A photographic message must meet all the design criteria discussed in the last chapter. Its parts must be in balance and in harmony with one another. In addition, its size and proportion should be pleasing.

RECORDING THE MESSAGE

Once the shot is composed, most photographic messages are produced using a two step process. The image or message is captured on film. Then, the film is used to reproduce the message, so that the audience can receive it, Fig. 21-8.

Recording the message requires two separate procedures, selecting the medium and exposing the film.

Selecting The Medium

The film used in producing photographs is grouped in several ways. These include type, format, size, and speed.

Types of Film. The two major types of film are black-and-white film and color film. **Panchromatic** is

Fig. 21-7. Notice how color adds interest to the photograph of the ''concept'' car. (Pontiac)

Fig. 21-8. A careful technical process is used to turn exposed film into photographic communication. (American Greetings Co.)

the most common form of black-and-white film. Panchromatic film reacts to all colors of visible light. It records those colors as shades of gray.

The typical panchromatic film has four layers, Fig. 21-9. There is a plastic film base that gives the film its structure. An emulsion of light-sensitive silver halide crystals coats one side of this base. This emulsion is then protected by a clear gelatin layer. Finally, the back of the film is coated with a antihalation coating. This layer absorbs light so that it does not reflect back through the film. Without this coating, photographs would be blurred by reflected light.

Black-and-white film is available only in negative form. **Negative** film produces a reverse image of the scene that is photographed. This means that the negative will be dark where the subject is light, and the negative will be light where the subject is dark. Later, it is used to make photographic prints, Fig. 21-10.

Color films have similar base, protective, and antihalation layers. However, the emulsion is made up of three layers of light-sensitive material. Each layer is sensitive to one of the three additive primary colors: red, green, and blue.

Color film is available in a positive (transparency) or a negative form. **Positive** transparencies, or slides (as small transparencies are called), produce the actual view, such that it can be viewed directly through a slide or movie projector. Color negatives are used to make color photographic prints.

Film Formats and Sizes. Each type of film comes in several formats and sizes. They are available in rolls and sheets. Roll film is the most commonly used format. It is loaded in a lighttight canister or cassette. The film is drawn from its container as it is exposed in a camera. Later, it is drawn back into the container and removed from the camera for developing. Both formats come in a variety of sizes. You can buy 4″ x 5″ sheet film or 35 mm (width) roll film, to name just two options.

Film Speed. Both panchromatic and color films are available in a number of speeds. This speed is given by a ASA/ISO number. The speed of the film indicates its sensitivity to light. Fast film is very light-sensitive, while a slow speed film takes a considerable amount of light to generate an image. Fast film is used to photograph objects and events in poorly lighted situations. It is also used in photographing fast moving objects. Slow film is used in areas that are well-lighted as well as for photographing outdoors.

Fig. 21-10. The negative (top) and positive (bottom) are of the same photograph. (Jack Klasey)

Fig. 21-9. These are the layers of black-and-white and color film. Color film has three separate emulsion layers. Black-and-white film has only one.

Technology: How it Works

INSTANT PHOTOGRAPHY: Photographic system where the picture is taken and the film is developed in one step, using special cameras and film.

The first instant photographic camera was developed by Edwin Land. It was commercially introduced in 1948. This process used film that had layers of light-sensitive emulsions and developers. The film used a wet process system. The exposed film was drawn between two rollers. The pressure of the rollers caused pods of chemicals in the film to burst. This started the developing of the photograph.

Modern instant films use a dry process, Fig. 1. This film consists of two plastic layers. On these layers are a series of coatings. These coatings are the image capturing, coloring, and developing chemicals. The negative portion of the film is made up of six layers. Three of these layers are silver halide compounds, sensitive to one specific color: blue, green, or red. The other three layers contain dyes used in developing the images. These produce cyan (blue), magenta (red), and yellow images. These are the same colors that are used to produce the color pictures found in this book.

Each of the silver halide layers acts as a light filter for its color. It captures that color and allows the other colors to pass through to the next layer. Thus, during exposure, the film produces three separate color images; one each for red, blue, and green. When the film is ejected from the camera, rollers burst the developer. The dry chemical develops the images in the three light-sensitive layers.

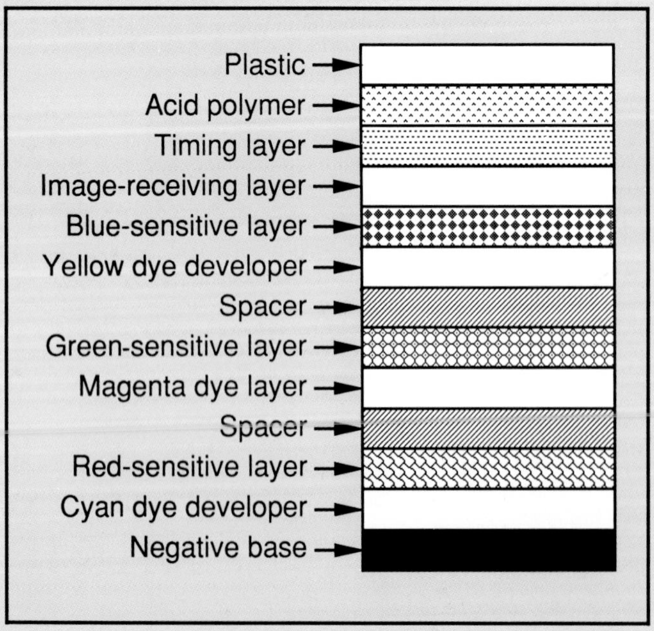

The roller pressure also releases the dyes from the adjacent layers. The dyes move freely up through the layers in the film. The dyes are captured by exposed portions of their respective layers. Exposed portions of the red layer capture the cyan dye. Unexposed portions allow it to move through. Partially exposed images capture some dye and allow some of the dye to pass through. The degree to which the layer is exposed will determine the amount of dye that the layer will capture and neutralize.

The uncaptured dye migrates to the image-receiving layer. There the final image is produced in a simple "print." If all three dyes are captured below any point of the print, it will appear as a white space. If no dyes are captured, yellow, cyan, and magenta dyes will all reach the image layer. This produces a black space. Mixing various amounts of the three dyes produces the range of colors between white and black.

Instant photography uses a special camera, Fig. 2. The camera exposes the film and applies enough pressure to burst the developer packet as the film is ejected from the camera. The picture develops shortly after it has been ejected.

Plastic
Acid polymer
Timing layer
Image-receiving layer
Blue-sensitive layer
Yellow dye developer
Spacer
Green-sensitive layer
Magenta dye layer
Spacer
Red-sensitive layer
Cyan dye developer
Negative base

Fig. 1. Shown are the multiple layers in instant photography film.

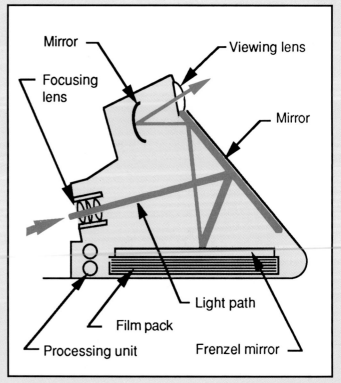

Fig. 2. This is a view inside a simple one-step camera.

Mirror
Viewing lens
Focusing lens
Mirror
Light path
Film pack
Processing unit
Frenzel mirror

The classification of fast film and slow film is somewhat subjective. However, films with ASA ratings of 25 to 100 are considered slow films. Fast films have ASA ratings of 200 and above.

Exposing Film

Film is almost always exposed in a camera. This technological device is, basically, a lighttight box used to expose a light-sensitive material. A camera has three main systems:

- Film exposing system.
- Viewing system.
- Film feeding system.

Film Exposing System. The basic function of a camera is to allow a controlled amount of light to strike the film. The key word is control. Control is necessary for good photographs. This control is done with three parts of the camera – the shutter, the diaphragm, and the lens. They are shown in Fig. 21-11.

The **shutter** is a device that opens and closes to permit or prevent light from entering the camera. The *length of time* the shutter is open controls the length of time light can strike the film. Many cameras have adjustable shutters. Speeds can range from 1 second or more, down to 1/1000 of a second.

The **diaphragm**, or aperture control, regulates the *amount* of light that can enter the camera at any given moment. Generally, it is made of a series of very thin metal leaves or plates. The leaves can be adjusted to

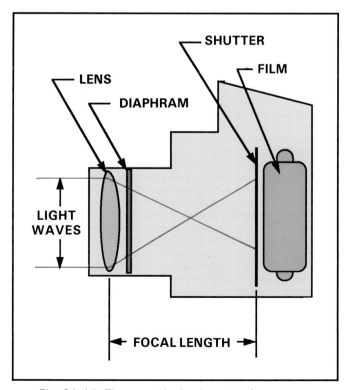

Fig. 21-11. These are the basic parts of a camera.

change the size of the hole (aperture) behind the camera lens. The size of the opening is identified by an **f-stop** number. The f-stop numbers are the *reverse* of the apertures – as the f-stop number increases, the size of the aperture decreases. As the f-stop number decreases, the size of the aperture increases. The amount of change is specific. It doubles or halves the amount of light from stop to stop. Thus, going from f/11 to f/16 will decrease the light by one-half. Moving from f/11 to f/8 will double the amount of light.

The f-stop setting also determines the **depth of field**. This is the range of distances in which the camera will capture objects in focus. Higher f-stops have a greater depth of field. For example, a high f-stop, f/16 or f/22, will allow objects at almost any distance from the camera to be in focus. A low f-stop, f/2.4, will focus on only a narrow range of distances. Objects outside this range (the depth of field) will be out of focus and appear fuzzy in the finished photograph. This is not necessarily bad. In some communications, the designer wants the point of interest to be in focus and the background fuzzy.

The purpose of the **lens** is to *focus* the light on the film. The lens also inverts the image. The focal length of the lens determines the size of the image. A typical 35 mm camera has a lens with a 50 mm focal length. This means that the front of the lens is 50 mm away from the film. Different lenses are used for varying effects. A 28 mm wide-angle lens produces a wide view for close objects. A 300 mm telephoto lens has a narrow angle of view and enlarges distant objects.

Quite often, the camera lens can be adjusted in or out, changing its focal length. This allows the object to be brought into focus on the film. Zoom lenses have a wide range of focal lengths available. For example, a 28-90 mm zoom lens can shoot everything from a 28 mm wide-angle shot to a 90 mm telephoto shot.

Many cameras, also, include a **light meter** to measure the amount of light available for the photo. This allows the operator to make the proper adjustments to the aperture opening and shutter speed. When the amount of light is low, the shutter must be slowed down and/or the aperture increased. Bright scenes require the opposite adjustments. The speed of the shutter can be increased, or the aperture can be closed down. Often, the operator does both.

A number of cameras on the market, today, are **automatic**. They use electronics to sense the light and then automatically set the f-stop, aperture opening, and/or the shutter speed. Also, many less expensive cameras have nonadjustable systems. They operate with a set f-stop, aperture opening, and shutter speed.

In all cameras, the actual photograph is taken when the shutter is momentarily opened. The film is exposed by light focused by the lens. The intensity of the light is controlled by the aperture.

Viewing System. A camera captures only the light reflected from the scene in front of the lens. Therefore, it is essential that the operator point the camera precisely at the desired scene. Also, it is important that the object is in focus. Both of these conditions can be achieved using the viewing system.

Cameras use one of two systems to select and adjust the view, the viewfinder system and the single lens reflex system, Fig. 21-12. In the viewfinder system, the operator looks through a viewing port on the back or top of the camera. The scene is viewed through a lens on the camera's front. The camera lens and the viewing lens are separate. The view the operator sees is approximately what the camera views. But, since the viewfinder lens and camera lens are separate, the view is not exactly what will be captured on film.

More accurate viewing is done using a single lens reflex system. This system uses a series of prisms and mirrors that allow the operator to look directly through the camera lens. The view seen through the single lens reflex is exactly the view captured on film.

Film Feeding System. Most cameras capture a series of photographs on a strip of film. This film is held in position behind the shutter. After each shot, the film is advanced so that the next area of unexposed film moves into position behind the shutter.

The exposed film must be protected from being exposed again. A second exposure to light will cause a double exposure. This is when two separate images are captured on the same piece of film. In most cases, this is unwanted. Occasionally photographers use double exposures to communicate certain ideas or feelings.

After all the film has been exposed, the film feeding system rewinds the film into a lighttight case, so that it can be removed from the camera. The film must be protected from light when it is removed otherwise it will be ruined.

Typically, film is available in rolls or cassettes, Fig. 21-13. Many of the more expensive camera use a 35 mm roll film. This film comes in lengths that allow 12, 20, 24 or 36 exposures (photographs). "Snapshot" cameras, generally, use film cassettes in the smaller 110 size. These hold 12 or 20 shots.

REPRODUCING THE MESSAGE

Communication technology is, often, used to produce messages for a large, diverse audience. Therefore, the message must be produced in quantity. Reproducing photographic communication media requires developing film and making prints.

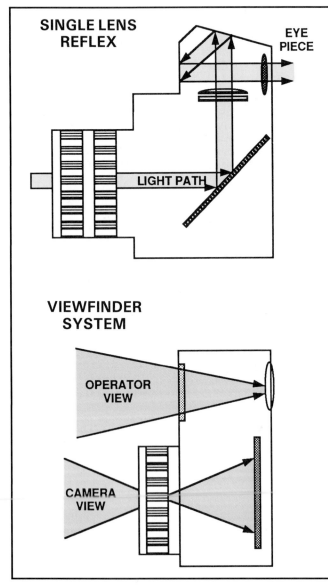

Fig. 21-12. Here are two viewing systems used in cameras. The single lens reflex system gives a more precise view of the image to be captured.

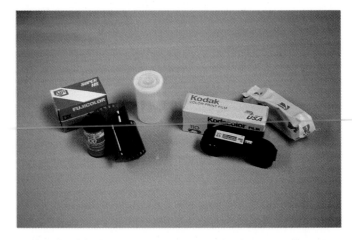

Fig. 21-13. Two common formats for film are rolls and cassettes.

Developing Film

Photographic communication is not possible until the captured image is retrieved. This retrieval process is called developing the film. Developing chemically treats the light-sensitive materials in the film emulsion to bring out the image. Black-and-white film is developed with a fairly simple process. Let's explore this activity, Fig. 21-14.

Developing film involves a series of chemical reactions. These reactions include:
- Developing.
- Stopping.
- Fixing.

The first step, **developing**, uses chemicals (developers) to alter the light-sensitive crystals in the emulsion. They are changed from silver halide, which cannot be seen, to black metallic silver that can be seen. The next step, **stopping**, stops the chemical action of the developer. It uses an acid solution, called **stop bath**, to neutralize the developer. **Fixing**, then, removes the unexposed silver halide crystals that remain in the film. This makes the image permanent. After these three steps are complete, the film is washed to remove any chemical residue. It is then dried, so it can be stored or used to make prints.

Safety With Darkroom Chemicals
Use care when mixing darkroom chemicals. They can cause skin and eye irritation. Also, use tongs and an apron when transferring prints from one solution to another. Wash your hands if they do come in contact with the chemicals.

Making Prints

Most types of black-and-white film produce negative (reverse) images. This negative is used to make photographic prints. Two basic techniques, contact printing and projection printing, are used to produce prints.

Contact printing is the simpler procedure. The negative is placed directly on top of a piece of light-sensitive photographic paper. The film and paper are held tightly under a sheet of glass. This assembly is exposed to a bright light. Then, the paper is removed and developed, employing the same steps used for developing film: developing, stopping, fixing, and washing. However, a different chemical is used for developing prints than is used for film.

It is hard to see how each negative will print out, so contact prints are widely used as proof prints. A contact print will show each picture as a small print, or photograph. This allows the photographer to select only the best negatives to make into larger prints.

Projection printing is done by projecting (shining) light through the negative onto a piece of photographic paper, Fig. 21-15. Projection printing requires an enlarger to adjust and control the print size and focus. The farther the enlarger lens is away from the paper, the larger the print will be. Producing a projection print uses the following steps:
- Cleaning and inserting the negative in the enlarger.
- Adjusting the enlarger for print size and focus.
- Exposing the photographic paper.
- Developing the print.
- Drying the print.

OTHER TYPES OF PHOTOGRAPHIC COMMUNICATION

Prints are one common use for photography. They can be used directly to communicate, or they may be integrated into a printed graphic message. This was discussed in Chapter 20.

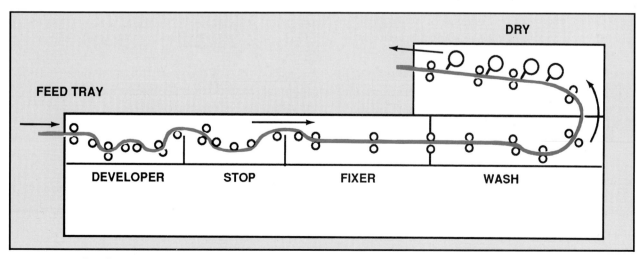

Fig. 21-14. The dark line shows the path of film in an automatic film developing machine.

FEED TRAY

DRY

DEVELOPER   STOP   FIXER   WASH

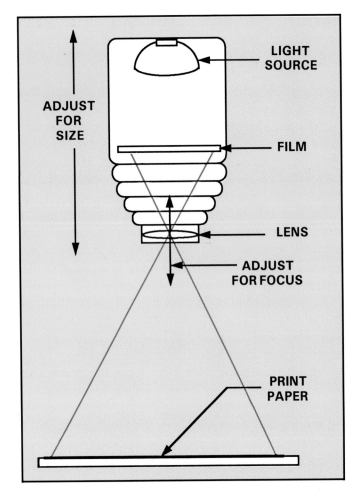

Fig. 21-15. This shows the basic operation of projection printing, using the enlarger.

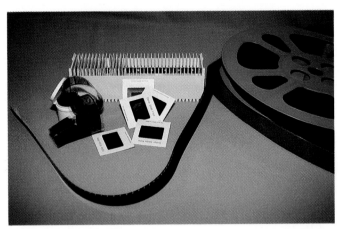

Fig. 21-16. Color transparencies, such as slides, filmstrips, and motion pictures, are used frequently in your class-rooms for graphic communication.

Additionally, color transparencies are used to reproduce color illustrations in books, magazines, and newspapers. Often, these transparencies are 35 mm slides. However, larger transparencies are produced on sheet film by cameras called view cameras. These transparencies, usually 4″ x 5″ or 8″ x 10″, are used in advertising and publishing. Most of the illustrations in this book were reproduced from transparencies.

SUMMARY

You see the world as a series of pictures. You look at something and capture the view. This gives meaning to the space around you. You often want to save these views for later use. One way to do this is with photography. You can use photography, film, and paper to record and store images.

In some cases you want to use photographs to communicate information, feelings, or ideas. Then, these photographs take on a new role. They become a communication medium. They are another technology that allows us to extend and share knowledge.

WORDS TO KNOW

All of the following words have been used in this chapter. Do you know their meanings?

Absorb
Amplitude
Automatic
Compose
Contact printing
Depth of field
Developing
Diaphragm
Direction
Distance

Photography is also used to produce transparencies. As you learned earlier, these are positives of the recorded scene. They show the image as the eye sees it. Three common uses of transparencies in communication are shown in Fig. 24-16.

- **Slides** are single transparencies designed to be viewed independently. They are cut from the film and mounted in frames. They can be arranged into sets called slide series. Later, they can be rearranged or supplemented with new slides to communicate a different message.
- **Filmstrips** are a series of transparencies designed to be viewed one at a time. Their sequence is rigid, and they communicate a specific message. The film is left in strip form as produced by a camera or filmstrip duplicator.
- **Motion pictures** are a series of transparencies shot over a span of time. They are, generally, shot at a rate of 24 frames (single transparencies) per second. This rate is fast enough that the human eye cannot distinguish each individual picture. When the transparencies are projected at the rate they were recorded, the series of still shots will give the illusion of movement.

Filmstrips
Fixing
f-stop
Lens
Light meter
Motion pictures
Negative
Panchromatic
Photograph
Photographic communication
Photography
Point of interest
Positive
Projection printing
Reflect
Scope
Shutter
Slides
Stop bath
Stopping
Transmit
Wavelength

TEST YOUR KNOWLEDGE

Write your answers on a separate piece of paper. Please do not write in this book.

1. Using photographs to convey information or feelings is called _____.
2. True or false. The height of a light wave is called the apogee.
3. List four types of electromagnetic waves.
4. The spot to which your eye is drawn in a photograph is called the _____.
5. True or false. Photographs used in communication are designed using the same criteria used in the printed message.
6. The criteria referred to in Question 5 are called _____.
7. Match the term on the right to the correct statement(s) on the left.

   _____ Controls the length of time light that can enter a camera.   A. Diaphragm.

   _____ Measures the amount of light that is available to expose film.   B. Shutter.
   C. Light meter.
   D. Lens.

_____ Controls the amount of light that can enter the camera.
_____ Focuses the light on the film.
_____ Its setting is measured by f-stop.
_____ Its speed can be adjusted from 1 second down to 1/1000 of a second in many cameras.

8. List the three main systems in a camera.

APPLYING YOUR KNOWLEDGE

1. Design and shoot a six photograph series. Have the series describe how to do something, or have the series communicate an idea about pollution, energy conservation, or another important issue. Develop a layout, similar to the one below, to describe the theme and each shot you would use. Sketch the location of the major elements as you would compose each photograph.

Theme or topic:		
Slide 1 (Description)	Slide 2 (Description)	Slide 3 (Description)
Sketch of the shot	Sketch of the shot	Sketch of the shot
Slide 4 (Description)	Slide 5 (Description)	Slide 6 (Description)
Sketch of the shot	Sketch of the shot	Sketch of the shot

2. Shoot and develop photographs that show (a) happiness, (b) sadness, (c) quietness, and (d) harmony with nature.

Telecommunications uses electronic devices to send messages over long distances. Integrated circuits are important tools in telecommunications. (Harris Corp.)

22
CHAPTER

TELECOMMUNICATION

After studying this chapter you will be able to:
- *Define telecommunication.*
- *List and describe the major types of telecommunication systems.*
- *Explain the differences between hard-wired and broadcast telecommunication systems.*
- *Diagram and explain a typical hard-wired telecommunication system.*
- *Diagram and describe the components of a broadcast telecommunication system.*
- *Explain the difference between frequency modulation and amplitude modulation.*
- *List and describe the types of scripts developed for broadcast communication.*
- *List and describe the work of creative, production, and performance workers in the broadcast industry.*
- *Describe how a television program is recorded and broadcast.*
- *Describe three mobile communication systems.*

Fig. 22-1. Telecommunication means communicating over distance. This is a satellite dish receiver, which aids in capturing messages sent from great distances. (Winnebago)

PHYSICS OF TELECOMMUNICATION

Important in the discussion of telecommunication are the principles of electricity and electromagnetism. These principles help you understand the technology you use daily.

ELECTRICAL PRINCIPLES

You have probably learned about the atom in your science classes. All matter is made up of atoms. Each atom has a nucleus, or center. This center is made up of positively charged particles, called protons, and neutral particles, called neutrons. A group of negatively charged particles orbit the nucleus. These particles, or bits of energy, are called electrons. The electrons are

Humans have always communicated their ideas and feelings. Early communication was person-to-person using a spoken language. Later, writing was developed to record and transmit information. Writing allowed us to express *and store* information, opinions, and concepts. However, this type of communication did not meet all human needs. People wanted to hear the human voice beyond the small limits of face-to-face communication. They wanted to communicate their thoughts and knowledge over great distances. Out of this desire came **telecommunication,** Fig. 22-1, meaning communication over distance. Telecommunication implies that there is a message and hardware (technology) to deliver it.

held in orbit by their attraction to the positively charged protons, Fig. 22-2.

In certain situations, electrons will travel from one atom to another. This movement is called electricity. It takes place, most often, in a metal called a **conductor.** When electrons are flowing in one direction along the conductor, it is called **direct current.** If the electrons flow in both directions along the conductor, reversing at regular intervals, it is called **alternating current.**

Movement of electrons in a conductor creates magnetic lines of force, Fig. 22-3. This force is known as an electromagnetic force. As these lines of force increase and decrease in strength, they can cause electrons to flow in an adjacent wire. This process is called **induction.** The principle of induction is commonly used to change sounds into electrical signals or to change electrical signals into sound. This process is used in microphones and speakers. These technological devices are examples of **transducers.** Transducers change energy of one form into energy of another form.

ELECTROMAGNETIC WAVES

Earlier in this book, you were introduced to basic electromagnetic wave theory. You will remember that two important characteristics of waves are frequency and amplitude, Fig. 22-4. **Frequency** is the number of cycles (complete wavelengths) that pass some point in one second. The number of cycles per second is measured in **hertz.** The basic units of measurement in telecommunication are **kilohertz,** or kHz (1000 cycles per second), and **megahertz,** or MHz (one million cycles per second).

Amplitude measures the strength of the wave. The higher the amplitude, the stronger the signal. Telecommunication uses changes in the amplitude and/or frequency of the waves to carry a communication message.

Within the electromagnetic spectrum is a series of frequencies we call **radio waves.** These frequencies extend from around 30 hertz to 300 gigahertz (300 billion

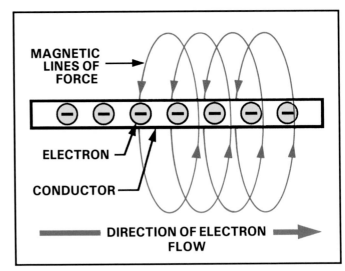

Fig. 22-3. Electron movement through a conductor is called electricity. This movement creates magnetic lines of force around the conductor.

cycles per second). At the low end of this range is sound you can hear. This range extends from 30 Hz to about 20 kHz. Below this range of frequencies is a series we call extremely low frequency (ELF). The series extends from about 10 Hz to 13.6 Hz. It is used for underwater communication. Naval commanders communicate with submarines using these frequencies.

Above audible sound is a series of frequencies known as **broadcast frequencies.** These frequencies are used for a wide range of communication systems. These systems include police and fire department radio, broadcast radio, cellular telephone, and television communication.

The Federal Communication Commission (FCC) assigns each type of communication to a range of fre-

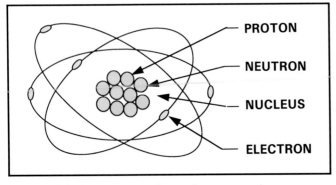

Fig. 22-2. Atoms are made up of protons and neutrons, which form the nucleus, and electrons, which orbit the nucleus.

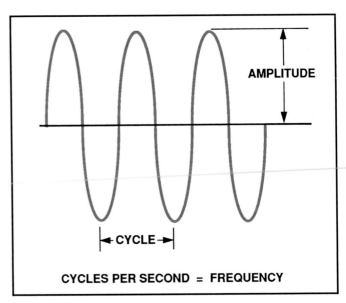

Fig. 22-4. The information in radio waves is coded through varying the frequency and amplitude of the wave.

quencies. For example, 160.215 to 161.565 kHz is assigned for railroad communication, while 50.0 to 54.0 MHz is assigned for 6-meter amateur radio. Frequency assignments are exclusive for each particular use nationwide.

TYPES OF TELECOMMUNICATION SYSTEMS

All of these frequencies are used to communicate data, information, or ideas. They are used in two major types of telecommunication systems. The types are hard-wired systems and broadcast systems, shown in Fig. 22-5.

HARD-WIRED SYSTEMS

You will remember that communication systems have three major parts: a sender, a communication channel, and a receiver. A typical **hard-wired system** is the telephone system, described in Chapter 7, Fig. 22-6. In this system, sound waves are changed into electrical impulses by the microphone in the mouthpiece. The frequency and duration of the electrical impulses are the coded message.

These electrical codes are, usually, conducted over a permanent waveguide that connect the sender and the receiver. This guide may be a copper wire or fiber-optic (glass fiber) conductor. In some cases, microwave radio signals take the place of a waveguide for a portion of the circuit. Microwaves are often used to send a message between major cities. Then, the signals are transferred back onto wire or cable.

Often, a special system, called **multiplexing,** is used to increase the capacity of the waveguide. This system allows several unrelated messages to travel down a single conductor at the same time. This can be done through time division multiplexing (TDM) or frequency division multiplexing (FDM).

Time division multiplexing divides time into very brief segments. Several messages are also divided into small discrete bits. Then, electronic equipment assigns each bit separate time slots. Next, the message bits are transmitted. The receiver collects all of the bits. It sorts the bits and assembles each message back into its original form.

Frequency division multiplexing uses a separate frequency to transmit each message. Several messages are transmitted at the same time over the same waveguide. They are blended together and channeled down the carrier. At the receiving end, the various frequencies are separated, and each message is delivered to a separate receiver.

Telephone messages are decoded by the speaker in the earpiece. There the electrical impulses are changed back into an audible sound.

BROADCAST SYSTEMS

Broadcast systems send radio waves through the air carrying the signal from the sender to the receiver. The transmitter (sender) changes sound into a signal that

Fig. 22-5. Telecommunication systems can be grouped as hard-wired or broadcast systems. Shown on the left is a telephone (hard-wired)unit. On the right is a remote unit for a television (broadcast) station. (Harper-Fritsh Studios, Gannet)

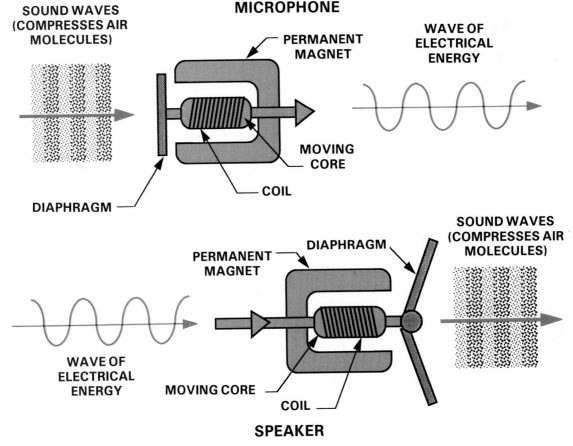

MICROPHONE

SOUND WAVES
(COMPRESSES AIR
MOLECULES)

PERMANENT
MAGNET

WAVE OF
ELECTRICAL
ENERGY

MOVING
CORE

COIL

DIAPHRAGM

PERMANENT
MAGNET

DIAPHRAGM

SOUND WAVES
(COMPRESSES AIR
MOLECULES)

WAVE OF
ELECTRICAL
ENERGY

MOVING CORE

COIL

SPEAKER

Fig. 22-6. Telephone systems use coded electrical messages to communicate the spoken word. Shown here are sound waves being transferred into electrical waves by a microphone. The electrical wave is then channeled to a receiver, where a speaker decodes the signal.

contains the message. This signal radiates into the atmosphere from an antenna. Another antenna attached to a receiver gathers the signal. The receiver separates the desired signal from other signals and changes it back into audible sound, Fig. 22-7.

Radio signals, generally, project in all directions from an antenna. Telephone microwave communication systems use directional antennas to focus the signal to receiving antennas.

Radio Broadcasting

Radio is one common broadcast system. Radio communication was the first widespread broadcast medium. Originally, it was called the *wireless* because there was no hard-wired connection between the sender (transmitter) and the receiver (radio set).

All radio broadcast systems use a **carrier frequency** that radiates from the transmitter. It is to this carrier frequency that you tune your radio to receive a station. Onto this frequency the code for the audible sound is imposed, Fig. 22-8. The earliest radios used **amplitude modulation** (AM) to code the carrier frequency. These systems merged the message onto the carrier wave by changing the strength (amplitude) of the carrier signal.

This type of broadcast radio is assigned the frequencies between 540 and 1600 kHz.

Later, radio broadcast systems that use **frequency modulation** (FM) were developed. These systems encode the message on the carrier wave by changing its frequency. The 200 separate FM radio broadcast frequencies range from 88.1 to 107.9 MHz.

Look back to Chapter 7 for an explanation on the operation of radio communication systems. The operation of radio transmitters and receivers are explained.

Television Systems

Television is another common broadcast system. Television broadcast systems are really two systems in one. These systems use a fairly broad range of frequencies within their broadcast band. Each channel is assigned a band 6 MHz in width, Fig. 22-9. One portion of the band is used by a radio-like system to send and receive the **audio** (sound) portion of the message. This portion uses frequency modulation to impose the audio message on the carrier wave.

The larger portion of the band is assigned to a second system that communicates the **video** (visual) part of the

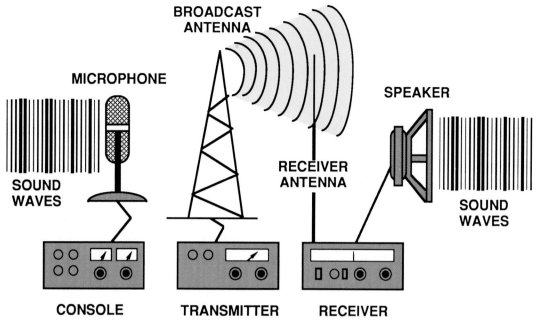

Fig. 22-7. A typical radio broadcast system.

program. This system uses amplitude modulation to send the picture portion of the message. At the ends of each channel are unused frequencies. This buffer zone keeps the signals from one channel from disturbing adjoining channels.

Television systems use a microphone to capture the sound while a camera generates the picture. The television receiver reproduces the program using a speaker for the sound and, usually, a cathode-ray (TV) tube to display the picture. The operation of these devices is discussed later in this chapter.

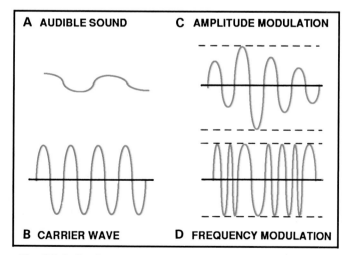

Fig. 22-8. Radio waves are changed to carry the message through amplitude or frequency modulation. Notice how the amplitudes of the sound wave (A) and the carrier wave (B) have been blended in amplitude modulation. The amplitude of the combined waveform (C) oscillates in a pattern similar to the sound wave. Notice in the frequency modulated waveform (D), how the frequency varies in a pattern with the initial sound wave.

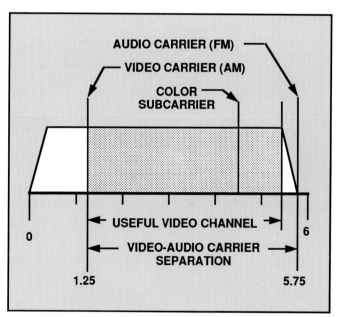

Fig. 22-9. This is a diagram of a television broadcast channel showing the audio and visual broadcast frequencies.

Two basic types of television stations exist. **VHF** (very high frequency) stations broadcast on channels 2-13. Each channel is assigned a specific 6 MHz range of frequencies. Channels 2-6 broadcast on frequencies between 54.0 and 88.0 MHz (megahertz). Channels 7 - 13 use the range between 174 and 216 MHz. The frequencies between these two broadcast ranges are used for FM radio, aircraft navigation and aircraft communication operations, weather satellites, and amateur (ham) radio.

The **UHF** (ultra high frequency) stations broadcast on channels 14-83. They use the frequency range from

470 to 890 MHz. Since VHF channels can broadcast a greater distance than UHF channels, they are, generally, assigned to the major television network outlets and large local stations. Public broadcasting and smaller independent stations are often found on the UHF channels.

The operation of television broadcast and receiver systems is discussed later in this chapter.

COMMUNICATING WITH TELECOMMUNICATION SYSTEMS

Earlier you learned that telecommunication includes a message and a delivery system. Communicating with telecommunication systems requires three distinct actions, Fig. 22-10:
- Designing the message.
- Producing the message.
- Delivering (broadcasting) the message.

DESIGNING COMMUNICATION MESSAGES

Many communication messages are not carefully designed. When you talk on the telephone you often "play it by ear." You have a basic message, or reason for the call. Yet, there is, generally, interaction between the caller and the receiver. The caller adjusts the message as he or she receives feedback from the receiver. The format and sequence of the message are not fixed.

This is not typical for broadcast telecommunications. Most broadcast messages are the result of focused design activities. These activities include:
- Identifying the audience.
- Selecting an approach.
- Developing the message.

To understand these steps, let's explore the development of a television program. A radio program follows a similar procedure, except without the use of visual impressions.

COMMUNICATING WITH TELECOMMUNICATION SYSTEMS

DESIGNING THE MESSAGE

PRODUCING THE MESSAGE

BROADCASTING THE MESSAGE

Fig. 22-10. Telecommunication messages move through three stages. They are first designed, then produced, and finally broadcast. (Ohio Art Co., Harris Corp., Ball State University)

Technology: How it Works

FIBER OPTICS: Channeling messages, in the form of light, through glass fibers.

Information has been transmitted with light for many years. One of the first methods used was smoke signals. More recently, flags and flashing lights have been used to convey information. All of these techniques, though, are limited to the line of sight between the sender and the receiver. Practical use of long-distance light communication came with the invention of optical cables.

Optical cables are channels that guide light waves, through internal reflection, over some distance. Internal reflection means that when the light waves strike the outer edge of the fiber they are reflected back toward the center. Optical communication of this type is called *guided optical transmission.* The development of glass fibers for guided optical transmission began in the 1960s.

A typical fiber-optic cable, as shown in Fig. 1, has three layers. The outside layer is a protective plastic coating. The middle layer is called *cladding.* It reflects the light waves back into the glass fiber. The inner layer is a strand of glass called the *core.* The individual core strands are as thin as a human hair. Typically they are about 0.0005 inch in diameter.

Several hundred of these strands are bundled into larger cables. Each of these fiber-optic waveguides can carry numerous messages. Billions of bits of information per second can move down an optical fiber. This is the same amount of information as is contained in thousands of independent telephone conversations. As you can see, optical fibers are capable of carrying encyclopedias of information per second.

Fiber-optic cables are rapidly replacing copper circuits in telephone systems. They are smaller in diameter, use less power, and are much more resistant to interference. During thunderstorms, you will hear a great deal of static on phone lines using copper cable. Phones linked with fiber-optic cable will remain unaffected by the flashes of lightning.

Fiber optics are finding uses in areas other than communication. In the medical field, fiber optics are being used in the endoscope. This device allows physicians to look inside the human body using a fiber-optic cable with a viewing lens at one end. In many cases, this can prevent unnecessary surgery.

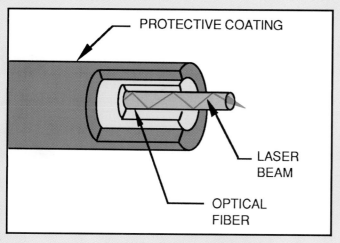

Fig. 1. Diagram of a fiber-optic cable.

PROTECTIVE COATING

LASER BEAM

OPTICAL FIBER

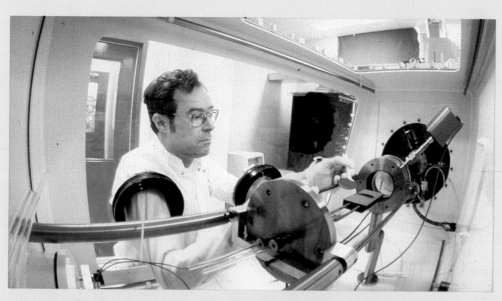

Fig. 2. A worker testing an optical fiber. (Northern Telephone Co.)

Identifying the Audience

Each television program is targeted to a specific audience. Saturday morning programming is largely devoted to children. Late night programming is given over to young adults. Afternoons are full of game shows and soap operas, which are popular with many retired people and others who are at home during daytime hours.

The first step in designing a television program or advertisement is to select the target audience. This includes determining who the audience is and what they like. Different people find different things important. Teenagers want to be part of the group. Therefore, advertisements and programming directed toward that age group will emphasize this factor. Advertisements, often, try to convince teenagers that wearing the "right" clothes or driving the "right" car will make them more popular.

Older people, generally, seek security and stability. Television messages directed at this group will focus on durability, soundness, and quality. Young executives have different goals. They may want to show that they are succeeding. Advertising directed toward them will emphasize status. The advertisements will try to communicate "You are what you drive!" or "You are what you wear!"

The aims of the communications are chosen through ample research. Actions that study audience needs and wants are called **audience assessment.** Audience assessments were discussed in Chapter 20.

Selecting the Approach

Once the audience has been identified and described, as in other communication media, an approach must be selected. In advertising, this may involve a catch phrase, such as "You're in good hands with Allstate" or "The only way to travel is Cadillac style."

Next, the tone of the message must be selected. The tone could use a serious or humorous angle. A regal (dignified) technique is another option. A contemporary (present day), historical, or futuristic theme could be selected. Whatever the style, the tone must appeal to the audience.

The tone must also fit the situation. A program covering the effects of air pollution would probably not have a humorous approach. On the other hand, a program on circus life might have humor as an important component.

Developing the Message

The ultimate success of a broadcast message is determined by the number of people that receive and act upon it. Therefore, the message must be carefully crafted. This most often requires a **script.** A script iden-tifies the characters, develops a situation, and communicates a story. Also, it provides the dialog for the characters and describes their movements on the stage.

Television is a visual and audio medium. Therefore, the script must be prepared with both sight and sound in mind. The script lays out the view to be shown and the words to be heard.

The script, like all communication media, is tailored to the audience. The complexity of a children's program is greatly different from the complexity of a PBS (Public Broadcasting System) documentary. Children have trouble following flashbacks and parallel story lines. They like simple, logical approaches to the issues and stories.

The previous knowledge level of the audience, also, must be considered. Local programming will assume the audience has an understanding of local politics and geographical features (such as towns or rivers). National programming may need to explain these details.

There are four common types of scripts used in television:

- **Full or detailed scripts** are complete scripts that contain (a) every word to be spoken, (b) sound effects and music information, (c) all major visual effects, and (d) production notes (timing, camera angles, etc.), Fig. 22-11. Generally, this script is written in a two column format. The left column contains video and production information. The right column contains audio information, including the dialog (words to be spoken) and sound effects (background sounds).

VIDEO	AUDIO
Show title	Music theme up then fade.
Host seen— head shoulders	Art has long been an important communication media. In recent years it has grown in popularity as the availability of reasonably priced, quality paintings has grown.
Background light as the camera dollies past host to pan art on back wall	In our city, this trend is easily seen with a visit to the Artists' Guild show at the convention center.
Zoom in on Watercolor #1	Artists using many media are exhibiting the results of their talent.

Fig. 22-11. An example of a full script format for a local public interest program.

Full scripts for commercials often develop from a storyboard, Fig. 22-12. A storyboard is a series of sketches or pictures that show all the scenes in the commercial. Under each sketch is its dialog and production notes.

- **Partial format scripts** contain complete scripts for the introduction and conclusion. The remaining parts of the show are simply outlined. This type of script is used for talk shows and sports events, where actions or responses give direction to the show. The script simply provides a skeleton around which the show develops. Often, it will contain time cues and other production notes.

- **Show format scripts** develop a list of the various film or show segments. Production notes and timing cues are included. This type of script is used for programs that follow a specific format for each show. Programs using this type of script would include morning and nighttime news programs, Fig. 22-13.

- **Fact or rundown scripts** develop a list of facts or characteristics. This type of script is used when the performer ad-libs a part in an advertisement, or a sportscaster uses statistics for broadcasting "color" commentary during a sportscast.

These scripts are designed to do three things for each program. They *establish the format, contain production direction, and communicate the content.*

PRODUCING COMMUNICATION MESSAGES

Once the design work is done, the stage is set for producing the message. Production steps include: casting, rehearsing, performing, and recording.

Casting and Rehearsing

Developing a television program is a controlled activity. Many different people are involved. These people fall into three major groups: creative, production, and performance.

Creative personnel develop the scripts and design the scenery. They plan the action and the "look" of the

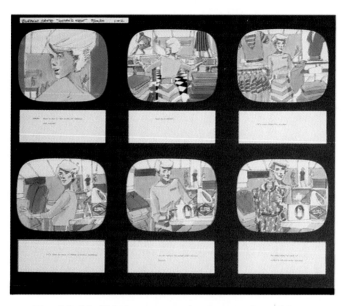

Fig. 22-12. This is an example of a storyboard for a commercial. (Clorox Co.)

Fig. 22-13. News programs often use show format scripts. (Gannett)

show. Directors, who identify camera angles, lighting, and performers' movements, are creative personnel.

Production personnel record the sights and sounds of the performance. They include the camera operators, the sound technicians, and the control booth personnel who capture the audio and video messages.

Performers deliver the message. They are the actors, announcers, newscasters, and musicians who use their talents to bring the program to life. The performers' voices, actions, and musical performances are used to attract the audience's attention to entertain, persuade, or inform them.

All of these groups are responsible to a **producer.** The producer oversees every element of the show. He or she helps develop the program idea, obtains financing, and controls the overall operation.

All these people are employed to deliver a service. The behind-the-camera personnel are hired like most other workers. They seek a job and are interviewed by the production company.

On-camera performers are employed through a process called **casting.** They are selected for their appearance and talent. They have to "look the part" of the character they will be playing. Then, they must be able to deliver a believable performance.

The production team and performers must rehearse. This involves the actors and musicians practicing their performance. Rehearsing also involves the camera and sound technicians running through their work, moving the camera and sound booms (microphones on long shafts) through their various stations.

Performing and Recording

The goal of a television project is to capture sound and sight on tape. This process uses two important transducers. A microphone changes audible sound into electrical signals. Likewise, the television camera changes light into electrical signals. Together, these coded messages are captured on video tape. Let's look at how these technological devices work.

Audio Recording. Recording sound is essentially the same for radio, audio recordings, or television. It usually involves a microphone, a mixer, and a recording device.

The microphone changes sound waves into electrical waves of a similar pattern. Each increase in the sound wave corresponds to an increase in the electrical wave. These waves form a continuous, flowing pattern that we call analog. The waveform moves upward and downward as the amplitude changes.

However, highly efficient communication cannot be done with analog waves. Analog signals are easily subject to internal distortion and external interference (static). Therefore, many newer systems convert the analog sound waves into digital signals. This is done by the process shown in Fig. 22-14.

Video Recording. Video recording uses a camera to capture the light reflecting from a scene. The lens in the camera focuses the light on the image plate in a vidicon tube. This image is changed into a pattern of electrical charges on the plate. Then, an electron gun shoots a beam of electrons at the image plate. The beam moves (scans) in a series of lines from left to right and top to bottom. The gun uses 525 lines (625 in Europe) per picture and scans 30 pictures (frames) per second.

The beam of electrons is reflected back from the plate. The intensity of the reflected electrons varies with the pattern of electrical charges on the plate. The reflected beam becomes a code for the picture. This code can be imposed on a carrier wave or recorded for future viewing.

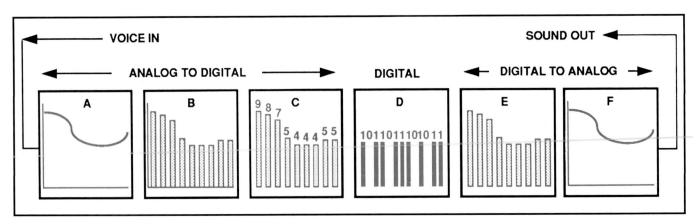

Fig. 22-14. Changing analog waves to digital signals. A–The system starts with an analog audio waveform that has been converted to an electrical wave. B–This wave form is sampled several thousand times a second, and the magnitude of the wave is determined for each sample (the bars). C–The magnitude of each sample is then given a numerical (digital) value that indicates its intensity. D–This number is changed into binary (computer) code. E–The code is transmitted to a receiver where it is converted back into digital code. F–The digital signal is changed into an analog wave that you can hear.

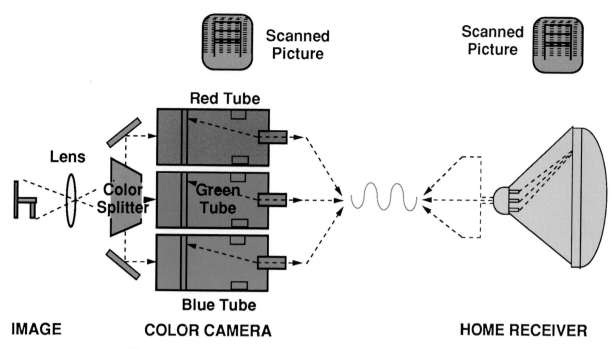

Fig. 22-15. This is a simplified drawing of a color television camera system.

Color television cameras use a three-tube system, Fig. 22-15. The beam of light is broken into red, green, and blue segments by a prism. Each beam is focused on a different tube. This action produces separate red, green, and blue signals. The signals are then transmitted to the television receiver, which uses three guns to project the three colors onto the tube screen, producing a picture.

Storing Signals. Often audio and video signals are recorded on tape or disc. Magnetic tape recording uses a specially coated plastic tape. The coating is a metal oxide material that is made up of needle-like crystals. The signal from some receptor (microphone, video recorder, etc.) is fed into an amplifier and then to a recording head. The electrical impulses from the signal generate magnetic forces in the recording head. These magnetic forces realign the metallic crystals in the tape. This new arrangement forms a code that corresponds with the sound or picture captured by the receptor. Later, this code can be used to reproduce the information.

Another recording system is the laser CD (compact disc). This system uses very tiny pits in the surface of a disc to record the digital code for sound or visual information. Later, a laser can "read" the code and reproduce the original signal.

BROADCASTING AND RECEIVING

All broadcast media (radio and television) use a carrier wave with a code imposed on the wave. This coded wave is sent through the air from a transmitter, or broadcast site, to receivers at distant locations. There are two different types of wave communication used in these broadcast systems.

Types of Wave Communication

Radio waves travel in straight lines. Two types of broadcast systems were developed that allow these waves to travel great distances, Fig. 22-16. The first system uses **direct waves.** Television and FM stations use direct waves. Direct wave transmitting and receiving antennas must have an open line of sight. Due to the curvature of the earth, these antennas must be placed within, roughly, 60 miles of each other. The earth will block the signal from reaching an antenna beyond this distance. This is why you have trouble receiving distant television and FM stations.

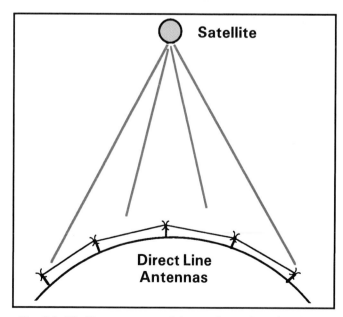

Fig. 22-16. The curvature of the earth requires that direct line antennas be placed within 60 miles of one another. Satellites present another way to allow direct waves to cross long distances.

One broadcast station and antenna may be used to cover a specific area. This type of system is used for some television and radio broadcasts. In other cases, a series of receivers and transmitters (relays) are used to transfer the waves between broadcast towers. This type of system is used for microwave telephone communication, Fig. 22-17.

Another way to overcome the problem caused by the curvature of the earth uses satellite communication systems, Fig. 22-18. The signal is transmitted to a satellite orbiting in space. One part of the satellite is a receiver, another part is a transmitter. The transmitter portion takes the message from the receiver and transmits it back to earth. This system allows a very large area of the earth to receive a message from one transmitter site. The area that is covered by a satellite is known as its *footprint*. Satellite communication systems are used by television "super stations" and networks like CNN. They have opened rural areas to more complete television programming, Fig. 22-19.

The second broadcast system uses **reflected waves.** In this system, the signal is bounced off the ionosphere (upper part of the atmosphere). The reflected wave is then received by a distant antenna. AM radio signals are transmitted through reflection. When atmospheric condition are just right, you may receive radio stations thousands of miles away.

Fig. 22-18. Communication satellites can send information in straight line waves over a large part of the earth. This photo shows a communication satellite being placed in orbit from the space shuttle. (NASA)

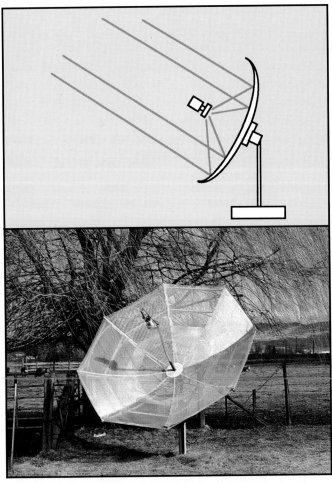

Fig. 22-19. Satellite dishes, like the dish pictured, allow many rural areas access to better communication.

Fig. 22-17. This shows a microwave tower, which transmits telephone messages in straight lines.

Broadcast System

Broadcast systems, as you have learned, have several major parts. The first is a **transmitter.** The transmitter imposes the message on the carrier. The resulting signal is radiated into the atmosphere by a broadcast antenna. This signal is captured by a receiving antenna. The signal is then fed into a **receiver,** where the coded message is separated from the carrier wave. The electrical impulses of the message are then converted by a transducer. Audio messages are converted to the sound you hear by speakers. Video messages are usually changed into pictures with cathode-ray tubes.

Fig. 22-20. Remote relay communication allows us to see live television pictures from almost any spot on the globe.

A fairly new broadcast system is the remote-link-system, Fig. 22-20. This system allows the signal to be produced at a remote location like at a sporting event, the site of a natural disaster, or the place of some other newsworthy incident. The on-site camera and transmitter generate the signal. The signal is transmitted from an antenna on a truck to a relay antenna. The relay antenna may be on a building or a satellite. Finally, the relay antenna relays the signal to the receiving antenna.

OTHER COMMUNICATION TECHNOLOGIES

Broadcast radio and television are not the only telecommunication technology available. Mobile radio and cellular communication systems are becoming a vital part of personal and business communication.

MOBILE COMMUNICATION SYSTEMS

There are three basic systems used in mobile communication. These, as shown in Fig. 22-21, are:
- **Simplex systems:** These systems use the same frequency, or channel, for both the base and mobile transmissions. The systems allow only one-way transmission at any given time. The mobile unit cannot break into a base transmission. It must wait until the base transmission is completed.

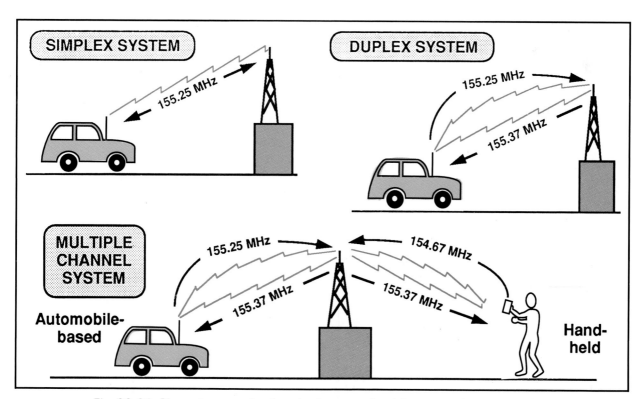

Fig. 22-21. Shown here are the three basic types of mobile communication systems.

- **Duplex systems:** These systems use two frequencies, or channels. Generally, the base unit transmits on one frequency and the mobile unit replies on another. The systems allow one mobile unit to transmit information to the base station while another unit is receiving a signal from the home base. Most of these systems, however, do not allow the mobile units to talk to one another. Their messages must be relayed through the base station.
- **Multiplex systems:** These systems use multiple frequencies to accommodate different types of units. The base may broadcast on one frequency. Vehicle units may use a second frequency and hand-held units a third frequency. These systems ensure that the low-power hand-held units can get through to the base station.

SATELLITE COMMUNICATION SYSTEMS

Satellite communication systems are another telecommunication system linking business together. Data, product orders, and other information are transmitted from one location to another. Fig. 22-22 shows a dish-antenna used to communicate parts orders from an automobile dealership to a factory warehouse.

SOUND RECORDING SYSTEMS

Still another system that impacts almost all of us, is sound recordings, Fig. 22-23. The fundamentals of this system were explained in the earlier parts of this chapter dealing with audio recording. Cassette tapes and CD recordings are the systems of choice today. They have, essentially, replaced the 45 rpm and 33 1/3 rpm records that your parents grew up with.

Fig. 22-22. The communication antenna on top of this automobile dealership connects the office with the parts warehouse.

Fig. 22-23. This audio technician is mixing (blending) sound from several microphones to make an audio recording. (Ball State University)

SUMMARY

Telecommunication is an important part of modern life. Telecommunication includes the telephone system, which carries personal and facsimile messages over hard-wired (copper wire or fiber-optic cable) circuits. The second type of telecommunication includes broadcast systems. These are, primarily, radio and television broadcasting. They also include mobile radio and cellular telephone communication.

Almost all telecommunication involves designing the message through audience assessment, script writing, message production, and message broadcasting. The main steps in broadcasting are encoding the message on a carrier frequency, transmitting the signal, receiving, and decoding the message.

WORDS TO KNOW

All of the following words have been used in this chapter. Do you know their meanings?

Alternating current
Amplitude
Amplitude modulation
Audience assessment
Audio
Broadcast frequency
Broadcast systems
Carrier frequency
Casting
Conductor
Creative personnel
Direct current

Direct wave
Duplex systems
Fact script
Frequency
Frequence division multiplexing
Frequency modulation
Full scripts
Hard-wired systems
Hertz
Induction
Kilohertz
Megahertz
Multiplex systems
Multiplexing
Partial format script
Performers
Producer
Production personnel
Radio wave
Receiver
Reflected wave
Script
Show format script
Simplex systems
Telecommunication
Time division multiplexing
Transducer
Transmitter
UHF
VHF
Video

TEST YOUR KNOWLEDGE

Write your answers on a separate piece of paper. Please do not write in this book.

1. What does telecommunication mean?
2. True or false. A wire that carries electricity is called a transducer.
3. Match:

   _____ Codes the message   A. Amplitude
   by changing the cycles   modulation.
   per second of the   B. Frequency
   carrier wave.   modulation.

   _____ Codes the message by
   changing the strength of
   the carrier wave.

   _____ Used to broadcast the audio
   portion of a television program.

   _____ Used to broadcast the video portion of a
   television program.

   _____ The oldest radio broadcast system.

4. True or false. Digital signals are less likely to be affected by distortion and interference than are analog signals.
5. Sending more than one message over a telephone waveguide is called _____.
6. List the two different types of telecommunication systems.
7. An action-type television show is likely to use _____ scripts.
8. Signals that have been converted in numerical values are called _____ signals.
9. True or false. Radio waves travel in straight lines.

APPLYING YOUR KNOWLEDGE

1. Select a product or an event and conduct an audience assessment for it. Record your data on a chart similar to the one below.

Product or event:
Intended audience:
Size of the audience:
What has high value to the audience?
What approaches are most likely to attract the audience's attention?

2. For the product or event selected above (a) select a theme or approach and (b) develop a full script or a storyboard for an advertisement to promote it.
3. Produce an advertisement using the script developed above by (a) selecting production personnel and performers, (b) rehearsing the production, and (c) recording the advertisement.

Section Six - Activities

ACTIVITY 6A - DESIGN PROBLEM

Background:

Communication technology is used to deliver information, project ideas, and generate feelings. Often, these goals are incorporated in advertising. Communication can promote a product or idea through delivering information and persuading people to act in a certain way.

Situation:

You are employed as an advertising designer for Breckinridge and Rice Agency. One of your clients wants to promote technological literacy as a public service effort.

Challenge:

Design a full page (6 3/4″ x 9 1/2″) magazine advertisement that promotes technological literacy. The ad should encourage students to select at least one class in technology education to help them understand technology as it impacts their lives. In this process, develop a theme, prepare a layout, specify the type size and style, ink colors, and the type and location of the photograph for your design.

Optional:

1. Print the advertisement as a flyer that can be sent home with students in your school.
2. Convert the layout to a poster, print copies, and post them on bulletin boards in your school.

ACTIVITY 6B - PRODUCTION PROBLEM

Background:

Photographs are used either to capture an event for historical purposes or to communicate information and feelings. The first use allows people to record family experiences, chronicle a trip, or capture a specific happening. The second use is communication. The use attempts to give directions, develop an attitude, or communicate a feeling.

Challenge:

You are a communication designer for a publisher of children's instructional books. They are developing a series of simple how-to-do-it booklets for children that are four to eight years old. Select a task that a child of this age must master. Design and produce a six-picture set that shows the child how to systematically complete the task.

Series title: _____

Shot #: _____

Description: _____

Fig. 1.

Equipment:
Layout sheets (Photocopy the form included with this activity.)
Pencils, magic markers, etc.
Camera and film (black-and-white print or color slide)
Darkroom equipment and chemicals
(Optional: The film can be sent to a commercial developer for processing.)

Procedure:

Designing The Product
1. Select a task.
2. List the steps that are needed to complete the task.
3. Group steps that can be shown in a single photograph (shot).
4. Develop a layout for each shot using the layout sheets, Fig. 1.

Producing The Product
1. Gather the items (props) needed for each shot.
2. Set up shot #1.
3. Shoot at least three photographs of the shot using different f-stops to "bracket" the exposures.

(Generally, one photograph is shot using the f-stop indicated by the light meter. Then, a second shot is exposed using one f-stop above and a third using one f-stop below the first setting.)
4. Repeat steps #2 and #3 until all six shots are taken.
5. Process the film.
   If black-and-white print film is used:
   a. Make a contact print of the strip of negatives.
   b. Select the best negative for each shot.
   c. Make a print of each of the six selected negatives.
   d. Mount the prints for display.

Optional:
1. For black-and-white prints: Make a story book by producing a printed narrative for each picture. Print the story for each photograph on a separate page, then mount the picture above the narrative. Design and produce a cover for your book.
2. For slides: Prepare a script, and record a narrative for the series. Produce a title and end slide for the series.

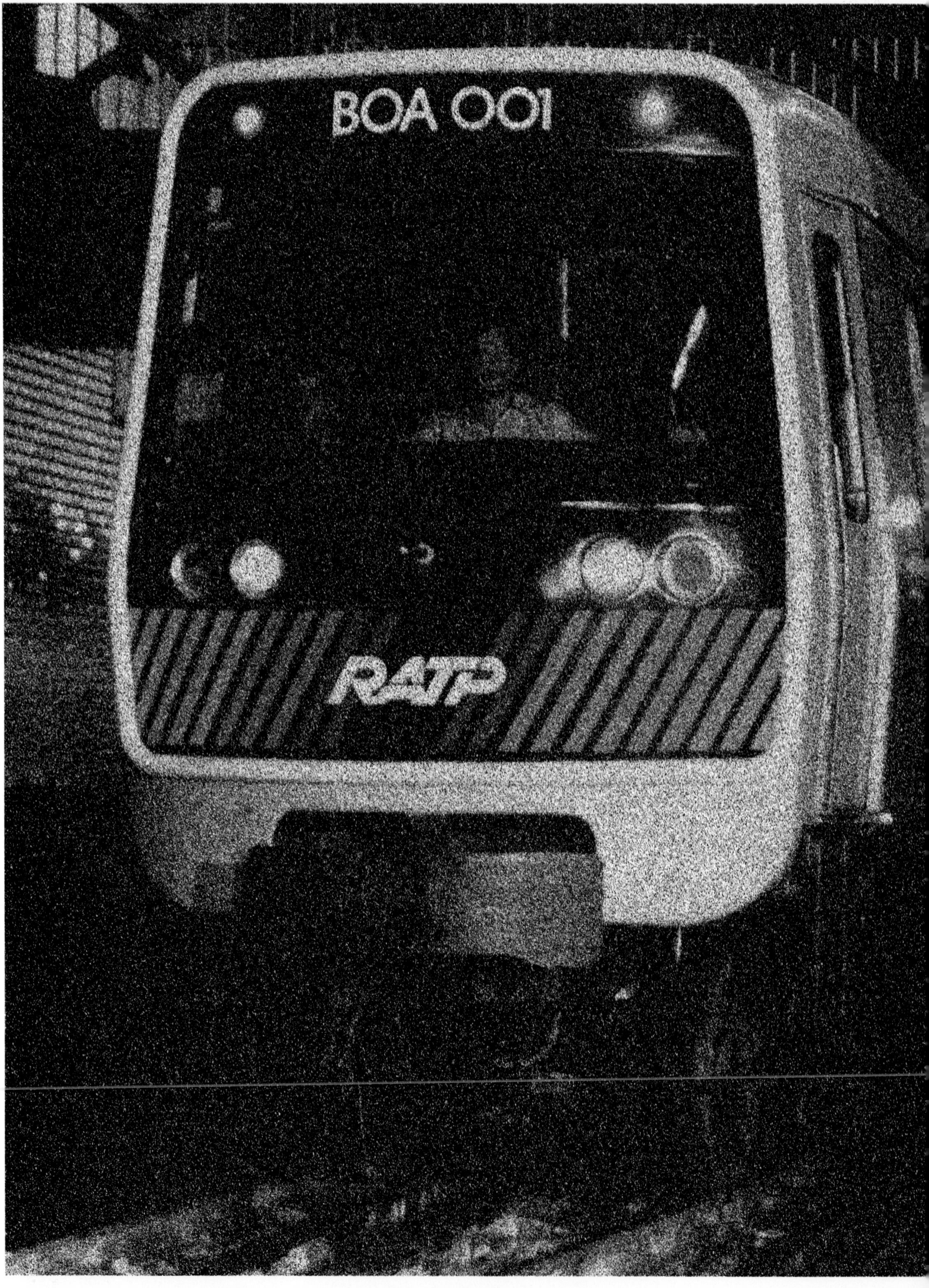
BOA 001
RATP

Applying Technology: Transporting People and Cargo

23. **Using Technology to Transport**

24. **Transportation Vehicles**

25. **Operating Transportation Systems**

Section 7 Activities

This is a model of the trains that run through the Eurotunnel. The Eurotunnel is 31 miles (51 km) long, connecting England to France. This transportation project brings together many types of construction, manufacturing, and communications knowledge. (Crown copyright reserved)

USING TECHNOLOGY TO TRANSPORT

After studying this chapter you will be able to:
- *Define transportation and transportation technology.*
- *List and describe the parts of a transportation system.*
- *Describe the four environments in which transportation systems operate.*
- *List the three components of a transportation system.*
- *Describe what is meant by place utility.*
- *Describe transportation pathways.*
- *List and describe the types of transportation vehicles.*
- *Describe the types and purpose of transportation support structures.*

The development of transportation and civilization are closely related, Fig. 23-1. Without transportation, humans are restricted to a very small area. This area is limited by the distance that a person can walk in a short period of time. This distance was less than 25 miles (40 km). In prehistoric times, a traveler had to make his or her journey out from home and back in one day. There were no places to stay along the way. Therefore, this area of travel was a very small circle. The domestication of animals enlarged this travel area several times. Now people could travel 40 to 50 miles (64 to 80 km) in a day. This advancement led people to design and build roads. As early as 30,000 B.C., established transportation routes existed. However, only the very brave ventured far from home. In fact, a restricted area of travel was the norm until recent times.

The development of sailing ships, followed much later by the railroad, further enlarged this area of travel. However, world-wide travel was available only to wealthy or adventurous people. Travel, as we know it today, is a very recent development. The advent of jet-powered aircraft after World War II made the far reaches of the globe available to many people. In one day, we can eat breakfast in New York, lunch in Los Angeles, and be back in New York for a late dinner.

IMPORTANCE OF TRANSPORTATION

Transportation is so important it has become a part of human culture. Try to imagine life without well-developed transportation systems. We think of transportation in the same light as food, clothing, and shelter. It has become a basic need. Transportation takes us to work, opens up areas for recreation, allows for easy shopping, and helps keep families in touch.

To help understand how important transportation is, consider the following:
- About 20 percent of the gross national product (the total of all goods and services sold in the U.S.) is related to transportation.
- There are over 110 million passenger cars in the country - nearly one for every two people.
- Nearly half the citizens of the country are licensed to drive an automobile.
- Almost 15 percent of personal spending goes toward transportation services.
- Nearly 64 percent of all petroleum is used for transportation, and 25 percent of all petroleum is consumed by personal cars.

TRANSPORTATION - A DEFINITION

Transportation is one of the basic areas of technological activity. People have moved themselves and their possessions from place to place since the start of

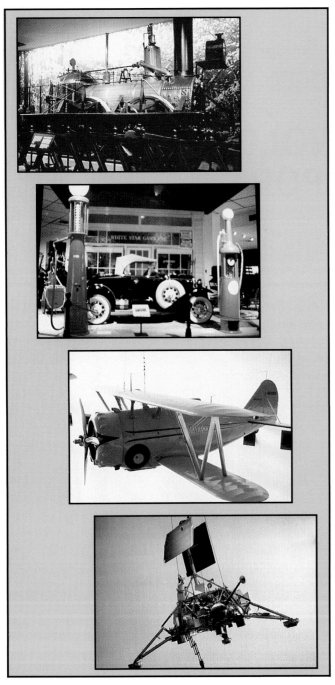

Fig. 23-1. Transportation systems have evolved to meet our changing needs. All the transportation vehicles shown above - trains, automobiles, airplanes, and spacecraft - were new transportation vehicles at one time.

civilization. We use transportation today, and will continue to use transportation systems in the future. Simply stated, **transportation** is *all acts that relocate humans or their possessions.* **Transportation technology** provides for this movement, *using technical means to extend human ability.* Transportation technology extends our ability beyond our own muscle power and our ability to walk.

Transportation is also an interaction between physical elements, people, and the environment, Fig. 23-2.

Transportation provides mobility for people and goods. Transportation uses such resources as materials, energy, money, and time. Transportation has a level of risk to the cargo and passengers that can result in damage, injury, or death. Finally, transportation has a societal impact through employment, pollution, and other factors.

TRANSPORTATION AS A SYSTEM

Transportation, like all other technologies, can be viewed as a system. It is a series of parts that are interrelated. The parts work together to meet a goal. Transportation uses people, artifacts, vehicles, pathways, energy, information, materials, finances, and time. These parts work together to relocate people and goods.

TYPES OF TRANSPORTATION SYSTEMS

There are four environments, or "modes," that people use for transportation. Transportation systems have been developed for land, water, air, and space, Fig. 23-3.

Land
Humans can move over land with ease. The earliest transportation systems were designed to move people over the land. **Land transportation** systems move people and goods *on the surface of the earth* from place to place. Land transportation systems have developed into three major types. These are:
- Highway systems: automobiles, buses, and trucks.
- Rail systems: freight, passenger, and mass transit systems.
- Continuous flow systems: pipelines, conveyors, and cables.

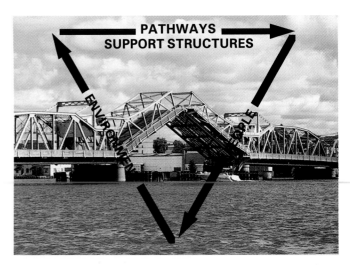

Fig. 23-2. Transportation systems are an interactive system of physical elements (vehicles, pathways, etc.), people, and the environment. (Bud Smith)

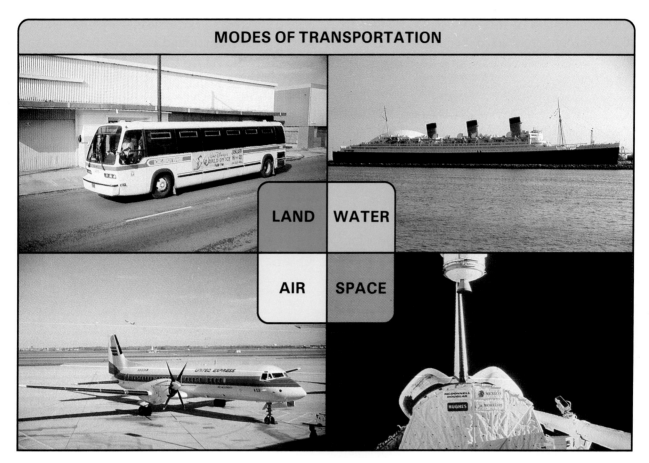

MODES OF TRANSPORTATION

LAND | WATER
AIR | SPACE

Fig. 23-3. Transportation systems operate on land, in water, and through air and space.
(MITS, United Airlines, NASA)

Water

The next system developed was water transportation. From the humble hollow-log canoe, water transportation has grown to be an important mode for moving people and cargo. **Water transportation** uses water to support the vehicle. Water transportation includes inland waterways (rivers and lakes) and ocean-going systems.

Air

Air transportation became practical in the twentieth century. The first successful flight was in 1903. The plane was built by Orville and Wilbur Wright. This flight happened at Kitty Hawk, North Carolina. In only ninety years, air travel has become a large industry. **Air transportation** uses airplanes and helicopters to lift passengers and cargo into the air so they can be moved from place to place. Today, air transportation includes commercial aviation (passengers and freight), and general aviation (private and corporate aircraft).

Space

The fourth system is **space transportation**. This mode can best be described as emerging. It started in

1957 with the launch of Sputnik, a Soviet satellite. Since 1957, space exploration has expanded our knowledge of the universe. Humans have traveled to the moon. The space shuttle now lifts satellites and scientific experiments into space.

On the horizon is personal space travel. Hypersonic aircraft will merge air travel and space travel technologies. This will allow people to travel anywhere on the globe in a matter of hours.

TRANSPORTATION SYSTEM COMPONENTS

Within each of the transportation systems are three major components. These, as shown in Fig. 23-4, are:
- Pathways.
- Vehicles.
- Support structures.

PATHWAYS

Transportation systems are designed to move people and cargo from one place to another. This movement provides **place utility**. People value being able to move things from one place to another. The things that are moved can be food, products, or people. We are willing to pay someone else to transport us or our goods from place to place.

To deliver place utility, a transportation system links distant locations with a network of pathways, Fig. 23-5. Pathways on land are readily visible. Land pathways are the streets, roads, highways, rail lines, and pipelines that people have built. These pathways form a network that connects most areas of the United States. Land pathways are the result of construction activities used to fulfill the needs of transportation.

Many water pathways are less visible. They include all navigable bodies of water. Navigable means that a body of water is deep enough for ships to use. Water pathways can be grouped into inland waterways and oceans. **Inland waterways** are rivers, lakes, and bays. Water pathways also include human-made canals that allow water transportation across areas that lack navigable rivers or lakes. Often, the canals use aqueducts to cross streams. Canals use locks to raise or lower water craft as the terrain changes, Fig. 23-6.

The oceans and seas provide a vast water pathway. The major land masses on earth are touched by an ocean or sea. Therefore, the continents can be easily served by an ocean transportation network.

Some pathways are hard to see. Air and space routes are totally invisible. They are defined by humans and appear only on maps. Airplanes and spacecraft can travel in almost any direction. Only human decisions establish the correct pathway. The pathway is chosen to ensure efficient and safe travel. Airplanes use pathways to take us easily over land and water barriers. Airplanes allow us to travel thousands of miles in a few hours. To cover these distances before airplanes or railroads took *months* of travel by wagon.

Fig. 23-5. Pathways include roads and rail lines (top); rivers, canals, and oceans (middle); and air and space routes (bottom).

Fig. 23-4. Transportation systems use pathways, vehicles, and support structures. This photo shows a guided-bus system. The pathway is the concrete guideway, and the vehicle is the bus. Support structures are the stations where passengers board the bus. (Daimler Benz)

Technology: How it Works

FLIGHT SIMULATOR: A computer-controlled device that allows pilots to practice flying an airplane without actually using an airplane.

The first time you rode a bicycle, you did not hop on and ride away. You had to learn how to balance the bicycle and how to pedal. Until you mastered these skills, you may have fallen over a few times. You made mistakes as you practiced how to ride. Making mistakes with a bicycle was not a big deal. You got up, dusted off, picked up your bike and tried again.

Pilots who are learning to fly also need to practice. However, making a mistake with an airplane can be dangerous and costly. In order to practice without risk, pilots use *simulators*. Simulators range from simple training devices for small aircraft to sophisticated units for large airliners, Fig. 1. We will discuss large simulators here.

A simulator mimics the sound, motion, and the view from an airplane in flight. Pilots can improve their skills and learn to fly new models of aircraft. A trainee pilot can develop the ability to take directions from flight controllers, and practice takeoffs, cruising, and landings. Emergency procedures can be practiced over and over. The safety of the pilot, the public, and a very expensive aircraft are not put at risk.

A simulator has several major systems. The first is the control system. The instructor uses the control system's computer to set up a desired flying condition. The second is a capsule motion system. This is a series of hydraulic cylinders that move the capsule. The movements of the capsule imitate the motion of an aircraft in a particular situation. The next part is an audio-visual system which uses projectors, screens, and speakers. This system conveys to the pilot what he or she would see and hear if he or she were flying a real airplane. Finally, the simulator has a set of controls and instruments just like those in a real aircraft. The pilot uses them to "fly" the simulator. They are connected to the control system computer.

The flight simulator uses a closed loop control system. The computer sets up a situation. It directs the audio-visual system to show the pilot what he or she would see and hear in a real situation. The computer also directs the motion system to provide the physical "feel" of the situation. The computer causes the instruments to show readings that would be expected in the situation. Then, when the pilot reacts, the computer makes the simulator "act" like an airplane. The computer will change the view, sounds, instrument readings, and motion to *simulate* the actions of an aircraft. If the pilot makes a mistake, the computer can simulate an emergency. The pilot can practice how to get the aircraft back to a safe condition. No one is injured, and the pilot can find out how to avoid a crisis.

Fig. 1. This simulator is used to train pilots to fly the Boeing 767 aircraft.

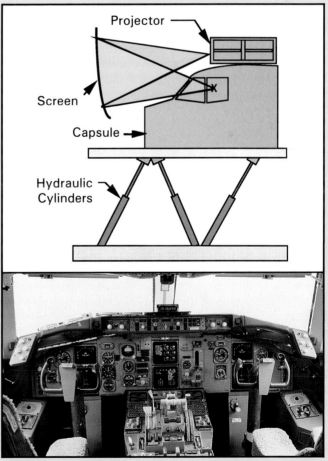

Fig. 2. The capsule is suspended on hydraulic cylinders and moves to mimic the motion of an aircraft. The capsule contains a complete set of instruments and controls (inset).

Fig. 23-6. Canals are constructed waterways. They include a waterway, locks (left) to raise or lower the ship, and aqueducts (right) to cross streams and gullies.

VEHICLES

All transportation systems, except for continuous flow systems, use vehicles. **Vehicles** are technological devices designed to carry people and cargo on a pathway. Vehicles are designed to contain and protect the cargo as it is moved from place to place. The demands on the vehicle change with the type of cargo, Fig. 23-7.

People need to be protected from the environment through which the vehicle travels. People need to travel in comfort. Passenger vehicles must provide proper seating, good lighting, and air conditioning. On long trips, food and rest rooms must be provided. Finally, passenger vehicles must provide safety in case of accident. Impact absorbing construction, seat belts, fire protection, and other safety measures must be designed and built into the vehicle.

Cargo vehicles are designed to protect the cargo from damage by motion and the outside environment. Cargo must be cushioned so that is does not become broken, dented, or scratched. Fumes, gases, or liquids that could affect the cargo must be kept away. Finally, the cargo must be protected from theft and vandalism.

SUPPORT STRUCTURES

All transportation systems use structures. Many transportation pathways use human-built structures. Rail lines, canals, and pipelines are constructed works.

More structures than just the pathway are needed. These structures are terminals. Terminals are where transportation activities begin and end. They are used to gather, load and unload passengers and goods, Fig. 23-8.

Fig. 23-7. Vehicles, like these aircraft, are designed differently for cargo (left) and people (right).
(Alaska Airlines, United Airlines)

Fig. 23-8. Transportation terminals are at the start and the end of transportation pathways. This photo shows an air terminal in Chicago.

Terminals provide for passenger comfort and cargo protection as they wait to be loaded onto transportation vehicles. Terminals also provide connections for various transportation systems. They may allow truck shipping to connect with air, rail, or ocean shipping. Other terminals connect automobile, rail, and bus systems with air transportation systems. These types of terminals allow for **intermodal shipping**. This means that people or cargo travel on two or more modes of transport before they reach their destination. An example of intermodal shipping is semi-truck trailers that are hauled on railcars.

Other structures are used to control transportation systems. They are communication towers, radar antenna, traffic signals, and signs. These structures help vehicle operators to stay on course and observe rules and regulations.

SUMMARY

Transportation is essential to modern-day life. We all use it and expect it to make our lives better. Each of us can choose from a variety of transportation systems to get to a place. These include land, water, and air transport. In the future we may also be using space as a transportation link.

All transportation systems have pathways and support structures. Most of them use vehicles to carry people and cargo from one point to another. In the next two chapters we will look at transportation vehicles and then look at operating transportation systems.

WORDS TO KNOW

All of the following words have been used in this chapter. Do you know their meanings?
Air transportation
Inland waterways
Intermodal shipping
Land transportation
Place utility
Space transportation
Transportation
Transportation technology
Vehicles
Water transportation

TEST YOUR KNOWLEDGE

Write your answers on a separate piece of paper. Please do not write in this book.
1. Define place utility.
2. List the three types of land transportation.
3. True or false. Walking to the store is an example of transportation technology.
4. Using more than one mode of transportation to ship a cargo is called _____.
5. List the three parts of a transportation system.

APPLYING YOUR KNOWLEDGE

1. Select a transportation system you use and describe it in terms of:
   (a) The pathway.
   (b) The vehicle.
   (c) The structures used.
   Enter your data on a chart similar to the one below.

TYPE OF TRANSPORTATION SYSTEM:		
	NAME	DESCRIPTION
PATHWAY:		
VEHICLE:		
STRUCTURE:		

2. Obtain a number of bus, rail, ship, or airline route maps and road maps. Choose a place that you would like to visit. Plan a trip that will use intermodal transportation. Make sure your route uses at least two different transportation modes.

Transportation vehicles have evolved from horsedrawn carts to high-performance jet airliners. (United Airlines)

24
CHAPTER

TRANSPORTATION VEHICLES

After studying this chapter you will be able to:
* *Describe a transportation vehicle.*
* *List the five systems present in a transportation vehicle.*
* *List the factors considered in developing a vehicle structure.*
* *Describe the types of land, water, and air transportation vehicles.*
* *Describe common propulsion systems used in transportation vehicles.*
* *Tell the difference between guidance and control in transportation systems.*
* *Describe how vehicle suspension systems operate.*
* *Describe how the five vehicle systems are applied to common land, water, and air transportation vehicles.*

Transportation is all the actions that move people or goods from one place to another. Most transportation actions use technology to make them more efficient.

This technology often takes the form of a vehicle, Fig. 24-1. A simple definition of a **vehicle** is a powered carrier that supports, protects, and moves cargo or people within a transportation system. Vehicles are used in all four environments of travel. They are used on land and in water, air, and space.

VEHICULAR SYSTEMS

All vehicles share some common systems. These are the structure, propulsion, suspension, guidance, and control systems, Fig. 24-2.

STRUCTURE SYSTEMS

All vehicles are designed to meet a common goal: to contain and move people and goods. People want to arrive safely and in comfort. Cargo must be protected

Fig. 24-1. Most transportation systems require a vehicle.

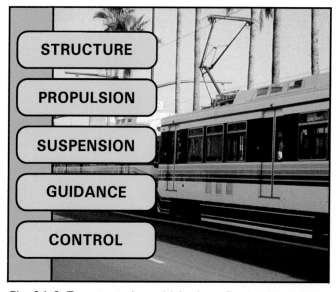

STRUCTURE

PROPULSION

SUSPENSION

GUIDANCE

CONTROL

Fig. 24-2. Transportation vehicles have five basic systems.

from the weather, damage, and theft. The structure of the vehicle helps to do these things. The structure is the physical frame and covering that provides spaces for people and cargo, power and control systems, and other devices.

PROPULSION SYSTEMS

Transportation vehicles are designed to move along a pathway. This pathway may be a highway, rail line, river, ocean, or air route. A vehicle must have a force to propel it from its starting point to its destination. This force is produced by the **propulsion system**. It uses energy to produce power for motion.

Propulsion systems range from the simple pedal, chain, and wheel system of a bicycle to complex heat engines, such as gasoline, diesel, and rocket engines, Fig. 24-3. The type of engine to be used in a vehicle is determined by several factors. These factors include:
- The environment where the vehicle travels.
- Fuel availability and cost.
- The forces that must be overcome, such as vehicle and cargo weight, rolling friction, and water or air resistance.

The engine must match the job. For example, using a jet engine to propel an automobile is overkill. The capabilities of a jet engine do not match the job of moving a car. Likewise, using a large diesel engine to power an airplane is not the best choice. The diesel engine would be too heavy. Therefore, there are many different sizes and many different types of engines used in transportation vehicles.

SUSPENSION SYSTEMS

All vehicles and cargo have weight. This weight must be supported as the vehicle moves along the pathway.

Proper support is produced by a **suspension system**. Suspension systems include:
- Wheels, axles, and springs on land vehicles.
- The wings on an airplane.
- The hull of a ship.

CONTROL SYSTEMS

A vehicle moves from its origin to its destination to relocate people and cargo. As a vehicle moves it requires control, Fig. 24-4. There are two types of con-

Fig. 24-4. Transportation vehicles require control in three degrees of freedom. The degrees are (1) forward - backward, (2) left - right, and (3) up - down. The steam locomotive shown in the photo has one degree of freedom.

Fig. 24-3. Propulsion systems range from simple to complex.

trol: **speed control** and **direction control**. The vehicle can be made to go faster through acceleration or slowed down by braking or coasting. How a vehicle changes its direction depends on its environment of travel. Land vehicles turn their wheels or follow a track. Ships move their rudder. Airplanes adjust ailerons, rudders, and flaps. Spacecraft use rocket thrusters.

The system of controls used will depend on the **degrees of freedom** a vehicles has. For example, rail vehicles have one degree of freedom. They move forward or backward by using speed control. Gravity and the rail eliminate the possibility of up-down and left-right movement.

Automobiles and ships have more freedom. They can move forward and backward, and can change their direction by turning left or right. They have two degrees of freedom.

Aircraft, spacecraft, and submarines have the most freedom–three degrees. They can move forward, backward, left, right, and change their altitude by moving up and down.

GUIDANCE SYSTEMS

Controlling any vehicle requires information. The operator must know his or her location, speed, and direction of travel. Information about traffic conditions and rules is also required. This information is provided by **guidance systems**.

Guidance information can be as simple as a speedometer reading or a traffic light. Guidance systems can be quite complex. Instrument landing systems (ILS) and land-based satellite tracking stations are examples of complex guidance systems.

LAND TRANSPORTATION VEHICLES

Land transportation includes all movement of people and goods on or under the surface of the earth. This includes highway, rail, material, and on-site transportation systems.

Highway systems use automobiles, trucks, and buses to move people and cargo. Rail systems move people and cargo from place to place. Material transportation systems include pipelines, conveyors, and lifts.

VEHICLE STRUCTURE

Each land transportation system requires special vehicles. The structure a vehicle has is based on its use. Passenger vehicles are different from cargo carrying vehicles. However, all the vehicles have three basic units, as shown in Fig. 24-5:
- Passenger or operator unit.
- Cargo unit.
- Power unit.

The size and location of these units or compartments will vary with the type of vehicle. Freight and some passenger rail systems place the power and operator units in one vehicle called a **locomotive**. The cargo and

Fig. 24-5. All vehicles have operator, power, and cargo or passenger units. (Freightliner Corp.)

passenger units are pulled by the locomotive. Mass transit systems have power units (electric motors) and passenger units in each car. The operator unit is located in the front car, Fig. 24-6.

Standard automobiles and delivery trucks have all three units in one vehicle. However, most long-distance trucks place the power and operator units in the **tractor**. The cargo unit is the one or more **trailers** attached to the tractor.

The passenger and operator unit must be designed for comfort and ease of operation. This requires the use of ergonomic principles. Ergonomics is the study of how people interact with the things they use. Ergonomic vehicle design requires that the seats adjust for people of different sizes. The instruments that provide guidance data must be easily seen, Fig. 24-7. The operating controls, such as for steering and braking, must be easy to reach and operate.

Fig. 24-7. This aircraft flight deck was designed using ergonomic principles for easy operation. Ergonomic principles are also used in land, water, and space vehicles. (Boeing)

The vehicle must also have an appropriate structural design. The vehicle must provide for safety, and for operator, passenger, and cargo protection. Most vehicles have a reinforced frame with a skin. The skin protects the vehicle interior from the outside environment. The frame supports the skin, and carries the weight of passengers and cargo. The frame also provides crash protection by absorbing the impact when vehicles are involved in accidents, Fig. 24-8.

PROPULSION

Most land transportation vehicles move along their pathway on rolling wheels. The rotation is produced by two systems: power generation and power transmission.

Fig. 24-6. The light rail vehicles (top) integrate operator and passenger areas. The freight rail system (bottom) uses a locomotive that combines operator and power units. The locomotive pulls railcars, the cargo units. (Long Beach Light Rail, Norfolk Southern)

Fig. 24-8. This vehicle is being crash tested to evaluate its structural design. (Arvin Industries)

Power Generation

The **power generation system** uses an engine as an energy converter. The engine produces the power needed to propel the vehicle. The most common engine in land vehicles is the **internal combustion engine**. This name means that fuel is burned *inside* the engine to convert energy from one form to another. The chemical energy in the fuel is first changed to heat energy. Then, the heat energy is converted to mechanical (rotating) energy by the piston and crankshaft to move the vehicle.

Land vehicles normally use either a four-stroke cycle gasoline engine (described in Chapter 7) or a diesel engine. Some rail vehicles use electric motors for propulsion. Electrically powered vehicles for highway use are on the horizon.

You will remember that the internal combustion gasoline engine uses four strokes to make a complete cycle. The first stroke is called the intake stroke. The intake valve opens and the piston moves down. A fuel/air mixture is drawn into the cylinder. When the piston is at the bottom of the intake stroke, the valve closes. The piston moves upward, beginning the compression stroke. The piston compresses the fuel/air mixture. When the piston is at the top of its travel, a spark plug ignites the fuel/air mixture. The burning fuel expands rapidly and drives the piston down. This is called the power stroke. At the end of the power stroke the exhaust valve opens. The piston moves upward to force burnt gases from the cylinder. The last stroke is called the exhaust stroke. The exhaust valve closes and the engine is ready to repeat the four strokes–intake, compression, power, and exhaust. You may want to look back to Fig. 7-22 to see a drawing of these four strokes.

Many large vehicles, such as buses, heavy trucks, and locomotives, are powered by diesel engines. The most common diesel engine uses a four-stroke cycle similar to a gasoline engine, Fig. 24-9. However, during the intake stroke only air is drawn in. Thus, only air is compressed. As the air is compressed it becomes very hot. When the piston is at the top of the compression stroke, fuel is injected directly into the cylinder. The fuel touches the hot air and burns rapidly. This causes the power stroke, which is followed by the exhaust stroke. These last two strokes are exactly like those of a gasoline engine.

Power Transmission

The power of an engine is worthless unless it can be controlled and directed to do work. This is the task of the **power transmission** system.

The transmission connects the engine with the drive wheels, Fig. 24-10. This is done in automobiles and

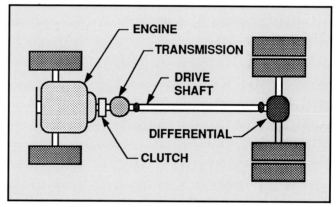

Fig. 24-10. The transmission joins the engine to the rest of the drivetrain.

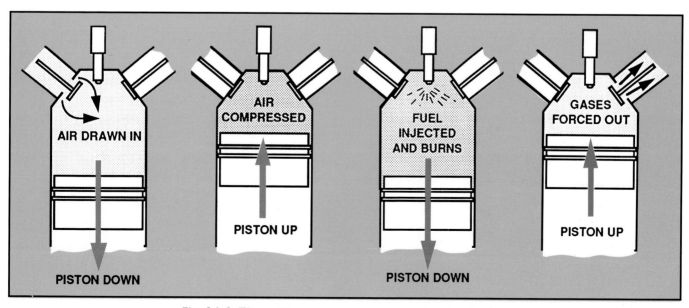

Fig. 24-9. These drawings show how a diesel engine works.

trucks by mechanical or fluid devices. Often a transmission has several input/output ratios, commonly called "speeds." A five-speed transmission has five ratios. The different ratios allow the power of the engine to be used efficiently. The transmission provides high torque (rotating force) at low speeds as the vehicle starts to move. Later, as gravity, rolling resistance, and air resistance are overcome, low torque-high speed outputs are used.

Mechanical transmissions (also called manual transmissions) have a clutch between the engine and the transmission. This allows the operator to disconnect the engine so that the transmission ratio can be changed. This action is called shifting gears. **Automatic transmissions** use valves to change hydraulic pressure so that the transmission shifts its input-output ratios.

When a vehicle makes a turn the wheels on the outside must be able to turn faster than those on the inside. The outside wheels turn faster because they have farther to travel. A device called a differential allows the wheels to turn at different speeds. The differential is a set of gears that independently drive each axle. The axles are connected to the vehicle's drive wheels. In rear-wheel drive automobiles and trucks, this device is separate from the transmission. Front wheel drive cars combine the transmission and differential into a single unit called a transaxle.

Diesel-Electric Power Units

Freight and passenger locomotives use a different type of power generation and transmission system. This type is called a diesel-electric system, Fig. 24-11. This system uses a very large (8 to 12 cylinder) diesel engine to convert the fuel to mechanical motion. This motion is used to drive an electric generator. The electricity is transmitted to electric motors that are geared to the drive wheels of the locomotive. This motor-gen-

erator unit is relatively light. Therefore, extra weight is added to the locomotive to give it better traction. This weight is called **ballast**. This extra weight is created by making the frame of the locomotive out of heavier steel plates than needed.

Electric Railways

Some rail vehicles are powered only by electricity. The vehicle has a moving connection that remains in contact with the electric conductor. Overhead wires or a third rail are used as the conductor. The electricity is conducted to motors that turn the vehicle's wheels.

SUSPENSION

Suspension systems keep the vehicle in contact with the road or rail. They also separate the passenger compartment from the drive system to increase passenger and operator comfort.

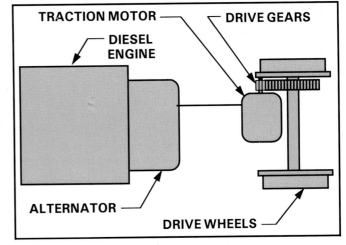

Fig. 24-11. This drawing shows a diesel-electric traction system used to power railroad locomotives.

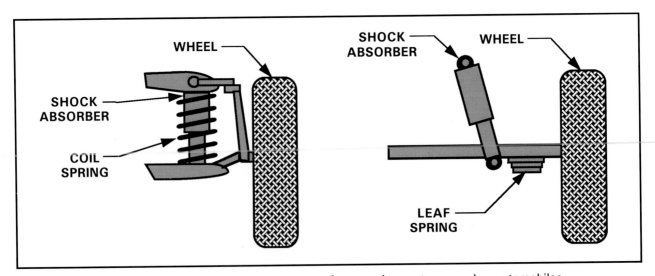

Fig. 24-12. This drawing shows two types of suspension systems used on automobiles.

The suspension system of a vehicle has three major parts:
- Wheels.
- Axles.
- Springs and shocks.

The wheels provide traction and roll along the road or rail. The wheel also spreads the weight of the vehicle onto the road or rail. A wheel can be used to absorb shock from bumps in a road or rail. Often rubber tires are attached to the wheels to absorb shock and increase traction. Most rail systems use steel wheels and depend on friction between the wheels and rails for traction.

Axles carry the load of the vehicle to the wheels. The axles support a set of springs. The springs absorb movements of the axles and wheels. However, springs give a very "bouncy" ride. To prevent this motion a set of shock absorbers is used. The shock absorbers are attached between the axle and the frame to dampen the spring action, Fig. 24-12. The result is a controlled, smooth ride.

A variation of the suspension system uses torsion bars. Instead of a spring, a steel bar absorbs the movement of the wheels and axles by twisting. The torsion bar untwists to release the absorbed energy. Shock absorbers are used to dampen the motion of the torsion bar.

GUIDANCE AND CONTROL

Most land transportation vehicles are controlled manually by an operator. The speed and direction of a vehicle can be changed by the operator. Throttles and accelerator pedals are used to control engine speed. Transmissions are shifted to increase the power being delivered to the drive wheels. Brakes can be applied to slow the vehicle. The wheels can be turned to take the vehicle in a new direction.

These actions require that the operator make decisions. These decisions are made on the basis of visual information and judgment. The operator receives information from signs and signals, Fig. 24-13. The operator also observes traffic conditions. Gages and instruments provide information on vehicle speed and operating conditions. However, all this information must be considered and acted upon. This is where operator judgment is required. Inexperienced operators and those influenced by drug and alcohol use may make poor judgments. Their actions may cause accidents and human injury or death.

WATER TRANSPORTATION VEHICLES

More than 70 percent of the earth's surface is covered by water. Thus, water transportation is an important form of travel. Water transportation includes all vehicles that carry passengers or cargo over or under water. This includes traffic on rivers, lakes, and along coastal waterways. This type of shipping is called **inland waterway** transportation. Water transportation also includes ships that sail the oceans and large inland lakes such as the Great Lakes in the U.S. This is called **maritime shipping**.

There are three kinds of water transportation vehicles (called vessels), Fig. 24-14. Vessels owned by private citizens for recreation are called **pleasure craft**. They allow people to participate in activities such as water skiing, fishing, sailing, and cruising. Large ships that are used for transporting people and cargo for a profit are called **commercial ships**. Some ships provide for the defense of a country. These vessels are called **military (or naval) ships** and are owned by the government of a country.

Fig. 24-13. Signs and signals, like highway signs (left) and rail signals (right), give operators guidance.

TYPES OF BOATS AND SHIPS

PLEASURE

COMMERCIAL

MILITARY

Fig. 24-14. There are three types of ships and boats.

VEHICLE STRUCTURE

Like land vehicles, ships can be designed to carry either cargo or passengers, Fig. 24-15. Cargo carrying ships are called **merchant ships**. Vessels that carry people are called **passenger ships** or passenger liners.

Nautical Terms

What a ship is used for will influence how it is designed. However, all ships have two basic parts:
- Hull–forms the shell that allows the ship to float and contain a load.
- Superstructure–the part of the ship above the deck that contains the bridge and crew or passenger accommodations.

Understanding the design and operation of ships requires a knowledge of some nautical terms. These, as shown in Fig. 24-16 are:
- Bow: the front of the ship.
- Stern: the back of the ship.
- Amidships: the middle of the ship.
- Port: the left side (as you look forward).
- Starboard: the right side (as you look forward).
- Topside: the outer surface of the ship above the waterline.
- Astern: behind the ship.
- Ahead: in front of the ship.
- Beam: the width of a ship at the widest point.
- Draft: the vertical distance from the waterline to the keel at amidships.
- Keel: a ridge down the center of the hull that helps the ship move in a straight line.
- Rudder: a flat paddle-like steering surface under the stern of the ship.
- Propeller: a set of blades that rotate and move the ship through the water.

MERCHANT SHIP

PASSENGER SHIP

Fig. 24-15. Commercial ships can be either merchant or passenger ships.

- Bridge: the control center where officers direct the ship's course and operation.
- Forward: toward the front of the ship.
- Aft: toward the rear of the ship.

Types of Ships

There are many types of ships in use today. Passengers are carried on ocean-going liners and cruise ships and, for short distances, on ferries. Cruise ships and ocean liners are miniature cities. They must supply all the needs of the passengers. This includes sleeping

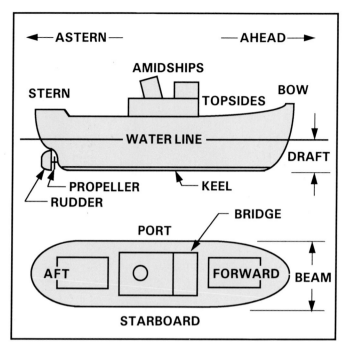

Fig. 24-16. This drawing shows some common terms related to boats and ships.

Tankers are used to move liquids across the oceans. Tankers carry petroleum and other chemicals in a series of large tanks. The liquid cargo can be pumped into and out of the tanks. Tankers have also been built to move gases.

Container ships are a new, fast way to ship large quantities of goods, Fig. 24-18. The shipments are loaded into large steel containers that resemble semi-truck trailers without wheels. The containers are loaded into the hold and stacked on the deck of the ship. The loading process that once took days is now done in hours.

Large ships sail the oceans while smaller ships can be used on lakes like the Great Lakes in North America.

quarters, kitchens, dining space, recreational areas, retail shops, and services, such as barber shops, beauty parlors, and cleaners.

Merchant ships are usually designed for one type of cargo. The types of cargo can include dry cargo, liquids, gases, and cargo containers, Fig. 24-17. **Dry cargo ships** are used to haul both crated and bulk cargo. Bulk cargo is the loose commodities that are loaded into the holds of ships. Typical bulk cargo includes grain, iron ore, and coal. Other dry cargo is contained in crates that are loaded into the various holds (compartments) of a ship. A final type of bulk cargo is land vehicles, such as cars and trucks. They are often transported on special vessels called RORO ships (for *R*oll *O*n-*R*oll *O*ff). The vehicles are driven onto the ship and off of the ship through large hatches (doors) in the hull.

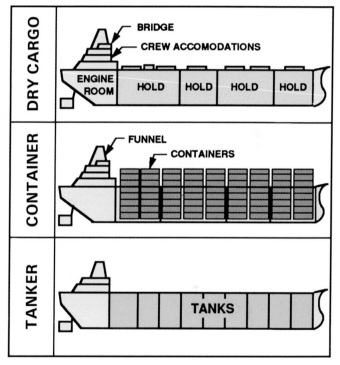

Fig. 24-17. This drawing shows the common types of merchant marine vessels.

Fig. 24-18. Containers are often shipped from inland points (left), then unloaded at the dock (center). Later, the containers are loaded onto container ships (right).

However, most shipping on inland waterways is done with tugboats and barges, Fig. 24-19. These are "water trains." The tugboat acts like the locomotive. It contains the power generation and transmission system and the operation controls. The barges are the waterway "rail cars." There are specifically designed containers for various types of cargo. The tug pushes or pulls a group of barges that are lashed together.

PROPULSION

Most commercial ships are propelled through the water by a propeller that is driven by an engine. This engine can be a steam turbine or diesel engine. Often the engine is located inside the ship or boat. However, small boats may be powered by a motor (internal combustion gasoline engine) attached to the stern of the boat. This type of power source is called an **outboard motor**, Fig. 24-20.

Some military ships and submarines are powered by nuclear reactors. Heat from the nuclear reactor is used to turn water into steam. This steam is used to turn a turbine. The turbine turns the propeller using a shaft.

The propeller (called a "prop") is a device with a group of blades that radiate out from the center. Propellers can have from two to six blades. The propeller attaches to a shaft at the center. Propellers range in size from 2-3 inches (51-76 mm) to over 30 feet (9 m) in diameter. Each blade of the propeller is shaped to "bite" into the water much like a window fan bites into the air. The rotation of the propeller forces the water past it. A law of physics tells us that for every action there is an equal and opposite reaction. Therefore, the action of

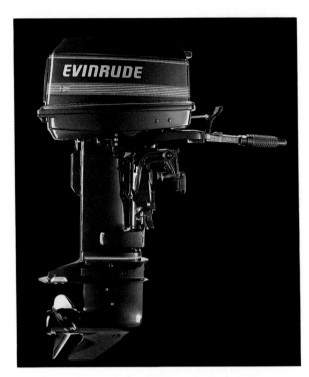

Fig. 24-20. Small boats can be powered by outboard engines. (OMC Power Products)

forcing the water through the spinning propeller causes an opposite reaction that pushes the boat forward.

However, not all boats are powered by engines. Some use wind power. One of the earliest sources of propulsion was the sail. A sail catches the wind and pushes the boat through the water, Fig. 24-21.

Fig. 24-19. Tugboats pull or push barges to carry large loads on inland waterways. (American Petroleum Institute)

Fig. 24-21. Sails are widely used on pleasure craft. Wind power was once the leading way to propel ships.

Maglev Train: A transportation vehicle that uses magnetism to suspend and propel itself along a guideway.

The rough ride and noise produced by traditional rail systems have been a drawback to their use for passenger travel. These drawbacks are caused by the steel wheels and rails used in most rail systems. Improvements in rails, wheels, and maintenance have made trains quieter. Some trains use rubber tires (such as monorails) or rubber inserts in a steel wheel. However, some noise is still produced because of the contact between the wheel and rail. In the 1960s a new land vehicle appeared that did not touch the rail. This type of vehicle is called a magnetically levitated train, commonly called a maglev. Maglevs use magnetic forces to support and move the vehicle along a special pathway called a guideway.

There are two types of maglevs: attraction and repulsion. An attraction maglev uses magnets pulling *toward* each other to support the train, Fig. 1. A repulsion maglev uses magnets pushing *away* from each other to support the train, Fig. 2. The guideway differs between the two types of maglev. Attraction maglevs wrap around the guideway. Repulsion maglevs sit in a trough-like guideway.

The maglev train includes a passenger compartment that provides a quiet and comfortable environment for the passengers. Beneath the passenger compartment is the suspension system. Attraction maglevs use electromagnets to support the train. The electromagnets pull

Fig. 2. Magnet *repulsion* can also be used to suspend a maglev. The MLU-002 maglev uses magnets pushing away from each other for suspension.
(Japan Railways Group)

toward a rail in the guideway, lifting the train. The amount of electricity is controlled so that the electromagnets and rail never touch. Repulsion maglevs use superconducting magnets to induce magnetic fields in coils in the guideway. The maglev floats on these magnetic fields.

Maglevs are propelled by a device called a linear induction motor, or LIM. A LIM works like a standard electric motor, using the principle that like poles of a magnet repel each other while unlike poles attract. A standard electric motor produces rotary motion. A LIM is laid out flat, in a straight line. Thus, a LIM produces *linear* or straight-line motion.

A LIM uses alternating electric current to create magnetic fields. Many times a second the direction of the electric current changes. When this happens, the lines of magnetic force collapse and change direction. This causes the rail to be attracted to the magnet and the vehicle moves along the rail. Then the current switches direction, causing the magnet to repel the rail. This cycle repeats, over and over, and pushes the maglev along the guideway.

Fig. 1. This is the Transrapid 06 maglev. It uses magnetic *attraction* (magnets pulling toward each other) to suspend the vehicle. (Transrapid International)

SUSPENSION

Designing water transportation vehicles is a unique challenge. They must be light enough to float. However, at the same time, a vessel must carry enough weight to make it economical to operate. A ship or barge floats due to a principle of physics called **buoyancy**. Buoyancy is the upward force exerted by a fluid on an object. A ship will be held up by a force equal to the weight of the water it displaces. This makes it easy to use a mathematical model to describe how an object will float. The formula we will use is:

$$\frac{\text{Submerged}}{\text{depth}} = \frac{\text{weight of object} + \text{weight of load}}{\text{density of fluid} \times \text{displaced fluid area}}$$

Let's look at an example, Fig. 24-22. A 10 ft. x 20 ft. raft is made from 12 inch thick planks of Douglas fir. The wood has a density of 28 lb. per cubic ft. Density is a measure of how much a material weighs per unit of volume. Thus, the weight of the raft is 5600 lb. [(10 x 20 x 1) x 28]. The raft will sink into the water until it displaces enough water to equal its weight. The depth can be calculated if you know the density of water: 62 lb. per cubic ft. for fresh water.

If we put the values into the equation: (the load is zero because the raft is empty).

$$\frac{5600 \text{ lb.} + 0}{62 \text{ lb./ft.}^3 \times 200 \text{ ft.}^2} = .45 \text{ ft.}$$

Multiply .45 ft. by 12 to get the answer in inches, and you get about 5.5 in.

The raft can carry a certain amount of cargo. Assume that you are willing to allow the top of the raft to be only 2 in. above the water. This will allow you to load additional weight. This can be calculated by determining the volume of the raft that can still go below the water. It is 5.5 in. into the water with its own weight. That leaves 6.5 in. above the surface. You want it to remain 2 in. above the surface after it is loaded. Therefore, it can sink an additional 4.5 inches. The additional wood will displace more water.

How much water will be displaced? To find this out, we need to determine the volume of the additional wood. Multiply the width, length, and thickness to learn the volume: 4.5/12 x 10 x 20 = 75 cubic ft. Multiply the volume by the density of water to find out the weight of the water displaced: 75 cubic ft. x 62 lb./cubic ft. The raft can hold 4650 lb.

You can increase the weight carrying power while using the same amount of material. A 20 ft. x 20 ft. barge with 10 ft. high sides and 2 in. thick walls can be built with the same amount of Douglas fir. The barge will displace a total of 248,000 lb. of water. It weighs 5600 lb. so it can safely carry 192,800 lb. and its sides will still be 2 ft. above the water.

Special Boats

Three special types of boats use unique suspension principles. The first is the hovercraft, Fig. 22-23. A **hovercraft** is suspended on a cushion of air. The air is forced by large fans into a cavity under the boat. As the air escapes this pocket, the boat is lifted above the water. Hovercraft are used over shallow water, swamps, marshy land, and where speed is important.

The second type of special boat is the hydrofoil. A **hydrofoil** has a normal hull and set of underwater "wings." These wings are called hydrofoils. A jet engine provides power for the boat. As the boat's speed in-

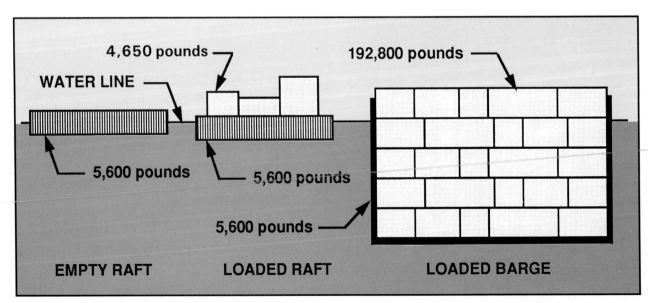

Fig. 24-22. Ships and barges suspend loads by displacing water. These drawings show how hollow hulls can increase load carrying capacity.

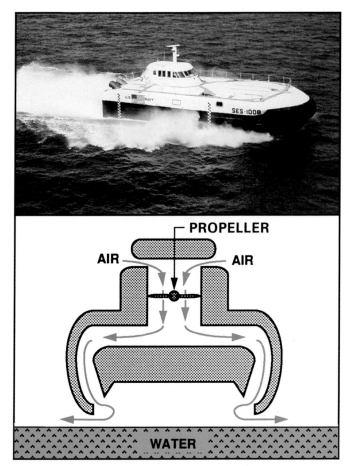

Fig. 24-23. Hovercraft are suspended above the water on a cushion of air.

creases, the water passing over the hydrofoils produces lift. The lift causes the hull to raise out of the water. The reduced friction between the water and the hull allows the boat to travel faster while using less fuel.

The third type of special boat is a **submersible** or submarine. This vessel can travel on the surface and underwater. Its buoyancy is adjusted by allowing water to enter or forcing it out of special tanks. As the tanks fill with water the vessel becomes heavier, or less buoyant. Therefore, the submarine sinks into the water. Compressed air can be used to force the water out of the tanks and increase the buoyancy. The submarine will then rise to the surface.

GUIDANCE AND CONTROL

The operator (called a "skipper" or "captain") of a ship is responsible for its course (path) and safety. He or she obtains guidance from a number of sources. A compass and the charts of rivers, harbors, and oceans are the basis for navigating ships of any kind. In or near a harbor, lighthouses and lighted buoys identify safe channels for navigation or mark hazards, Fig. 24-24. Flags and radio broadcasts communicate weather con-

ditions. Special electronic systems help to pinpoint the ship's location and indicate the depth of the waterway.

This information is used as the operator makes judgments about the best route. The path a ship follows is controlled in a number of ways, Fig. 24-25. Ships that have their own power source are generally guided with a **rudder**. This is a large flat plate at the stern of the ship. When it is turned away from the ship's course, it will deflect the water passing under the hull. This deflection forces the ship into a turn. Unlike most land vehicles, the back of the ship changes its path and causes the vehicle to turn.

Some large ships have two propellers or "twin screws." Increasing the speed of one propeller will create additional force on that side of the ship. The ship will then turn toward the side with the slower propeller.

Another way large ships can be turned, especially in docking, is with **bow thrusters**. This device is a propeller that is mounted at a right angle to the keel. The blades of the propeller can be adjusted to provide thrust in either direction. The ship will turn opposite the thrust.

Sailboats are controlled by a combination of rudder and sail. The rudder uses the force of the water to turn the boat. Sails use forces produced by the wind to aid the rudder in guiding the boat.

AIR TRANSPORTATION VEHICLES

The newest transportation vehicle that is in widescale use is the airplane. In the past 40 years the airplane has changed from a vehicle for the wealthy to an everyday transportation device.

Fig. 24-24. Safe channels are marked by lighted buoys and lighthouses.

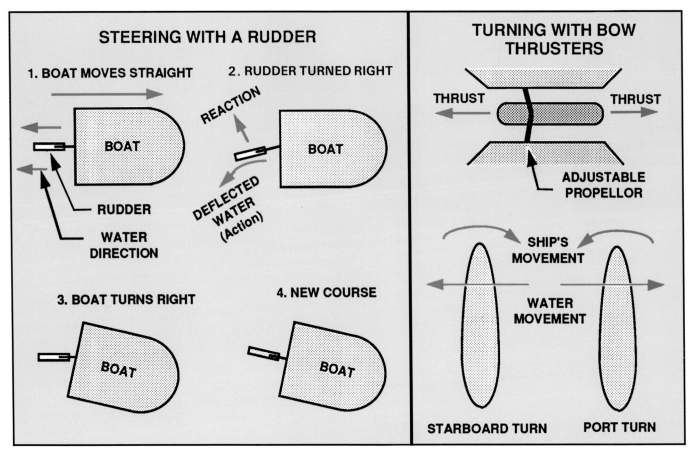

Fig. 24-25. Boats and ships use rudders to change their direction. Some large ships have bow thrusters to make small changes while docking.

Airplanes can be divided into two groups: general and commercial. **General aviation** is travel for pleasure or business on an aircraft owned by a person or a business. The airplane is not available for the use of the general public.

Commercial aviation includes the airplanes used by businesses that make money by transporting people and cargo. These companies include scheduled airlines, commuter airlines, air freight carriers, and overnight package companies.

VEHICLE STRUCTURE

There are three major types of air transportation vehicles, Fig. 24-26. They are lighter-than-air, fixed wing, and rotary wing aircraft.

Lighter-than-air Vehicles

Lighter-than-air vehicles use either a light gas (such as helium) or hot air to produce lift. These were the first air vehicles to be used by people. The earliest was a hot

Fig. 24-26. There are three types of aircraft: lighter-than-air (left), fixed wing (center), and rotary wing (right). (Goodyear, Grumman Corp., Bell Helicopter)

air balloon developed in France in 1783. This balloon, like present-day hot air balloons, was basically a fabric bag with an opening in the bottom. This opening allows warm air to enter the balloon and displace heavier, cold air. Since warm air is less dense than cold air, it rises to the top of the bag. When enough warm air has built up in the balloon, it rises off the ground.

More advanced lighter-than-air vehicles are blimps and dirigibles. A **blimp** is a non-rigid aircraft, meaning it has no frame. Its shape is determined by the envelope that is filled with a light gas, usually helium. Slung beneath the envelope is an operator and passenger compartment. Attached to the compartment is one or more engines to give the blimp forward motion.

Dirigibles are rigid airships with a metal frame that is covered with a skin of fabric. They used hydrogen gas to give maximum lift. Dirigibles were used for trans-Atlantic passenger service in the 1930s. However, since hydrogen is highly flammable, and after several disastrous accidents occurred, they ceased to be used.

Fixed Wing Aircraft

The first successful fixed wing aircraft was developed by the Wright Brothers. The *Flyer* was first flown in 1903. Later, Charles Lindberg made his historic trans-Atlantic flight in a fixed wing aircraft called the Spirit of St. Louis, Fig. 24-27.

Today most passenger and cargo aircraft are **fixed wing aircraft**. They all use similar structures. The flight crew, passenger, and cargo units are contained in a body called the **fuselage**. The lift necessary to fly is provided by one or more wings attached to the fuselage. A **tail assembly** provides steering capability for the aircraft.

Rotary Wing Aircraft

The helicopter is the most common **rotary wing aircraft** in use today. The craft has a body that contains the operating and cargo unit and encloses the engine. Above the engine is a set of blades. The blades are adjustable to provide both lift and forward and backward motion. The tail has a second, smaller rotor that keeps the body from spinning in response to the motion of the main rotor.

Helicopters are widely used by the military. In civilian life they are used for emergency transportation, law enforcement, and specialized applications. These special jobs include communication tower erection, high-rise building construction, flying to remote areas, and aerial logging.

PROPULSION

Aircraft use two major types of propulsion systems: propellers or jet engines. Smaller aircraft use propellers

Fig. 24-27. These historic aircraft, the Wright Flyer (top) and The Spirit of St. Louis (bottom), are on display at the Smithsonian Institution.

attached to internal combustion engines. The engine operates on the same principle as the automotive internal combustion engine. Some large aircraft use a variation of the jet engine to turn the propeller. This type of engine is called a turboprop.

Jet Engines

Most business and commercial aircraft are powered by one of three types of **jet engines**: turbojet, turbofan, and turbo prop, 24-28.

The first type to be used was the **turbojet** engine. This engine was developed during World War II. It operates by:

- Drawing air into the front of the engine.
- Compressing the air at the front section of the engine.
- Feeding the compressed air into a combustion chamber.
- Mixing fuel with the compressed air.
- Allowing the fuel/air mixture to ignite and rapidly burn.
- Causing the rapidly expanding hot gases to exit the rear of the engine.

The exiting hot gases serve two functions. The gases turn a turbine that operates the engine compressition

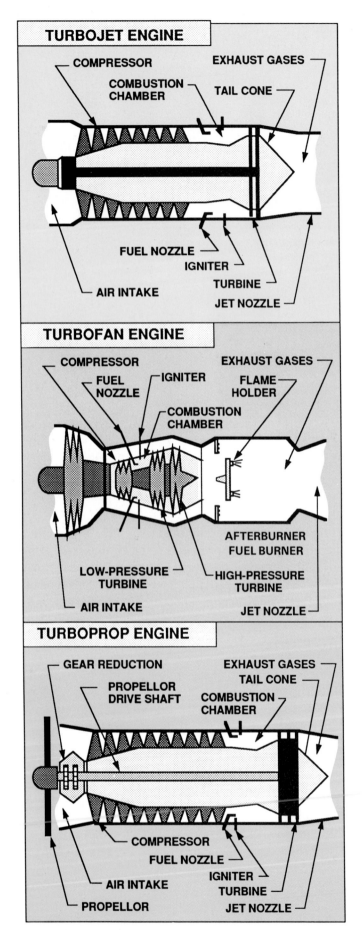

TURBOJET ENGINE

COMPRESSOR
COMBUSTION CHAMBER
EXHAUST GASES
TAIL CONE
FUEL NOZZLE
IGNITER
TURBINE
JET NOZZLE
AIR INTAKE

TURBOFAN ENGINE

COMPRESSOR
FUEL NOZZLE
IGNITER
EXHAUST GASES
FLAME HOLDER
COMBUSTION CHAMBER
AFTERBURNER FUEL BURNER
HIGH-PRESSURE TURBINE
JET NOZZLE
LOW-PRESSURE TURBINE
AIR INTAKE

TURBOPROP ENGINE

GEAR REDUCTION
PROPELLOR DRIVE SHAFT
EXHAUST GASES
TAIL CONE
COMBUSTION CHAMBER
COMPRESSOR
FUEL NOZZLE
IGNITER
TURBINE
JET NOZZLE
AIR INTAKE
PROPELLOR

Fig. 24-28. The are three types of jet engines in use today.

section. The gases also produce the thrust to move the aircraft.

The turbojet engine operates at high speeds and is used in military aircraft. Early commercial jet airliners also used these engines.

The **turbofan** engine is the engine of choice for most commercial aircraft in use today. The turbofan operates at lower speeds than a turbojet engine. Also, a turbofan uses less fuel to produce the same power. In this engine the turbine drives a fan at the front of the engine. The fan compresses the incoming air. The compressed air is then divided into two streams. One stream of air enters the compressor section. In this section the air is compressed further, fuel is injected, and the fuel-air mixture is ignited. The other stream of air flows around the combustion chamber. This stream is used to cool the engine and reduce noise. In the rear section of the engine, the exhaust gases and cool air mix. If necessary, additional fuel can be injected and ignited to provide additional thrust. This arrangement is called an afterburner.

A variation of the jet engine is the **turboprop** engine. It operates in the same manner as a turbojet engine. However, the turbine also drives a propeller that provides the thrust to move the aircraft. Turboprop engines operate more efficiently at low speeds than turbojet or turbofan engines. Therefore, they are widely used on commuter aircraft.

A new type of jet engine is being developed. It is called a propfan. Like a turboprop, it uses a jet engine. The propfan differs from a turboprop in that two propellers are driven. The propellers rotate in opposite directions. Propfan engines promise to be fuel efficient while operating at high speeds.

SUSPENSION

Air transportation vehicles are suspended in the atmosphere. This requires a knowledge of the principles of physics. All air transportation vehicles depend on the fact that an area with less dense air will allow heavier air to force the vehicle up. This force must be greater than the force of gravity that draws the craft toward the earth.

Lighter-than-air Vehicles

Lighter-than-air vehicles use air weight differences to cause the vehicle to rise and be suspended in the atmosphere. There are two ways to cause this difference in weight.

The first uses a closed envelope filled with a light gas. As you learned earlier, blimps use helium and dirigibles used hydrogen to generate lift. The vehicle is made to rise and lower much like a submarine. There are air tanks or bags inside the vehicle. The tanks are filled with outside air to make the aircraft heavier. Therefore, the aircraft will descend. Removing the air from the tanks or bags will make the craft light and it will rise.

The second system uses circulating, warm air in an open-ended envelope to cause the craft to ascend and descend. This craft is called a hot-air balloon, Fig. 24-29. The balloon has a burner suspended below its opening. To cause the balloon to rise the burner is ignited. The flame heats air to about 212°F (100°C) and it rises into the balloon. The warm, rising air displaces colder, heavier air. The warmer air is less dense and makes the balloon lighter-than-the air it displaces in the atmosphere. Therefore, it rises until its displaced air is equivalent to its weight. As the balloon rises it enters air that is less dense (lighter).

Also, the air in the balloon cools and becomes heavier. Therefore, the burner must continue to operate to keep the balloon flying. To descend to earth the burner is turned off. During flight the burner is used intermittently (turned on and off) to keep the balloon at a constant height. Running the burner all the time may cause the balloon to rise too high. Not running the burner often enough will allow the balloon to descend too low over the earth.

Fixed Wing Aircraft

The most common aircraft is the fixed wing airplane. An airplane has four major forces that affect its ability to fly. These forces, as shown in Fig. 24-30, are:
- Thrust: The force that causes the aircraft to move forward.
- Lift: The force that holds or lifts the craft in the air.

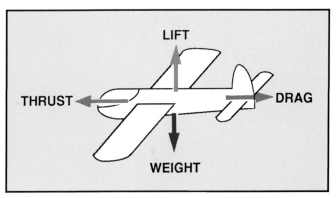

Fig. 24-30. There are four forces that affect flight.

- Drag: The air resistance force that opposes the vehicle's forward motion.
- Weight: The pull of gravity that causes the craft to descend.

Critical for all flight is **lift**. It is generated by air flowing over the wing of the aircraft. The **wing** is shaped to form an **airfoil**, Fig. 24-31. The wing separates the air into two streams. The airfoil shape causes the upper stream to move farther than the lower stream. Therefore, the upper stream speeds up. This increased speed causes a decrease in pressure. The high pressure air below the wing forces the wing up. This gives the plane the required lift.

The greater the slope of the upper surface of the wing, the greater the lift. However, drag is also increased because of the larger front profile of the wing. An airplane needs greater lift during take-off and landing. Devices called flaps on the leading and trailing edges of the wings are extended, Fig. 24-32. The flaps increase the lift, and are also used to slow the aircraft during landings.

Helicopter Suspension

The blades on a helicopter operate like any other airfoil. As they rotate, they generate lift. The angle of

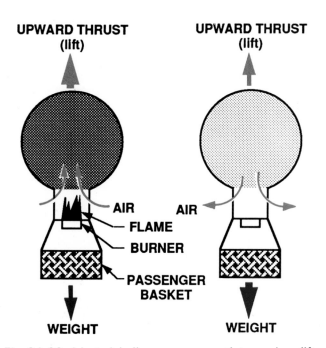

Fig. 24-29. A hot-air balloon uses warm air to produce lift.

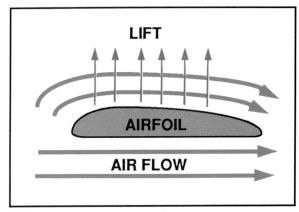

Fig. 24-31. Air flows a greater distance over the top of an airfoil and produces lift.

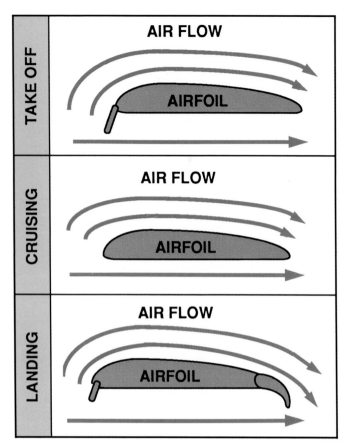

AIR FLOW

AIRFOIL

AIR FLOW

AIRFOIL

AIR FLOW

AIRFOIL

Fig. 24-32. These drawings show the airflow over the wing during takeoff, cruising, and landing. Devices called flaps are extended during takeoff and landing. The flaps increase the lift.

the blades is increased to generate lift. This will cause the helicopter to rise vertically. When the proper altitude is reached the pitch (angle) of the blades is changed. The pitch is adjusted so that the lift equals the weight. When the weight equals the lift, the helicopter will *hover*. Reducing the pitch will allow the helicopter to descend.

Forward flight is accomplished using a complex mechanism. This mechanism changes the pitch of the blades and tilts the blades as they rotate. In forward flight, the blade is tilted slightly forward. Also, the pitch of the blade is increased in the rear part of their travel. The pitch is reduced as they rotate in the front part of their travel. This causes the craft to have more lift in the back than in the front. Therefore, the helicopter leans slightly forward. The combination of the lift and thrust of the rotors causes the helicopter to move forward.

Backward flight is accomplished by reversing the tilt. The rotors are pitched the most in the front part of their travel. This causes the craft to tilt slightly backwards and travel in reverse. Helicopters can move left and right in the same way, by tilting the rotor and adjusting the pitch.

GUIDANCE AND CONTROL

Guidance for the pilot of a commercial aircraft comes in many forms. Ground personnel help the pilot bring the aircraft into the terminal safely, Fig. 24-33. Control tower personnel are in radio contact as the airplane taxies from the terminal and is cleared for take off. Once in the air, regional air traffic controllers take over. They monitor and direct its progress across their region, and hand the craft off to controllers in adjoining regional centers. General aviation aircraft do not normally use regional air traffic control centers. They use visual flight rules to govern their movements.

A number of instruments help the pilots monitor and control the aircraft, Fig. 24-34. These instruments are called **avionics**. This term is derived from *AVI*ation electr*ONICS*. It includes all electronic instruments and systems that provides navigation and operating data to the pilot. The amount of avionics an airplane carries depends on its size and cost. Large passenger airliners have a wide variety of avionics. A small general aviation airplane will have very basic equipment.

Avionics include a number of different functions and systems. They include:

- Communication systems: Short distance radio systems operate on frequencies between 118 and 135.975 MHz. These radios allow pilots to communicate with air traffic controllers and other people on the ground.
- Automatic pilot: Electronic and computer-based systems that monitor an aircraft's position and adjust the control surfaces to keep it on course.

Fig. 24-33. This ramp attendant is helping a pilot guide an aircraft to the jetway, a passenger walkway.

Fig. 24-34. This photo shows a passenger jet cockpit. Note the number of instruments. (Alaska Airlines)

Fig. 24-35. This photo shows a jet landing over runway approach lighting. (United Airlines)

- Instrument Landing System: This system (called ILS) helps airplanes to land in bad weather. ILS uses a series of radio beams to help the pilot guide the airplane to the runway. A wide, vertical beam helps the pilot stay on the center of the runway. A wide horizontal radio beam tells the pilot if the plane is above or below the correct approach path. Vertical beams serve as distance markers to tell the pilot how far away he is from the runway. Pilots in aircraft without ILS equipment use approach lighting to guide the plane in for landing, Fig. 24-35.
- Weather radar: Provides the pilots with up-to-date weather conditions in front of the aircraft.
- Navigation systems: Instruments used to guide the plane along its course and indicates air speed.
- Engine and flight instruments: These instruments give the pilots information about how the engine(s) are operating. Flight instruments tell the direction (in three degrees of freedom) the airplane is flying.

The information provided by avionics allows the pilot to properly control the flight of the aircraft.

SPACE TRANSPORTATION VEHICLES

Space travel may be the transportation system of the future. However, today it is in limited use. Space travel is restricted to conducting scientific experiments and placing communication, weather, and surveillance satellites into orbit.

TYPES OF SPACE TRAVEL

Space travel can be classified in two ways:
- Manned or unmanned.
- Earth-orbit or outer space.

Manned or Unmanned Flight

The first space flights were **unmanned**. These flights used rockets to place a payload into orbit. Typically, the payload was either a scientific experiment or a communication satellite, Fig. 24-36. Today, unmanned space vehicles continue to launch these and other payloads.

Fig. 24-36. Satellites are a typical payload for unmanned space flights.

A **manned space flight** carries human beings into space and returns them safely to the earth. The first manned space flight was flown on April 12, 1961, when a Soviet cosmonaut named Yuri Gagarin made one orbit of the earth. Manned space flight in the United States also started in 1961. Astronaut Alan Shepard completed a sub-orbital flight less than one month after Gagarin's flight. This was followed by John Glenn's first orbital flight in the Mercury program. The Mercury program was followed by the Gemini and Apollo programs that ended with the first man on the moon, Fig. 24-37. All these programs used capsules attached to the nose of a rocket to place humans into orbit or into outer space.

The latest manned flights are carried out by the Space Shuttle. The craft rockets into orbit, where its crew carries out experiments and places satellites into orbit, Fig. 24-38. At the end of the mission the shuttle returns to earth. The shuttle glides in for a landing like an airplane. After landing, the shuttle is prepared for its next flight.

Earth Orbit

The two types of space travel are earth orbit and outer space travel. **Earth orbit** travel is fulfilled by communication satellites and the space shuttle. They travel at about 18,000 miles per hour in an elliptical orbit around the earth. This orbit comes *closest* to earth at a point called the **perigee**. The *farthest* distance away from earth is called the **apogee**.

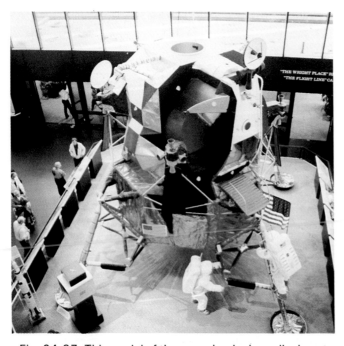

Fig. 24-37. This model of the moon lander is on display at the Smithsonian Institution.

Satellites can be placed into two types of orbits. One type allows the satellite to circle the earth, viewing a complete path around the globe. This type of orbit can be used for weather monitoring, geological and agricultural surveys, military surveillance, and other uses.

The second type of orbit is called **geosynchronous**. In this type of orbit the satellite travels the same speed the earth is turning. Therefore, the satellite stays over a single point on the globe. This type of orbit is used for communication and weather satellites.

Outer space travel places the spacecraft in earth orbit first. Then the speed of the spacecraft is increased to over 25,000 mph. At this speed the spacecraft is "thrown" into a path that will cause it to move out of the earth's gravitational field. The spacecraft can then travel to distant planets and out of the solar system. The Voyager probe is an example of this type of spacecraft. It moved out of earth orbit and traveled past a number of planets. Now it is traveling out of the solar system into what is called deep space.

VEHICLE STRUCTURE

A space vehicle and its launch systems have three major parts. These parts are: a rocket engine to place the vehicle or payload into space, a operating section, and a cargo and/or passenger compartment. As you have learned, early space programs integrated these components into a single launch vehicle or rocket. The engine was at the rear. The operating controls were placed above and around the engine. The cargo or passenger capsule formed the nose of the rocket.

The space shuttle is a newer type of space vehicle. It uses two distinct systems:
- Two solid fuel rockets strapped to the external fuel tank.
- The shuttle orbiter.

The solid fuel rockets lift the orbiter off the launch pad and give it initial acceleration. After they exhaust their fuel, they fall away. The solid fuel rockets parachute into the ocean and are recovered to be used again. At this point the three engines in the orbiter take over. They use the fuel in the external tank to place the shuttle into orbit. As it enters orbit, the external fuel tank falls away and burns up as it falls toward the earth.

The shuttle is about the size of a small airliner (such as a McDonnell Douglas MD-80)–122 feet (37 m) long with a 78 foot (24 m) wingspan. It is operated by its crew from the flight deck at the front of the orbiter. The middle of the shuttle is a large cargo bay. It can carry nearly 30 tons (27 metric tons) into orbit and almost 15 tons (13.5 metric tons) on re-entry.

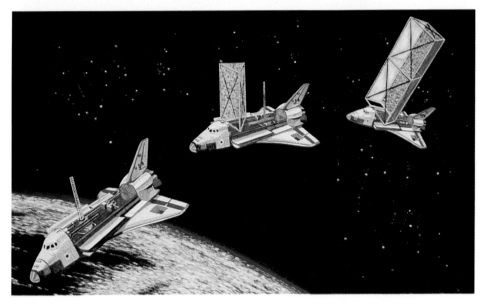

Fig. 24-38. This illustration is an artist's representation of a space shuttle experiment. (NASA)

PROPULSION

Space transportation depends on rocket propulsion systems. The rocket is based on Newton's third law of motion: for every action there is an equal and opposite reaction. Thus, rocket engines can be called *reaction* engines. The rocket applies this principle by:
- Burning fuel inside the engine.
- Allowing pressure from the exhaust gases to build up.
- Directing the pressurized gases out of an opening at one end of the engine.

This action produces motion that is in the opposite direction from the exiting gases. Have you ever blown up a balloon and allowed it to fly around as the air escapes? If so, you have seen the principle of a rocket engine.

The fuel used in rockets is either liquid or solid. Early rockets were developed in China and India before 1800. They used gunpowder as a solid fuel. The first modern liquid fuel rocket was designed and built by Robert Goddard in 1926.

Today's **liquid fuel rockets** have two tanks. One contains the fuel or propellant. The other contains oxygen– the oxidizer. The fuel and oxidizer are fed into the combustion changer. There they combine and burn to generate the thrust needed to propel the rocket and its payload into orbit.

Liquid fuel rockets have several distinct advantages. First, the amount thrust can be controlled by the amount of fuel and oxidizer fed into the engine. Second, a liquid fuel rocket engine can be used intermittently. It can be started and stopped several times during the flight. Finally, the rocket engine can be recovered and reused after the space flight.

Solid fuel rockets use a powder or a sponge-like mixture of fuel and oxidizer. Once the mixture is ignited, it burns without outside control. Therefore, the thrust cannot be changed. Also, the mixture will burn completely; it cannot be stopped once it is started.

AREAS OF OPERATION

Space craft and satellites operate in several regions. The lowest is called the **troposphere** which includes the first six miles (9.7 km) of space above the earth. General aviation and commuter aircraft operate in this region.

Above this region is the **stratosphere**. This region extends from 7 to 22 miles (11 to 35 km). Commercial and military jet aircraft operate in the lower part of this region. The upper part of the stratosphere is called the **ozone layer**. This layer absorbs much of the sun's ultraviolet radiation. Evidence of damage to this layer has caused great concern about global warming and the health of the planet.

The next layer is the **mesosphere**. This region extends from 22 to 50 miles (35 to 80 km) above the earth. The thermosphere or **ionosphere** lies just above the mesosphere. It ranges from 50 to 62 miles (80 to 99 km) above the earth. Many satellites operate in this layer of the atmosphere. The last layer is called the exosphere and blends directly into outer space.

SUMMARY

Transportation has been present as long as humans have lived on earth. It involves all actions that move people and goods from one place to another. Today there are four types of transportation: land, water, air, and space. Almost all these systems use vehicles to move the people or cargo. Each of these vehicles have structural, propulsion, suspension, guidance, and control systems.

WORDS TO KNOW

All of the following words have been used in this chapter. Do you know their meanings?

Airfoil
Apogee
Automatic transmission
Avionics
Ballast
Blimp
Bow thrusters
Buoyancy
Commercial aviation
Commercial ships
Container ships
Degrees of freedom
Direction control
Dirigibles
Dry cargo ships
Earth orbit
Fixed wing aircraft
Fuselage
General aviation
Geosynchronous orbit
Guidance system
Hovercraft
Hydrofoil
Inland waterway
Internal combustion engine
Ionosphere
Jet engine
Lift
Lighter-than-air vehicles
Liquid fuel rockets
Locomotive
Manned space flight
Maritime shipping
Mechanical transmission
Merchant ships
Mesosphere
Military ships
Outboard motor
Ozone layer
Passenger ships
Perigee
Pleasure craft
Power generation system
Power transmission
Propulsion system
Rotary wing aircraft
Rudder
Solid fuel rockets
Speed control

Stratosphere
Submersible
Suspension system
Tail assembly
Tankers
Tractor
Trailer
Troposphere
Turbofan
Turbojet
Turboprop
Unmanned flight
Vehicle
Wing

TEST YOUR KNOWLEDGE

Write your answers on a separate piece of paper. Please do not write in this book.

1. What is meant by degrees of freedom in vehicle control?
2. True or false. A tugboat serves the same purpose in the water as a locomotive does on land.
3. Match the vehicle on the left with the number of degrees of freedom it has:

    _____Automobile.   A. One degree
    _____Submarine.   of freedom.
    _____Train.   B. Two degrees
    _____Airplane.   of freedom.
    _____Elevator.   C. Three degrees
                            of freedom.

4. Ships that carry cargo in sealed steel boxes are called _____ ships.
5. True or false. Rocket engines burn kerosene.
6. List three types of power sources for boats and ships.
7. A boat that rides on a cushion of air is called a _____.
8. True or false. Hydrofoils and airplanes are suspended using the same principle of physics.
9. List the three types of jet engines.
10. A "stationary" satellite is said to be in a _____ orbit.
11. List the two liquids that liquid fuel rockets use.

APPLYING YOUR KNOWLEDGE

1. Design a land vehicle that is powered by a rubber band that has all five vehicle systems. Analyze the vehicle and explain the features of each system.
2. Build and fly either a model airplane or a model rocket.
3. Design a cargo container for a raw egg that can withstand the impact of a 15 ft. (4.6 m) fall onto the floor.

25

OPERATING TRANSPORTATION SYSTEMS

After studying this chapter you will be able to:
- *Discuss how the speed of transportation has increased over time.*
- *Describe the factors to be considered when developing a transportation system.*
- *Explain the differences between personal and commercial transportation systems.*
- *Discuss how transportation modes interface into a system.*
- *List and describe the common elements of all transportation systems.*
- *Describe transportation routes.*
- *Understand transportation schedules.*
- *Discuss the parts of transportation terminals.*
- *Describe the types of and impacts of regulations on transportation systems.*
- *Tell the difference between domestic and international transportation.*

Walking – 1 mph
Wagon, Sailing ship – 2.5 mph
Early steamship – 10 mph
Automobile, Train – 50 mph
Propeller airplane – 200 mph
Jet airplane – 600 mph
Orbiting satellite – 18,000 mph

Fig. 25-1. The speed of transportation has increased throughout history. (NASA)

For each of us transportation is an important part of everyday life. When we walk we are using a common form of transportation. We may also use many different kinds of transportation technology. We use vehicles and systems to improve our ability to move from one place to another.

Technology has increased the speed of moving cargo and relocating people. Throughout history there has been a great change in the speed of travel, Fig. 25-1. In early times, we depended on our ability to walk. A human can walk at an average of one mile per hour (1.6 km/h) over a great distance. Later, animals pulled wagons and sails captured wind to push boats. These advancements increased the travel speed by nearly three times. The development of canals and steamships also made travel faster. The advent of the railroad multiplied this speed by twenty fold to 50 mph (80 km/h). Today, jet aircraft move people at about 600 mph (966 km/h), and satellites move at 18,000 mph (28, 968 km/h).

TYPES OF TRANSPORTATION

Most people combine two types of transportation. These, as shown in Fig. 25-2, are:
- Personal transportation.
- Commercial transportation.

Personal transportation is travel using a vehicle owned by one person. The most common personal transportation vehicle is the automobile. Other typical personal vehicles are bicycles, motorcycles, mopeds, and skateboards.

Personal transportation vehicles are very flexible. They allows us to travel when we want and can go almost anyplace. However, personal transportation (except for bicycles) is not always energy efficient. A car burns the same amount of fuel if one or five people ride in it. Quite often, we drive around alone in our cars. The passenger seats are empty and, therefore, are underutilized.

Most personal transportation systems use public routes. These pathways are built using public funds generated by taxes and fees. Most often these are the streets, roads, and highways that criss cross the nation.

Commercial transportation includes all the enterprises that move people and goods for money. These are the land, water, and air carriers that operate locally, nationally, and internationally. These carriers include taxi, bus, truck, rail, ship, and air transport companies. Most of these are profit-centered enterprises that use private funds to finance their initial operation. These companies are owned by people who expect to make a profit on their investment. However, some commercial transportation companies are government owned. These companies provide transportation services where there is little chance for a private company to make a profit. An example of this is public transit companies. They operate city and regional bus, rapid transit, and commuter rail services.

COMPONENTS OF A TRANSPORTATION SYSTEM

All commercial transportation systems share some common elements, Fig. 25-3. They all have:
- A vehicle.
- A route the vehicle travels from the origin to the destination.
- An established schedule for the movement of people and goods.
- Terminals at the origin and destination points of the system.

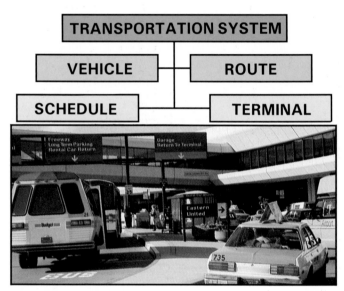

Fig. 25-3. All transportation systems have vehicles, routes, schedules, and terminals.

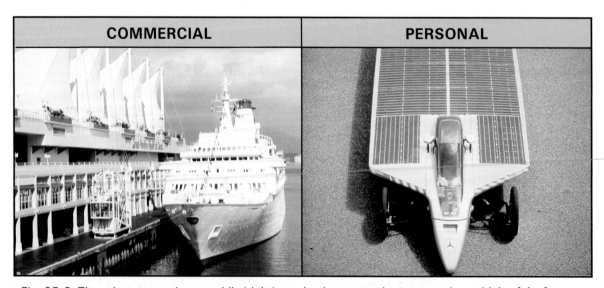

Fig. 25-2. The solar-powered automobile (right) may be the personal transportation vehicle of the future. The cruise ship (left) is an example of a commercial vehicle. (Daimler-Benz)

VEHICLES

You studied vehicles in Chapter 24. You learned that vehicles are powered carriers that support, move, and protect people and cargo. You also learned that all vehicles have structure, propulsion, suspension, guidance, and control systems.

TRANSPORTATION ROUTES

All transportation vehicles move from a point of origin to a destination. The path a vehicle follows is called a **route**. Personal transportation vehicles follow individual routes. These routes depend on the destination and the purpose of the trip. For example, a person hurrying to get to work may follow the most direct route or the one having the shortest travel time. However, a person on vacation may take a less direct, but more scenic route.

Commercial transportation vehicles often travel on *specific routes*. This means a vehicle follows one path from its origin to its destination, Fig. 25-4.

Typically, transportation systems are made up of many individual routes. These routes are designed to collect vehicle traffic. To understand this concept, consider the road system shown in Fig. 25-5. These roads are typical of those found in the midwestern U.S. In this part of the country there are often roads on each

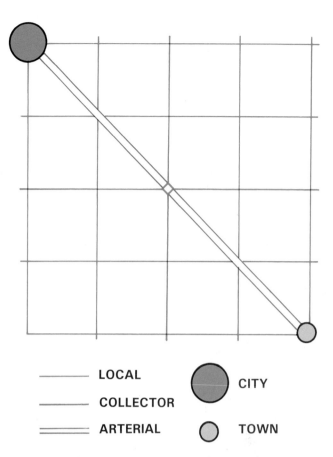

———————	LOCAL	⬤	CITY
———————	COLLECTOR		
═══════	ARTERIAL	◯	TOWN

Fig. 25-5. This map shows how the various parts of a highway transportation system feed into each other to provide local access and connections.

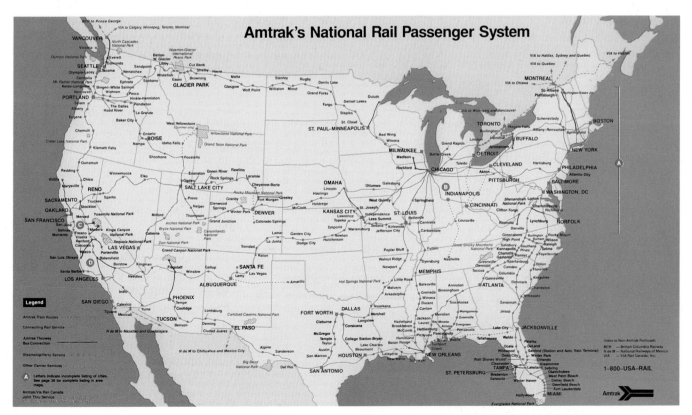

Fig. 25-4. This is a partial route map for a major railway. (Amtrak)

section line. A section is a one mile square piece of land. The smallest of these paths give access to individual parcels of land, homes, and farms. These paths may be dirt, gravel or paved roads. This series of local access roads feed into collector roads. Collector roads are larger pathways designed to carry more traffic. The collector roads feed into arterial roads. These are high speed highways that connect cities and towns. Some highways are *limited access* highways. On this type of highway, vehicles can only enter and leave at interchange points. Thus, the access to some land along these highways is limited. Larger U.S. highways and the interstate highway system are good examples of arterial highways.

Hub-and-Spoke Systems

Airlines use a route pattern called a **hub-and-spoke** system, Fig. 25-6. However, these routes are less apparent because they are pathways through the sky. A hub-and-spoke system is made up of small local airports and large regional airports. The local airports are on the end of two-way routes that radiate from the regional airport. Their routes appear to be spokes attached to a hub of a wheel.

The local airport serves as collecting and dispersing points for the system. The passengers normally arrive at a local airport using one or more transportation modes. They may arrive by personal car, taxi, shuttle van, bus, or train.

At scheduled times, flights leave the local airport and travel to the hub airport. Small- and medium-sized jet aircraft are used, or these routes are flown by small, commuter airlines.

The passengers arriving at the hub airport can transfer to other flights that travel to their destination. The simplest hub-and-spoke systems use one hub. For example, a wave of aircraft may arrive from points to the east and south of the hub. The same airplanes will then be used for flights to the west and north. Later, the flow is reversed with arrivals from the west and north and departures that travel to the east and south.

More complex systems tie two or more hubs together. Some passengers use one of the hub airports to travel within the region of the country it serves. Other passengers use both hubs. They fly into one hub then take a flight that connects it to a second hub across the country. The final flight segment will take them from the second hub to their destination. This system also allows the hub city to be served from a number of directions.

Often the passenger that transfers at a hub never leaves the airport. For example: O'Hare International Airport in Chicago is a major hub terminal. Chicago is also a major commercial and industrial city. Some passengers arriving at O'Hare will stay in Chicago for personal or business reasons. However, more than half of the 60+ million passengers that use the airport each year never step foot outside of the airport. They simply arrive on one flight and leave on another.

Airlines are not the only transportation systems using hub-and-spoke type systems. Trucking lines use small trucks to pick up and deliver freight. These vehicles work from a local terminal that collects and dispenses cargo for a city or local region. The freight may then move to a regional terminal where it is sorted for long distance transfer on large cross-country trucks. At

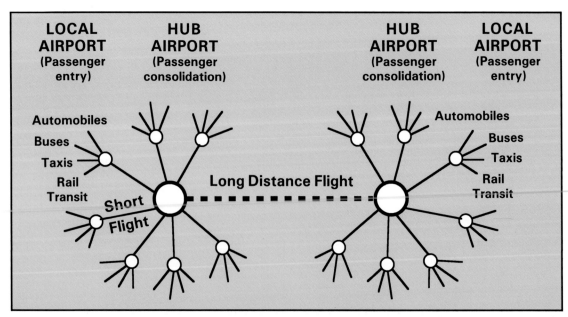

Fig. 25-6. This is a model of a typical hub-and-spoke airline system.

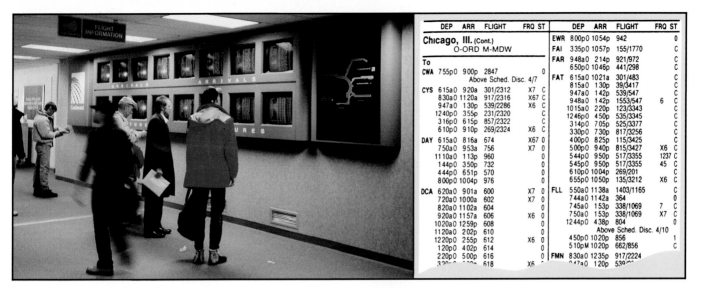

DEP	ARR	FLIGHT	FRQ	ST		DEP	ARR	FLIGHT	FRQ	ST
Chicago, Ill. (Cont.)						EWR	800p0	1054p	942		0
O-ORD M-MDW						FAI	335p0	1057p	155/1770		C
To						FAR	948a0	214p	921/972		C
CWA	755p0	900p	2847		0		650p0	1046p	441/298		C
		Above Sched. Disc. 4/7				FAT	615a0	1021a	301/483		C
CYS	615a0	920a	301/2312	X7	C		815a0	130p	39/3417		C
	830a0	1120a	917/2316	X67	C		947a0	142p	539/547		C
	947a0	130p	539/2286	X6	C		948a0	142p	1553/547	6	C
	1240p0	355p	231/2320		C		1015a0	220p	123/3343		C
	316p0	615p	857/2322		C		1246p0	450p	535/3345		C
	610p0	910p	269/2324	X6	C		314p0	705p	525/3377		C
DAY	615a0	816a	674	X67	0		330p0	730p	817/3256		C
	750a0	953a	756	X7	0		400p0	825p	115/3425		C
	1110a0	113p	960		0		500p0	940p	815/3427	X6	C
	144p0	350p	732		0		544p0	950p	517/3355	1237	C
	444p0	651p	570		0		545p0	950p	517/3355	45	C
	800p0	1004p	976		0		610p0	1004p	269/201		C
DCA	620a0	901a	600	X7	0		655p0	1050p	135/3212	X6	C
	720a0	1000a	602	X7	0	FLL	550a0	1138a	1403/1165		C
	820a0	1102a	604		0		744a0	1142a	364		0
	920a0	1157a	606	X6	0		745a0	153a	338/1069	7	C
	1020a0	1259p	608		0		750a0	153a	338/1069	X7	C
	1120a0	202p	610		0		1244p0	438p	804		0
	1220p0	255p	612	X6	0				Above Sched. Disc. 4/10		
	120p0	402p	614		0		450p0	1020p	856		1
	220p0	500p	616		0		510pM	1020p	662/856		C
	320p0	618		X6		FMN	830a0	1235p	917/2224		C
							947a0	120p	539/...		

Fig. 25-7. The schedule may be displayed on television monitors in a terminal (left) and contained in a printed schedule (right). The lower view is a portion of a flight schedule showing flights from Chicago to Wausau, WI (CWA), Dayton, OH (DAY), Washington, D.C. - National Airport (DCA), Denver (DEN), Newark (EWR), Fairbanks (FAI), Fargo, ND (FAR), Fresno, CA, (FAT) Fort Lauderdale (FLL) and Farmington, NM (FMN). (United Airlines, Bruce Kincheloe)

the other end the freight is separated into smaller shipments. They are delivered to local terminals and on to their final destination.

Parcel services like Federal Express and United Parcel Service also gather and group shipments using the hub-and-spoke principle. Likewise, many railroads and interstate bus systems use this principle.

TRANSPORTATION SCHEDULES

People travel for various reasons. They may be going to attend a conference, to visit other people, to participate in a business meeting, or to watch a sports event. Therefore, the arrival time is very important. Travelers will select a transportation mode by how long the trip takes and the expected arrival time.

This information is presented in transportation schedules. **Schedules** list the departure and arrival times for each trip. These schedules are available for all commercial transportation systems. These include freight hauling and passenger travel on bus, train, or airline systems.

Quite often the information is presented in printed form. Schedules for passenger travel are often displayed on television monitors or on schedule boards at the terminal, Fig. 25-7.

TRANSPORTATION TERMINALS

Terminals are important parts of commercial transportation systems. **Terminals** are where passengers and cargo are loaded onto and unloaded from vehicles. These structures are most often located at the origin and the destination of a transportation route.

Cargo systems have terminals that provide storage for goods awaiting shipment. They also have methods

Fig. 25-8. Two types of terminals are shown. At the left is a water transportation terminal which includes warehouses, storage yards, and a dock for berthing and loading/unloading ships. Right is a rail terminal. It includes a rail yard for sorting and storing rail cars, warehouses for product storage, and a loading/unloading area. (Norfolk Southern)

to load cargo onto transportation vehicles, Fig. 25-8. Finally, they have spaces for vehicles to stop as they are loaded and unloaded. These places may be a dock for ships, an apron for aircraft, or a loading dock on a warehouse for trucks and railcars.

Passenger terminals are more complex than cargo terminals. People need and demand more attention and comfort. Many terminals resemble small cities. They include at least six different areas, Fig. 25-9:

1. Passenger arrival areas that allow vehicles from other modes to unload passengers. Also, parking lots and bus shuttle services are provided at many large terminals.
2. Processing areas where passengers may purchase or validate tickets and check in luggage.
3. Passenger movement systems to allow passengers to move from ticket counters to loading areas. Often moving sidewalks, escalators, and automatic people movers are used.
4. Passenger service areas that provide food, gifts, and reading materials. Also, rest rooms and VIP (very important person) or frequent traveler lounges are available.
5. Passenger waiting areas where people may sit while waiting to board their vehicle.
6. Passenger boarding areas and mechanisms that allow people to board the vehicle.

Behind the scenes in all passenger terminals are many functions. Baggage is sorted for departing travel.

Baggage is also returned to arriving passengers. System managers direct the arrival, loading, unloading, and departure of vehicles. Operating crews have areas to prepare for trips and rest between assignments.

TRANSPORTING PEOPLE AND CARGO

You have been introduced to transportation vehicles, routes, schedules, and terminals. These things exist for only one reason: moving people and cargo from one point to another. This task requires three major actions as shown in Fig. 25-10. These are loading, moving, and unloading.

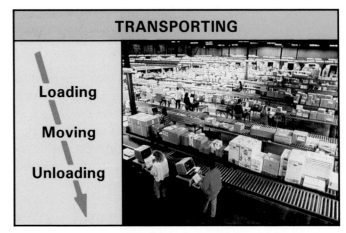

Fig. 25-10. Transportation involves loading, moving, and unloading people and goods.

Fig. 25-9. Passenger terminals serve a number of purposes.

Technology: How it Works

CANAL LOCK: A device that allows ships to change elevation as they travel through a canal.

Canals are a form of inland waterway transportation. Canals must have three features. First, there must be an adequate source of water to keep the canal full. The canal can be fed from a river or a reservoir. Second, there must be a way to cross ravines and streams. A "water bridge", called an *aqueduct*, is used. Third, there must be a method for raising or lowering a ship to different canal levels as the elevation of the land changes. This need is filled by a device called a lock, Fig. 1.

A canal lock is essentially a trough with a gate at each end. The operation of the lock is simple. Fig. 2 shows how a boat moves from a higher elevation to a lower elevation in the lock.

First, the lock is filled to the upstream water level. Then the upstream gate is opened and the boat enters the lock. Second, the upstream gate is closed to seal the lock. Third, sluice gates are opened in or around the lower gates. The sluice gates allow water to flow out of the lock until the level of water inside the lock is equal to the downstream level. Fourth, the downstream gates are opened, letting the boat sail out of the lock.

To move a boat upstream, the procedure is reversed. The boat enters the lock when the water level equals the downstream level. Then the lock is closed and the sluice gates are shut. The lock fills, lifting the boat to the upstream level. The upstream gate is opened and the boat sails out of the lock.

The St. Lawrence Seaway is an inland waterway that has several canal sections. The seaway rises more than 180 ft. (54 m) from Montreal to Lake Ontario through two lakes, three separate canals, and seven locks. Once the seaway reaches the Great Lakes, ships pass through Lake Ontario to the Welland Ship Canal. This canal has eight locks to raise ships to the Lake Erie level. The difference between the two lakes is about 330 ft. (100 m). The channel from Lake Erie to Lake Huron and onto Lake Michigan is the same elevation. No locks are needed. However, five more locks are required to raise ships 22 ft. (7 m) up to the level of Lake Superior.

Fig. 1. These large gates are part of a lock on the Illinois River. (Jack Klasey)

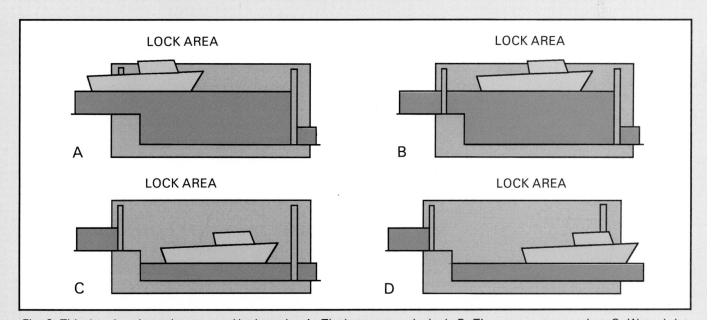

Fig. 2. This drawing shows how a canal lock works. A–The boat enters the lock. B–The upper gates are shut. C–Water is let out of the lock until the boat is even with the river downstream. D–The lower gates open and the boat leaves the lock.

| PERFORMING ROUTINE MAINTENANCE | SERVICING THE AIRCRAFT | LOADING PASSENGERS, LUGGAGE, AND CARGO |

Fig. 25-11. During the loading phase of a trip, passengers and cargo are loaded, the vehicle is serviced, and routine maintenance is performed. (United Airlines)

LOADING

Loading involves placing cargo onto or allowing people to board a vehicle. This involves checking bills of lading (shipping tickets) for cargo or passenger tickets. Then, the cargo must be loaded or the people allowed to board the vehicle. During loading, cargo must be secured so that it does not shift during the trip. Unwanted movement can cause damage to the freight.

People must be securely seated before a land or air trip can be started. Air transportation regulations, especially, require that all passengers be properly seated. They must have their seat belts securely fastened before the plane can leave the terminal. Likewise, seat belt use in automobiles is required in many states. Bus and rail transportation does not normally have this requirement.

At the same time that the people and cargo are loaded, a number of other tasks are completed, Fig. 25-11.

Fig. 25-12. This railroad employee is using computer controls to monitor part of a rail system. The computer allows him to signal engineers and conductors on trains. He can also open and close switches to reroute the train onto specific tracks. (Norfolk Southern)

Vehicles receive scheduled service. Fuel is added, oil levels are checked, drinking water and food is delivered, and wastes are removed. Also, routine maintenance and minor repairs may be completed.

MOVING

Moving people and cargo involves driving a vehicle, piloting a ship, or flying an airplane. However, this is only part of the task. It also involves monitoring and controlling the progress of the vehicle. This can be done by air traffic controllers talking to the pilot of an aircraft. It also involves railroad employees monitoring the progress of a train and switching it onto appropriate tracks, Fig. 25-12.

UNLOADING

Unloading passengers and cargo is the opposite of loading them. The passengers are allowed to get off the vehicle. Cargo is untied and lifted out of the vehicle.

Both people and cargo may enter the transportation system again by being loaded onto or allowed to board another vehicle. This may be a vehicle of the same or a different mode. For example, airline passengers may get off one plane and board another one at a hub airport. This is called "making a connection". Passengers may also leave the airport and board a bus or train, or rent an automobile.

MAINTAINING TRANSPORTATION SYSTEMS

Transportation systems are made up of a group of technological devices and structures. Each of these require maintenance. Typically, the roadways, rail lines, navigational structures, and terminals are maintained using construction knowledge. It was this knowledge that was used to build the components and, therefore, is needed to maintain and repair them.

STRUCTURE REPAIRS

Roadway and railroad track repair ensures that support structures meet the requirements of the system, Fig. 25-13. This may involve patching holes and resurfacing roads, or replacing rails and cross ties. Signals and signs must also be periodically checked, repaired, or replaced. Likewise, painted surfaces and stripes must be repainted.

Terminals also require repair of their functional parts. Their roofs will need to be periodically repaired. Exterior surfaces will need painting or residing. Entrances may need new steps and sidewalks. All these repairs ensure that the building can fulfill its function: to protect people and cargo.

However, appearance is also important to passengers. They want and expect a clean, bright, well maintained terminal. Therefore, walls may be repainted or wall coverings replaced. New carpeting and lighting fixtures may be installed. Seating may be repaired or replaced.

VEHICLE REPAIRS

Everyone wants to travel in comfort and in safety. This requires service and repair of vehicles and other equipment, Fig. 25-14. You'll remember from Chapter 18 that **maintenance** is the routine tasks performed to

Fig. 25-13. Transportation support structures must be kept in good repair. This highway is being repaved so it can last for many more years.

Fig. 25-14. Transportation vehicles must be maintained continuously and repaired when needed. (Alaska Airlines).

keep a vehicle operating properly. Generally, when a vehicle (bus, airplane, or train) reaches a terminal it receives daily service. This involves checking lubricants, adjusting controls and gages, and other simple tasks. More extensive servicing, including engine overhauls, are carried out on a fixed schedule. Look at an owner's manual for an automobile. It details the servicing required for each time or mileage anniversary of the vehicle.

Repairing is diagnosing technical problems, replacing worn or damaged parts, and testing the repaired vehicle. Minor repairs and parts replacement may take place in the field (at the terminal). Extensive repairs are generally completed at a vehicle repair and servicing center.

REGULATING TRANSPORTATION SYSTEMS

Transportation is considered essential for the general welfare of a country and its people. Therefore, it is subject to **regulation**, so that all people will be served. This regulation takes place on two levels. The first is **domestic transportation**. This type of transportation occurs within the geographic boundaries of one country. The other level is **international transportation**, which moves passengers and cargo between two nations.

In the United States, domestic transportation is regulated by local, state and Federal agencies. Cities may control the number of taxi cabs and the area they serve. Buses and rapid transit lines are often owned and operated by city or county governments. Therefore, these governments regulate routes and the level of bus service.

State agencies are responsible for building and maintaining the highway system. Their actions regulate their use, set speed limits, and control other operating conditions.

Much of the traffic in the U.S. and Canada crosses state or provincial lines. This type of traffic involves **interstate commerce**. This means that *business dealings* extend across state or provincial lines.

Therefore, in the U.S., the federal government plays an active regulatory role over much of the transportation industry. This is done by a number of agencies, including:

- Department of Transportation: coordinates federal programs and policies to promote fast and safe transportation. Two of the most important parts of the department are:
  - Federal Aviation Administration: Regulates airspace, air safety, and provides navigational aids. The FAA licenses pilots and determines the airworthiness of aircraft. The FAA also regulates the economics of air transportation.
  - Federal Highway Administration: Sets highway construction standards for federally financed roads and regulates the safety of interstate trucking.
- Interstate Commerce Commission: Regulates surface and inland waterway transportation between the states.
- Federal Maritime Commission: Regulates and promotes ocean shipping.

In addition, the National Transportation Safety Board promotes safe transportation and investigates serious accidents. The Environmental Protection Agency (EPA) sets controls on auto emissions, traffic control plans, and other transportation issues.

International transportation involves diplomacy. Diplomacy is the negotiations between countries that end up with agreements. These agreements may establish transportation routes between two countries, set **fares** (the cost of a ticket), and approve schedules. These actions often consider the needs of the passengers, but may attempt to protect one country's transportation companies.

SUMMARY

Transportation is vital to the welfare of any country. Moving people and cargo may be completed by personal, privately owned vehicles. However, they serve only the owner and cannot move large quantities of goods or large numbers of people. Therefore, commercial transportation systems have developed. These systems include vehicles, routes, schedules, and terminals. Each of these are regulated by various local, state, and Federal agencies. Their goal is to encourage the development of an integrated transportation system which serves the individuals and society as a whole.

WORDS TO KNOW

All of the following words have been used in this chapter. Do you know their meanings?

Commercial transportation
Domestic transportation
Fares
Hub-and-spoke system
International transportation
Interstate commerce
Maintenance
Personal transportation
Regulation
Repairing
Route
Schedule
Terminal

TEST YOUR KNOWLEDGE

Write your answers on a separate piece of paper. Please do not write in this book.

1. List the six major areas contained in passenger terminals.
2. True or false. As transportation systems have developed, the speed of travel has increased.
3. Match the type of transportation system on the left with the correct classification on the right:

    _____ United Airlines.   A. Personal.
    _____ Bicycle.   B. Commercial
    _____ Greyhound   (for profit).
            Bus System.   C. Commercial
    _____ Chicago Transit   (public).
            Authority buses.
    _____ Washington D.C. Metro
            Rail system.
    _____ Airport shuttle vans.
    _____ Family automobile.

4. The pathways within a transportation system are called _____.
5. True or false. Commercial transportation systems maintain specific schedules for the arrival and departure of vehicles.

6. Controls placed on transportation systems by governmental agencies are called _____.

7. The published cost of a ticket is called the _____.

APPLYING YOUR KNOWLEDGE

1. Obtain a rail, bus, or airline schedule. Use the schedule to plan a trip that will take you from your city to a major destination. List the departure time, arrival time, and travel time for each segment of the trip and for the entire trip.

2. Obtain a map and plot the routes for the trip you developed above.

3. Visit a local bus terminal or airport. Make a rough floor plan of the terminal. Indicate the various passenger areas.

4. Develop a drawing for a new terminal to replace the one you visited.

ACTIVITY 7A - DESIGN PROBLEM

Background:

Most transportation systems use a vehicle to move people and cargo to their destination. The vehicle must have a propulsion system to cause it to move along its pathway.

Situation:

You are employed as a designer for the Technology Kits Company. This company markets simple kits that are fun to assemble and teach basic principles of technology.

Challenge:

You have been chosen to design a transportation vehicle kit. You may:

1. Use air, land, or water as the medium of travel for the vehicle.
2. Use a rubber band, mousetrap, or a spring as the power source.
3. Use any material that would be commonly available in a craft store, such as tongue depressors, balsa strips, wheels, and dowels.

Your supervisor wants a working model, a parts list for the kit, and a set of assembly directions for the purchaser.

Optional:

- Design a package for the product.
- Design a catalog advertisement for the kit.

ACTIVITY 7B - FABRICATION PROBLEM

Background:

The internal combustion engine is the most popular source of power for land vehicles. These engines use petroleum-based fuels which, in the near future, will be in short supply. Internal combustion engines also produce emissions that pollute the environment. Developing alternative fuel systems for land vehicles is a major challenge that faces designers and engineers.

Challenge:

Build a land vehicle that can be powered by a propeller, Fig 7-A. Develop an electric power system for the vehicle. Then, test the performance of the vehicle. Finally, change the number of blades on the propeller.

Equipment:

One piece 1 1/2″ x 1 1/2″ x 11 1/2″ wood (pine, fir, spruce)
Four wheels (available from many hobby stores).
Two pieces 1/8″ diameter x 3″ welding rod (for use as the axles)
Two pieces 1 1/2″ long of soda straw
One low voltage electric motor (available from a hobby or electronics store).
One battery that matches the motor voltage.
12″ electrical hook-up wire
One each 4-blade and 3-blade propeller (available from a hobby store)

Procedure:
Building the Prototype

1. Obtain a block of wood that is 1 1/2″ x 1 1/2″ x 11 1/2″ long.
2. Lay out the block including:
   a. A taper from 3/4″ at one end to 1 1/2″ at the other end.

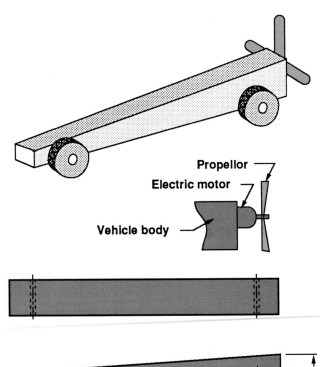

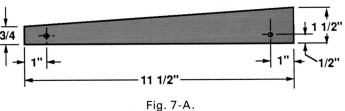

Fig. 7-A.

b. Axle holes that are 1″ from each end and 1/2″ above the bottom of the vehicle block.

3. Cut the block along the taper.

CAUTION: Follow all safety rules demonstrated by your teacher!

4. Drill two holes for the axles.

NOTE: The diameter of the holes should be the same as the outside diameter of the soda straw axle bearings.

5. Measure the diameter of the electric motor.

6. Drill a hole the same diameter of the electric motor into the tall end of the body.

NOTE: The depth of the hole should allow at least 1/2″ of the motor to extend out of the block. The vertical location hole is determined by the diameter of the propeller and the depth of the hole. The motor should be as high in the vehicle as possible.

7. Drill a hole from the top of the vehicle into the motor hole.

NOTE: This hole will allow the motor wires to be drawn out of the mounting hole.

8. Cut the axles to length.

9. Mount the axles and wheels onto the body.

Installing the Propulsion System

1. Feed the motor wires into the motor hole and out the wire escape hole.
2. Press the motor into the body.
3. Attach one motor wire to the battery.
4. Mount the battery onto the vehicle.
5. Attach the 3-blade propeller to the motor.

CAUTION: The propeller will now rotate when the motor spins. Be sure to keep your hands out of the way.

Testing the Vehicle

1. Attach the second wire to the motor.
2. Place the vehicle on the floor.
3. Test the distance it will travel in a given period of time.
4. Disconnect one wire from the motor.
5. Change the propeller to the 4-blade model.
6. Repeat steps 1 through 4.

Optional:

If the propellers are made of stamped metal, change the pitch (angle) of the blades. Test the propeller again to check the performance.

Applying Technology: Using Energy

SIEMENS

Without energy, technology cannot function.

26
CHAPTER

ENERGY: FOUNDATION OF TECHNOLOGY

After studying this chapter you will be able to:
- *Define energy.*
- *Describe the difference between potential and kinetic energy.*
- *Explain the differences between energy, work, and power.*
- *List and describe the six important forms of energy.*
- *Describe the importance of energy as an input to all technology systems.*
- *Describe exhaustible, renewable, and inexhaustible energy sources.*
- *Define biomass, biogas, and biofuels.*
- *Explain how energy technology can make our lives better, and how energy technology can cause damage.*

We use the word *energy* in different ways. We say "I don't have the energy to mow the lawn." We are worried about energy dependence on foreign petroleum. We hear people talking about energy conservation. However, not everyone using the word, energy, knows exactly what it means.

Energy comes from the Greek word *energeia* which means work. As time passed, the word came to describe the force that makes things move. Today, **energy** is defined as *the ability to do work*. This ability includes a broad spectrum of acts. Energy is used in simple human tasks, like walking. Energy can be obtained from petroleum, which is then used to power a ship across the ocean, Fig. 26-1. Energy can be used to provide motion in vehicles or machines. It can be used to produce heat or light. Energy is fundamental to our communication technologies. It is used in manufacturing products and constructing structures. Energy is everywhere and used by all of us.

ENERGY IS USED FOR...

HUMAN ACTIVITIES

TECHNOLOGICAL SYSTEMS

Fig. 26-1. Energy is used in all actions, from walking to powering complex technological devices.

TYPES OF ENERGY

There are two types of energy. Energy can be either associated with a force doing the work or associated with a force that has the *ability* of doing work. Energy that is involved in moving something is called **kinetic energy.** It is the energy in motion. A hammer striking a nail is an example of a technological act that uses kinetic energy. A sail capturing the wind to power a boat uses kinetic energy. A river carrying a boat or turning a water wheel are still other examples of kinetic energy.

However, not all energy is being used at any given time. Some energy is stored for later use. Energy, in this condition, has the ability, or potential, of doing work when it is needed. For example, water stored behind a hydroelectric dam possesses energy. It will release this energy to turn a turbine when it flows through the power generating plant. This stored energy is called **potential energy,** Fig. 26-2. A flashlight battery and a gallon of gasoline also contain potential energy.

ENERGY, WORK, AND POWER

It takes energy to do work. For example, you must eat well if you plan to run a marathon or hoe a garden. When you do one of these things, you might say, "I really worked hard!" But, what does this mean? In scientific terms, **work** is *applying a force that moves a mass a distance in the direction of the applied force.* You may have lifted boards off the floor and placed them on a table. This is work. The boards had mass and they were moved some distance.

MEASURING WORK

We measure work by combining the weight moved and the distance it was moved, Fig. 26-3. The result is a

Fig. 26-2. The coal being mined in this photo has potential energy. The energy will be released when the coal is burned. (AMAX)

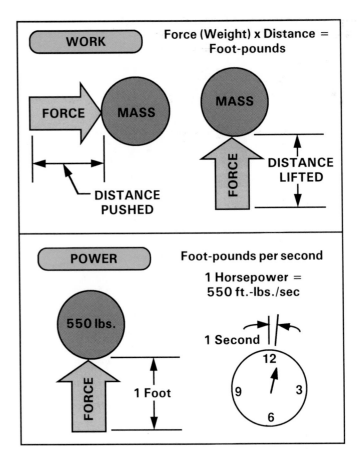

Fig. 26-3. Work is done when a force moves a mass over a distance. Work done per unit time is power.

measurement called **foot-pounds (ft.-lbs.).** This figure will tell you the amount of energy needed to move an object from one location to another.

Suppose you weigh 140 pounds. You plan on walking across a room that is 40 feet wide. You need 5600 ft.-lbs. of energy to complete the task. Likewise, lifting a 20 pound weight off the floor and placing it on top of a 36 in. high table requires 60 ft.-lbs. of energy. The amount of work completed can be measured with the formula:

Work (in foot-pounds) = Force or Weight (in pounds) x Distance (in feet)

In the metric system, work is measured in newtons per meter, or **joules (J).** The force or weight is measured in newtons, and the distance is measured in meters. The metric work formula is:

Work (in joules) = Force or Weight (in newtons) x Distance (in meters)

MEASURING POWER

Work is done in a context of time. Measuring the *rate at which work is done* gives you a term called **power** Fig. 26-3. Power can be calculated by dividing the work done by the time taken.

Power (in ft.-lbs./second) = Work done (in ft.-lbs.) / Time (in seconds)

The metric version is:

Power (in watts) = Work done (in joules) / Time (in seconds)

Two common power measurements are the horsepower and the kilowatt hour. **Horsepower** is used to describe the power output of many mechanical systems. One horsepower is the force that is needed to move 500 pounds one foot (500 ft.-lbs.) in one second. The factor of time is important to power. A motor that lifts 500 pounds in one minute can be smaller than one that lifts 500 pounds the same distance in one second. Likewise, the engine that moves a car from 0 to 60 mph in 7 seconds must be more powerful than one that does the same job in 9 seconds.

The term horsepower is used in several different ways. The theoretical, or *indicated horsepower,* is the rated horsepower of an engine or motor. This number suggests the maximum power that can be expected from the device under ideal operating conditions. Most often, this amount of power is not available from the device.

The *brake horsepower* is the power delivered at the rear of an engine operating under normal conditions. *Drawbar horsepower* is the power delivered to the hitch of tractors. *Frictional horsepower* is the power needed to overcome the internal friction of the technological device.

One **watt** is equal to one joule of work per second. One **kilowatt hour** is the work that 1000 watts will complete in one hour.

In electrical apparatus, the power consumed is determined by the resistance of the device. Its wattage rating is the product of the electrical current flowing through it and the voltage drop across it. The formula can be expressed as follows:

Power (in watts) = I (current in amperes) x E (electromotive force in volts)

FORMS OF ENERGY

Energy is everywhere we look. It is in the fires that burn coal and wood. Energy is in sunlight, wind, and moving water. In fact, we could not exist without the aid of energy.

The hundreds of examples of energy can be grouped into six major forms. These are, as seen in Fig. 26-4:

- Mechanical.
- Radiant.
- Chemical.
- Thermal.
- Electrical.
- Nuclear.

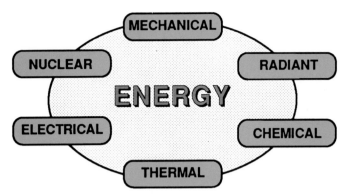

Fig. 26-4. There are six important forms of energy.

These energy sources do work that ends up as motion, light, or heat. They are used to power manufacturing machines, light buildings, propel vehicles, and produce communication messages.

MECHANICAL ENERGY

Most of us are familiar with **mechanical energy.** Often, it is produced by motion of technological devices. We associate machines with mechanical energy. This is correct, but it does not include all types of mechanical energy. Wind and moving water have motion and are, thus, sources of mechanical energy, Fig. 26-5.

Fig. 26-5. This waterfall is a natural source of mechanical energy.

Technology: How it Works

WIND TUNNEL: A device used to test the aerodynamics of vehicles and structures under controlled conditions.

Vehicles are affected by wind resistance as they move people and cargo. This wind resistance has an impact on the operating efficiency of the vehicle. In designing these vehicles, engineers often subject models to tests to maximize the efficiency of the design. One important test instrument in this quest is the wind tunnel.

Originally, this device was used solely to test airfoils during aircraft design activities. However, the uses of wind tunnels have been expanded. With current concern for fuel efficiency, wind tunnels are now used extensively to ensure that vehicles offer the least amount of wind resistance. Wind tunnels are also used to test the wind patterns over and around buildings as well as various other structures, Fig. 1.

Fig. 1. A wind tunnel is frequently used to test airflow patterns in industrial areas. (Colorado State University)

The wind tunnel is designed to pass high-speed air over a full-size or scale model of a vehicle or structure. An important design requirement is that a smooth and uniform flow of air be produced in the tunnel. To accomplish this, a number of wind tunnel designs have been produced. One of these is shown below, Fig. 2. Diagramed is a closed loop wind tunnel.

Most wind tunnels have a fan or turbine that develops an airflow in a large duct. The diameter of the duct increases as the air travels away from the fan. This reduces the airspeed as well as reducing frictional losses. The tunnels have mitered corners. This addition reduces the wind loss as the airflow changes directions. Also, in high-speed tunnel models, the air passes through cooling tubes to remove heat that the air gains while passing through the fan or turbine.

The air produced by the fan has a swirling pattern. This airflow creates unreliable test results. Therefore, the airflow passes through a wind smoothing unit. This unit is a series of tubes that remove the swirls and direct the air in a straight line.

As the air reaches the test chamber, the diameter of the tunnel shrinks rapidly. This causes the speed of the airflow to increase. The amount of the decrease in the tunnel's diameter controls the final airspeed in the test chamber.

Models in the chamber are carefully tested. Anemometers test the airspeed. Smoke might be introduced to visually observe the flow patterns of the air as it passes structures or vehicles. The vehicle itself might be attached to instruments to measure the lift and drag it develops as air passes.

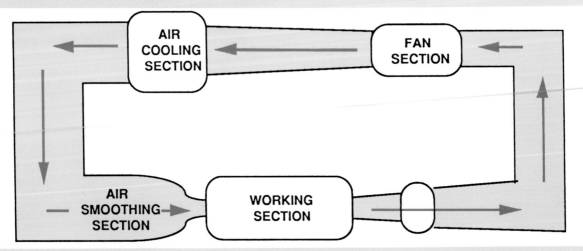

Fig. 2. This diagram shows the path the airflow follows in a common wind tunnel design.

RADIANT ENERGY

Radiant energy is the energy in the form of electromagnetic waves. You learned about these waves in Chapter 22. They extend from sound waves (long waves) to gamma rays (short waves). Cool objects give off longer waves than hot objects. Low frequency waves contain less energy than high frequency waves. Therefore, ultrasonic (sound wave) machining is slower than laser (concentrated light) machining.

The main source of radiant energy is the sun. Radiant energy may also be emitted by objects heated with a flame, or from a lightbulb in a lamp. Sometimes radiant energy is called light energy. This is not completely correct because many waves with wavelengths longer or shorter than the wavelengths of light possess radiant energy.

CHEMICAL ENERGY

Chemical energy is energy that is stored within a chemical substance. Typical sources of chemical energy are the fuels we use to power our technological machines. The most common are petroleum, natural gas, and coal. Wood, grains (such as corn), and biomass (organic garbage) are less-frequently used sources of chemical energy.

Chemical energy is released when a substance is put through a chemical reaction. This may be done by rapid oxidation (burning) or other chemical actions, such as digestion and reduction.

THERMAL ENERGY

Thermal energy is another name for heat energy. Thermal energy cannot be seen directly. But, you can see its effects by watching the heated air waves above a road on a very hot day. Thermal energy is, generally, felt. The energy strikes a surface, like your skin, and elevates its temperature.

Thermal energy is created by the internal movement of atoms in a substance. These particles are always in motion. If the atoms move or vibrate rapidly, they give off heat, or thermal energy. The faster they move, the more heat they give off.

Heat energy is widely used in technological devices. It provides the energy for our heating systems and some electrical generating plants.

ELECTRICAL ENERGY

Electrical energy is associated with electrons moving along a conductor. This conductor may be a wire in human-developed electrical system. The conductor could also be the air, as with lightning. Lightning is a natural source of electrical energy. Electrical energy is used as a basic source for other forms of energy. It is often converted into heat energy, to warm buildings, and into light energy, to illuminate our homes.

NUCLEAR ENERGY

Nuclear energy is associated with the internal bonds of atoms. When atoms are split, they release vast quantities of energy. This process is called **fission**. Likewise, combining two atoms into a new, larger atom releases large amounts of energy. This process is called **fusion**.

ENERGY IS INTERRELATED

All these forms of energy are related to one another. Radiant energy can be used to produce heat. If you have ever been sunburned, you have experienced this relationship. The *radiant* energy of the sun *heated* your skin until it burned.

A fire causes fuel to undergo a chemical action. For example, coal may be changed into carbon dioxide and water. In the process of this chemical action, heat is given off, Fig. 26-6.

The mechanical motion of an electrical generator causes magnetic lines of force to cut across any nearby conductor. This process induces an electrical current in the conductor.

Fig. 26-6. This person is burning fuel (wood) to generate thermal (heat) energy to cook food.

SOURCES OF ENERGY

Energy is a basic input to all technological systems. All energy comes in one of three basic types of resources. As shown in Fig. 26-7, these are:

- Exhaustible resources.
- Renewable resources.
- Inexhaustible resources.

EXHAUSTIBLE RESOURCES

Exhaustible energy resources are those materials that cannot be replaced. Once they are used up, we will no longer have that source. The most common exhaustible resources are petroleum, natural gas, and coal. These resources are called **fossil fuels.** They originated from living matter. Millions of years ago plant and animal matter were buried under the earth. Over time, this matter was subjected to pressure, and it decayed. This resulted in deposits of solid (coal and peat), liquid (petroleum), and gaseous (natural gas) fuels. These deposits have been found in many locations on the earth. Chapter 14 described how the deposits are located and the fuels extracted.

Uranium is another exhaustible energy source. It is an element that developed when the solar system came into being. Uranium is a radioactive mineral that is used in nuclear power plants.

RENEWABLE RESOURCES

Renewable energy resources are biological materials that can be grown and harvested. Their supply is directly affected by human propagation, growing, and harvesting activities. These activities can be improved by practices known as **bio-related technology** or **biotechnology.** These activities improve the types and the quantities of resources that are grown. One activity is gene modification, which changes the structure of the organism. The new plant or animal may have improved growth characteristics or reduced susceptibility to disease and parasites. Another activity involves careful management and land use practices to ensure a steady supply of the renewable resources.

The most common renewable energy resources are wood and grains. They can be burned directly to generate thermal energy. Corn is often converted to alcohol (ethanol), which then can be used as a fuel.

Organic matter such as garbage, sewage, straw, animal waste, and other waste can be an energy resource. They are often referred to as **biomass** resources. The *bio* means they have a biological, or living, origin. The resources can be traced back to plant or animal matter.

Fig. 26-7. The three types of energy resources. A – Oil - an exhaustible resource. B – Wood - a renewable resource. C – Wind - an inexhaustible resource. (Marathon, U.S. Department of Energy)

These organic materials can be burned directly as **biofuels.** Also, these materials can be converted into methane, a highly flammable gas. This process generates a **biogas,** which can replace some exhaustible fuel resources.

INEXHAUSTIBLE RESOURCES

Inexhaustible energy resources are part of the **solar weather system** that exists on earth. This natural cycle starts with solar energy. About one-third of the solar energy that reaches the earth's atmosphere is reflected back into space. The other two-thirds enters the atmosphere. Much of this solar energy is absorbed by the atmosphere.

About 23 percent of this energy powers what can be called a **water cycle,** Fig. 26-8. A very small portion of the earth's water is in rivers and lakes. The vast majority is in the oceans that cover much of the globe. The solar energy causes the water in the oceans to heat and evaporate. The warm water vapor rises into the atmosphere and forms clouds. The clouds rise and are carried inland by the wind. As the clouds travel upward, the water vapor cools. This cooling effect condenses the water vapor into droplets. The droplets fall to the earth in the form of rain or snow. Much of the water runs off the land and collects in rivers. From there the water flows into the ocean where it begins the cycle once again. Not all the water follows this exact pattern.

Some quantity of water is used by plants in their respiratory cycle. Some water ends up in lakes, from which it evaporates into the air to join clouds. Other portions flow underground into rivers and oceans.

Solar energy also heats the land, but the heating effects on the oceans and the land are very different. Since different amounts of solar energy strike various areas of the globe, temperature differences are created. Warmer air rises and is replaced by cooler air. This air movement is called wind.

The *water cycle, winds,* and *direct solar energy* become inexhaustible energy resources. They produce energy through hydroelectric power generators (water), wind generators, and solar converters.

Another inexhaustible energy resource is geothermal energy. The geysers found in Yellowstone National Park and other locations are examples of this energy resource, Fig. 26-9. This energy source comes up from the earth (geo) in the form of heat (thermal). It uses water that has been heated by the hot core of the earth. Geothermal energy is, generally, tapped by wells and used to heat buildings or power electrical generators.

ENERGY TECHNOLOGY: HELP OR HARM

As you learned, the world is a large energy system. Fuels and biomatter burn. This energy produces the power that drives our society. Yet, it also places

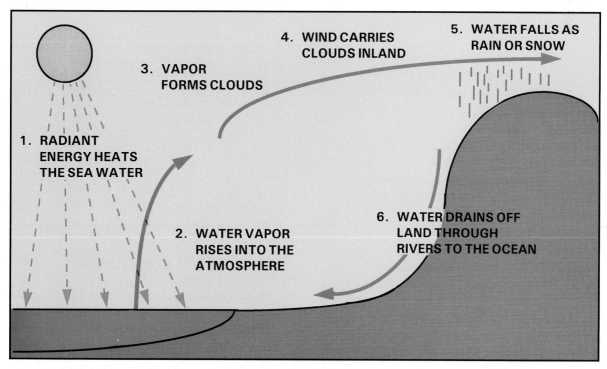

Fig. 26-8. The sun's radiant energy powers the earth's water cycle. The water cycle provides an inexhaustible energy source.

Fig. 26-9. This geyser is produced by geothermal energy.

into steam. The steam is passed through turbines, which turn generators (mechanical energy) to produce electricity (electrical energy). The electricity is later used to power motors (mechanical energy), to heat buildings (thermal energy), and to light rooms (radiant energy). This is an example of the type of energy conversion that is the focus of the next chapter in this book.

WORDS TO KNOW

All of the following words have been used in this chapter. Do you know their meanings?

Biofuel
Biogas
Biomass
Biotechnology
Chemical energy
Electrical energy
Energy
Exhaustible energy resources
Fission
Foot-pound
Fossil fuel
Fusion
Horsepower
Inexhaustible energy resources
Joule
Kilowatt hour
Kinetic energy
Mechanical energy
Nuclear energy
Potential energy
Power
Radiant energy
Renewable energy resources
Solar weather system
Thermal energy
Water cycle
Watt
Work

dangerous pollutants into the air, land, and water. Nuclear power provides a much longer-lasting source of energy. These power plants do not leak the toxic gases into the atmosphere that are produced by chemical-consuming power plants. Yet, the waste from nuclear power is more dangerous and longer-lasting. People must look at both the good and bad effects of technology. Properly used, energy allows us to produce the goods needed to survive and grow. Improperly used, technology can waste energy resources and create human suffering.

SUMMARY

Energy is a basic need for all technological activities. It is the foundation for power generation and work. Energy takes the forms of mechanical, thermal, electrical, chemical, radiant, and nuclear energy. These energy forms are derived from exhaustible, renewable, and inexhaustible resources. A major challenge facing society is to shift our use of exhaustible sources to renewable and inexhaustible sources.

Science tells us that energy can neither be created nor destroyed. However, much of human effort is devoted to converting energy in one form to another. We use burning fuels (thermal energy) to change water

TEST YOUR KNOWLEDGE

Write your answers on a separate piece of paper. Please do not write in this book.
1. Define energy.
2. True or false. Work measures the force of moving an object some distance over a period of time.
3. List the two types of energy.
4. True or false. Power is expressed in foot-pounds per second.

5. Two measurements of power are _____ and _____ .
6. List the six forms of energy.
7. Match each resource on the left with its correct classification on the right.

_____ Petroleum.   A. Exhaustible.

_____ Wind.   B. Renewable.

_____ Wood.   C. Inexhaustible.

_____ Falling water.

_____ Coal.

_____ Biomass materials.

_____ Natural gas.

_____ Sunshine.

_____ Corn.

APPLYING YOUR KNOWLEDGE

1. Construct a simple device to change wind energy into rotating mechanical motion.
2. Identify a renewable energy resource. Describe it in terms of the factors shown in the chart to the right. (Prepare a similar chart for your chosen resource.)

Energy resource:
Exhaustible resources it can replace:
Present location of the resource:
Advantages of using the resource:
Problems associated with producing the resource:
Problems associated with using the resource:

3. List and describe the types of energy resources you use in one day. Write how you could reduce your use of each of these resources.

This automobile uses natural gas as its fuel. Fueling stations, such as this one in Denver, Colo., are being built to serve the needs of cars that burn alternative fuels. (The Coastal Corp.)

ENERGY CONVERSION SYSTEMS

After studying this chapter you will be able to:
- *Describe early devices that converted inexhaustible energy into mechanical motion.*
- *Describe a common geothermal energy conversion system.*
- *List and describe the main ways solar energy is converted into other forms of energy.*
- *Explain the differences between passive and active solar conversion systems.*
- *Explain the operation of a common biomass converter.*
- *Describe heat engines in terms of energy conversion.*
- *Explain the differences between internal and external combustion engines.*
- *Describe common ways to heat homes and buildings.*
- *Diagram and describe the major parts of an electric energy generation and conversion system.*
- *Describe the common energy input systems for electric generation plants.*
- *Explain how an electric generator operates.*

THE ROLE OF AN ENERGY CONVERTER

Science tells us that *energy can neither be created nor destroyed.* However, a great deal of human action is devoted to converting energy from one form into another form. For example, we burn fuels to change water into steam, which contains energy in the form of heat. The steam may then be passed through devices used to warm rooms or dry lumber.

As you can see in this example, energy converters should be viewed as part of a larger system. An energy converter is a unique device. A converter has energy as its input and its output. Mechanical energy is the input to a turbine in a hydroelectric generator. Electricity, (electrical energy), is the output. This same electricity can be the *input* to several other energy converters. Incandescent lamps convert electrical energy into light (radiant energy). Motors convert electrical energy into rotary motion (mechanical energy). Resistance heaters change electrical energy into heat (thermal energy).

Your body is an energy converter. It converts food (your fuel) into energy, which moves muscles allowing you to walk, talk, and see. Likewise, an automobile engine is an energy converter. The engine converts the potential energy in gasoline into heat energy to produce mechanical motion. Energy converters power our factories, propel our transportation vehicles, heat and light our homes, and help produce our communication messages, Fig. 27-1.

Humans have developed hundreds of energy converters to meet their needs. In this chapter you will explore four broad categories of energy conversion systems:
- Inexhaustible energy converters.
- Renewable energy converters.
- Thermal energy converters.
- Electrical energy converters.

INEXHAUSTIBLE ENERGY CONVERTERS

The earliest energy conversion technologies were designed to power simple devices. These devices fall into a category that mechanical engineers call **prime movers.** A prime mover is any device that changes a natural source of energy into mechanical power.

Fig. 27-1. Energy is the foundation for technology. Energy powers our machines (top) and carries us across long distances (bottom).

Most early prime movers used inexhaustible energy sources. Specifically, they used wind and water power. Almost all societies used energy converters in transportation. Wind and flowing water helped move their boats.

However, there were several uses for energy converters. On land, two important technological devices were developed to harness these forces. They were windmills and waterwheels.

These two devices convert natural mechanical energy (flowing air or running water) into controlled mechanical energy (rotary motion). For example, they can produce the motion needed to power a water pump or an electric generator.

Other important converters use solar, geothermal, and ocean energy. These converters can be used to produce energy needed to heat and light our homes or power other technological devices.

WIND ENERGY CONVERSION

The sun is the original source of most of the energy on the earth. This energy is stored in growing plants and animals and in decayed organic matter–peat, coal, natural gas, and petroleum.

The sun, also, causes the winds to blow all over the earth. These air currents are produced by unequal heating of the earth's surface. Each day the sun's rays heat the land and water masses they touch. But, not all areas are touched at the same time or with equal energy. Polar areas receive less solar energy than do areas near the equator. Areas under cloud cover receive less solar heating than do areas in direct sunlight.

The heat from the land and water warms the air above it. The warm air rises, and cooler air moves in to replace it. This movement produces air currents we call wind.

The air above the hot areas near the equator is always rising. The cooler polar air moves toward the equator. In addition, air above water heats and cools more rapidly than air over land. Thus, during the morning hours, the air above the water warms quickly and rises. Cool air from the land moves in to replace it. In the evenings, the air above the land stays warmer longer than the air above the water. The cooler sea air moves inland to replace the rising land air.

Early humans designed technological devices to use these air currents. An early use of wind power was the

Fig. 27-2. Sails were one of the earliest technological devices to harness wind power.

ship's sail, Fig. 27-2. This device was developed in Egypt around 12,000 years ago. Sails remained the primary power for ships until the development of the steam engine in the late 1700s.

Wind power, the principle of the sail, was adapted to land applications with the development of the **windmill.** Its first use was, probably, in the Middle East around 200 B.C., Fig. 27-3. These mills were used for grinding grains into flour. This small start developed over the years leading to today's windmills and turbines. The modern windmill is primarily used to pump water for livestock on large western cattle ranches. The wind turbine is used to power electric generators.

Windmill and wind turbine designs can be grouped into two classes: vertical axis and horizontal axis, Fig. 27-4. The horizontal axis design has one or more blades connected to a horizontal shaft. The wind flows over the blades, causing them to turn. To see this action, place a household cooling fan in front of a blast of air. The blades will turn even though the fan's power is off.

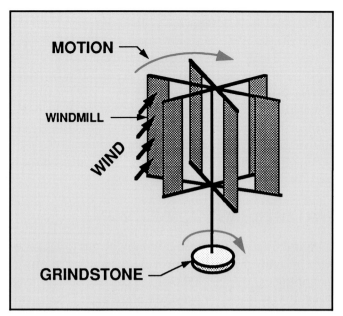

Fig. 27-3. This drawing is of a design thought to be used in ancient windmills. The windmill harnessed the wind for ease in grinding.

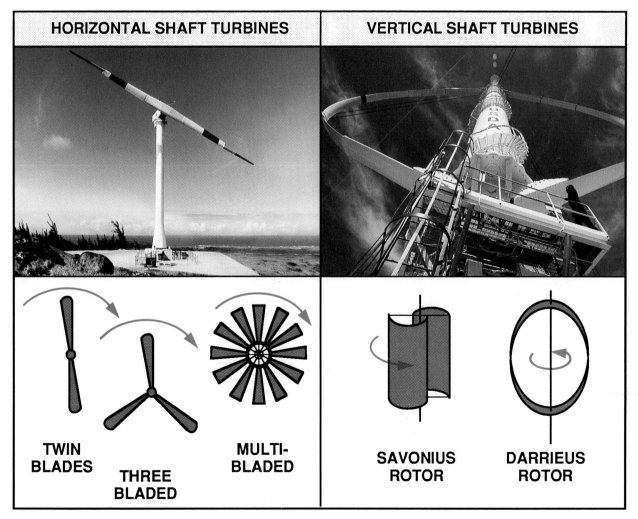

Fig. 27-4. Shown are the two classes of wind turbines. On the left are examples of wind turbines with their axes parallel with the air flow (horizontal shaft). On the right are examples with their axes at right angles to the wind (vertical shaft). (U.S. Department of Energy)

Vertical shaft designs have blades arranged around the shaft. As the wind blows past the blades, torque (turning force) is generated, rotating the shaft. The most common vertical shaft devices are the Darrieus and Savonius wind turbines.

Currently, a great deal of experimentation is being done to develop efficient wind turbines that can power electric generators. Large numbers of these devices are grouped together in "wind farms" located in various parts in the western United States, Fig. 27-5.

WATER ENERGY CONVERSION

Windmills are powered by a moving gas (air). Likewise, moving *liquids* can provide the energy to power technological devices. One of the earliest devices used to capture this energy was the **waterwheel.** This device is, essentially, a series of paddles that extend outward from a shaft. The flowing water drives the paddles causing the wheel to rotate.

Fig. 27-5. A view of a windmill farm in California.

Waterwheels powered the first factories of the industrial revolution. The wheels were produced in two basic designs, undershot and overshot, Fig. 27-6. The undershot waterwheel is powered by water rushing under it. The overshot wheel is powered by water falling onto it from an overhead trough or pipe.

A modification of the waterwheel is the water turbine. The water turbine is a series of blades arranged around a shaft. As water passes through the turbine at a high speed, the blades spin the shaft. Water turbines are used widely to power electric generators in hydroelectric power plants.

The steam turbine is a very similar device, only it uses steam (hot water vapor) to drive the turbine. Coal, oil, natural gas, and nuclear power plants use steam turbines to drive their generators.

SOLAR ENERGY CONVERSION

A third inexhaustible energy converter is the **solar converter.** The solar converter uses the constant energy source of the sun. The sun generates 3.8×10^{20} MW (380 million million million megawatts) of power through internal nuclear fusion.

The need to conserve exhaustible energy resources has brought solar energy into consideration as a replacement resource. Most solar energy conversion systems have two major parts–a collection system and a storage system.

Solar energy is very intense before it reaches the earth. The energy could produce temperatures approaching 6000° K. However, as it reaches the earth's protective ozone layer, much of this energy is absorbed and heats our atmosphere. Additional energy is absorbed by water vapor in the air. Some solar energy reaches the earth and is available as an inexhaustible energy source. The amount of this energy that is available depends on the inclination (height above the

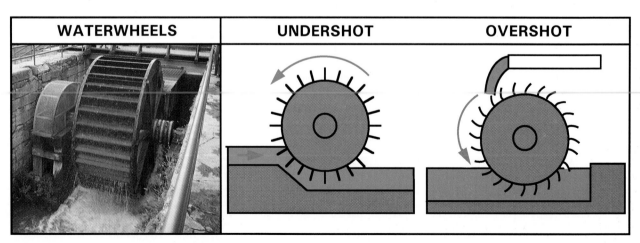

Fig. 27-6. Shown here are the two common types of waterwheels.

horizon) of the sun and the atmospheric conditions (cloud cover) over the earth. The term **insolation** is used to describe the solar energy available in a specific location at any given time. Insolation varies with the seasons and the weather. The maximum insolation on a clear, summer day is about 1000 MW per square kilometer (0.38 sq. mile).

Solar collectors depend on the principle that black surfaces absorb most of the solar energy that strikes them. This causes black surfaces to gather heat when they are exposed to sunlight.

Typical solar collectors can be grouped into two categories–passive collectors and active collectors.

Passive Collectors

Passive collectors directly collect, store, and distribute the heat they convert from solar energy. Actually, an entire house is a solar collector in a passive solar collection system. The building sits quietly in position as the sun heats it. This can be done in three ways as shown in Fig. 27-7. A **direct gain system** allows the radiant energy to enter the home through windows, heating inside surfaces.

An **indirect gain system** uses a black concrete or masonry wall (Trombe wall) that has glass panels in front of it, Fig. 27-7(B) . The wall has openings at its bottom and top. The wall heats up as the sunlight strikes its surface. In turn, the air between the wall and the glass panels becomes heated by energy radiating from the wall. The warm air rises and flows into the building through the openings at the top of the wall. This creates natural convection currents that draw cooler, heavier air into the openings at the bottom. This new air, in turn, is heated and rises.

The Trombe wall also retains a great deal of heat. Consequently, after the sun sets, the wall continues to radiate the heat and warms the air between the wall and the glass panels.

Adobe homes in the southwest part of the United States use a similar solar heating principle. The solar energy that strikes the adobe brick warms the surface. During the day the energy slowly penetrates the thick wall. This penetration takes about twelve hours. In the evening, the heat finally reaches the inside of the dwelling and provides warmth during the cool nights. By morning the wall is cool, and it insulates the rooms from the daytime heat. The twelve hour lag between heating and cooling makes a very effective daytime cooling and nighttime heating system for areas that have hot days and cool nights.

An **isolated system** uses solar collectors or greenhouses that are separate from the house. These collectors are built below the level of the house. The heat

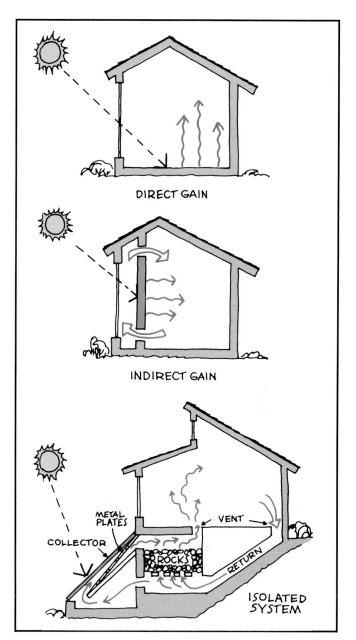

Fig. 27-7. Three types of passive solar systems that can be used to heat a home. A–Demonstrates the simplest system, direct gain. B–Shows how a Trombe wall, from the indirect system, works. C–Diagrams one type of isolated system. (PPG Industries)

generated in the collector can be channeled directly to heat the home. Additional heat is stored in a thermal mass (rock bed) for night and cloudy day heating.

Active Collectors

Active collectors use pumps to circulate the water that collects, stores, and distributes the heat that they convert from solar energy. These systems are used in many areas to provide hot water and room heat (space heating). There are two major types of active collector systems–direct and indirect. The typical **indirect active system** has a series of collectors. Each collector has a

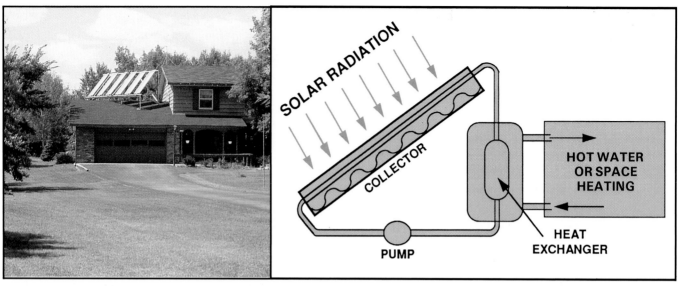

Fig. 27-8. The solar panels on the building at the left power a system similar to the one in the diagram. (U.S. Department of Energy)

black surface to absorb solar energy. Above or below this surface is a network of tubes or pipes. Water is circulated through these channels. As the fluid passes a warm black surface, it absorbs heat. Then, the warm water is pumped to a heat exchanger, where it can heat water for domestic use or provide thermal energy for a heating system, Fig. 27-8.

Direct active systems do not have a heat exchanger. The water that is circulated in the system is used as domestic hot water. The water flows directly to household faucets, washing machines, showers, etc.

A more recent use of active solar converters is in electric power generation. The converters produce steam that drives the turbines in the generation plant, Fig. 27-9.

Another type of solar converter is the **photovoltaic (solar) cell.** The cell can be a small device that powers a pocket calculator or part of an array of units that provide electricity for such devices as satellites and solar-powered vehicles, Fig. 27-10.

The cells are made of certain semiconductor materials like crystalline silicon. The cell is impacted by small bundles of light energy called photons. These photons strike the cell causing electrons to dislodge from the silicon wafer. The electrons move in one direction

Fig. 27-9. This large bank of solar collectors collects solar energy to produce steam for an electrical generating plant located in the Mojave desert in California.

Fig. 27-10. This experimental electric-powered car uses photovoltaic cells to convert solar energy into electricity. (Daimler-Benz)

across the wafer. A wire attached to the wafer provides a path for the electrons to enter an electrical circuit. A second wire allows electrons from the circuit to return to the wafer. This current flow supplies the power for each device.

GEOTHERMAL ENERGY CONVERSION

Geothermal energy is heat originating in the molten core of the earth. This energy can be found at great depths all over the planet. However, at certain locations it reaches the surface. The energy appears as volcanoes, hot springs, and geysers, Fig. 27-11. This energy is tapped in a number of ways. Electricity is readily produced using geothermal energy to form steam that drives a steam turbine powered generator.

Other applications use geothermal energy for direct heating. Geothermal heat pumps use the constant 55° F temperature of ground water to heat homes. Often the water is pumped from a well that extends into an underground aquifer. The water enters the heat pump, which removes heat from the water and transfers it to the dwelling. The cool water is returned to the aquifer through a second well. In Klamath Falls, Oregon, a

Fig. 27-11. The Old Faithful geyser in Yellowstone National Park is produced by geothermal energy.

number of buildings are heated and a section of highway is de-iced with hot water from geothermal wells. In Sandy, Utah, 14.6 acres of greenhouses are heated with geothermal energy.

OCEAN ENERGY CONVERSION

Ocean energy is an inexhaustible source that has only recently been considered a major source of energy for the coming generations. The oceans cover over 70 percent of the globe. They contain two important sources of energy, thermal and mechanical (wave and tide motion).

Ocean Thermal Energy Conversion (OTEC)

Ocean thermal energy conversion systems use the differences in temperature between the various depths of the ocean. The basic system has three steps. First, warm ocean water is used to evaporate a *working fluid*. Second, the vapors are fed into a turbine that turns an electrical generator. Finally, cold ocean water is used to condense the vapors to complete the energy conversion cycle. The process requires water with at least a 38° F difference. This occurs only at the equator.

Ocean Mechanical Energy Conversion

Ocean mechanical energy conversion devices use the mechanical energy in the oceans to generate power. Two sources of mechanical energy are tapped–wave energy and tidal energy.

Wave Energy Conversion. Presently, only two small wave generation plants exist in the world. However, several devices are being developed that show commercial promise. The first is a *mechanical surface follower,* which is used as a navigational aid. This device is one of the simplest designs. It is a buoy that floats in the water. Inside the device is a mechanism that uses the up and down movement of the buoy to ring a bell or blow a whistle.

The second system is a *pressure-activated device.* This is also a buoy that is used for navigational aid. The device uses the bobbing action created by the waves to compress air in a cylinder. As the water rises, it compresses the air. When the buoy falls, the compressed air is released, powering a small generator. The resulting electricity can power a navigational light.

Tidal Energy Conversion. Tidal energy devices use the difference between the height of the ocean at high tide and low tide to generate power. As the ocean rises, water is allowed to flow over a dam into a basin. As the tide recedes, the water flows back through turbines. This generates electricity, Fig. 27-12.

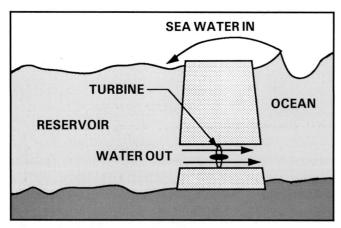

Fig. 27-12. This diagram demonstrates a tidal power ocean energy conversion system.

A system in France creates power from water flow in both directions. The rising tide drives the turbine in one direction. Later, the falling tide powers it in the other direction.

RENEWABLE ENERGY CONVERTERS

Our early ancestors depended heavily upon renewable energy resources. They burned wood and cattle dung to heat their homes and cook their food. In 1850, these energy sources provided 90 percent of our energy needs. In many parts of the world, these resources are still very important. However, in the United States today, they supply a very small percentage of our energy needs.

One source of renewable energy that is being actively considered as an alternate energy supply is **biomass resources.** Biomass resources are all the vegetable wastes and animal wastes that are generated through biological actions. Most biofuels come from three sources:
- Forest products industry–sawdust, bark, wood shavings, scrap lumber, paper, etc.
- Agriculture and food processing–corncobs, nut shells, fruit pits, grain hulls, sugar cane bagasse, manure, etc.
- Municipal waste–sewage and solid waste (garbage).

Bio energy conversion is completed using one of two basic processes. These are thermochemical conversion and biochemical conversion.

THERMOCHEMICAL CONVERSION

Thermochemical conversion produces a chemical reaction by applying heat. The most common method is *direct combustion.* The biofuel is burned to produce heat for buildings or to produce steam to power electric generation plants. This system is widely used in the forest products industry. Burning mill waste produces the steam that heats buildings, dries lumber, and operates the processing equipment.

A second thermochemical process is **pyrolysis.** In this process, the material is heated in the absence of oxygen. The heat causes the biofuel to form liquids, solids, or gases. The solids are carbon and ash. The liquids are much like petroleum and require further processing. The gases are flammable hydrocarbons. All these material can be directly or indirectly used as fuels. Other names used for this process are **liquidification** and **gasification.**

A third process is **liquefaction.** In this process, the biofuel is heated at moderate temperatures under high pressure. During heating, steam and carbon monoxide, or hydrogen and carbon monoxide, are present. A chemical action takes place that converts the material into an oil that has more oxygen than petroleum. This oil requires extensive refining to develop usable fuels.

BIOCHEMICAL CONVERSION

Biochemical processes use chemical reactions caused by fungi, enzymes, or other microorganisms. The two common biochemical conversion processes are anaerobic digestion and fermentation.

Anaerobic digestion is a controlled decaying process that takes place without oxygen. The material used is agriculture waste, manure, algae, seaweed (kelp), municipal solid waste, and paper. The reaction produces methane (a flammable gas) as the bio-materials decay, Fig. 27-13.

Fermentation is a very old process that uses yeast (a living organism) to decompose the material. The yeast changes carbohydrates into ethyl alcohol (ethanol). Grain, particularly corn, is often used for this process. The ethanol can be directly burned or can be mixed with gasoline as an automobile fuel.

THERMAL ENERGY CONVERTERS

Heat and thermal energy have had important parts in history. The industrial revolution that took place in Europe and America in the 18th century was greatly dependent on heat engines. These engines have been replaced by electrical motors in most industrial applications. Transportation is one exception.

However, we still depend on thermal energy. The comfort of the home you live in depends on heating and cooling systems. Many industrial processes use heat to cook, cure, or dry materials and products. Next, you will explore two major applications of thermal energy–heat engines and space heating.

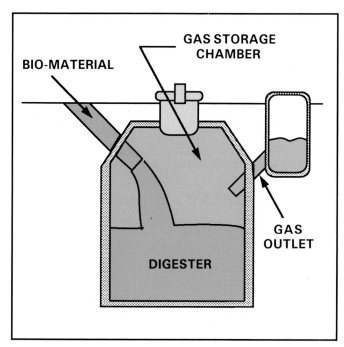

Fig. 27-13. This is a diagram of a biomass converter. The product is methane, a highly flammable gas.

HEAT ENGINES

Today, most of our transportation systems are based on fossil fuel-powered engines. These technological devices burn fuel to produce heat. In turn, the heat is converted into mechanical energy. All heat engines can be classified as either internal combustion engines or external combustion engines.

These classifications are based on the location of the thermal energy source. Internal combustion engines burn the fuel within the engine; external combustion engines burn the fuel away from the engine.

Internal Combustion Engines

You were introduced to a number of **internal combustion engines** in earlier chapters of this book. They included the gasoline and diesel engines that are widely used in land and water transportation vehicles, Fig. 27-14. Jet and rocket engines that were introduced in Chapter 24 are also internal combustion engines.

To create power, all of these engines use expanding gases produced by burning fuel. They change heat energy into mechanical motion. Let's review the common gasoline-powered internal combustion engine. You learned that the engine operates on a cycle that has four strokes: intake, compression, power, and exhaust.

During the intake stroke, the piston moves downward to create a partial vacuum. Then, a fuel and air mixture is introduced into the cylinder.

The intake stroke is followed by the compression stroke. During this stroke, the piston moves upward, compressing the fuel-air mixture into the small cavity at the top of the cylinder.

Then, the power stroke starts with an electrical spark produced by the spark plug. This action ignites the compressed fuel-air mixture, which expands. The resulting gases force the piston downward in a powerful movement.

The final stroke is the exhaust stroke. During this stroke, the piston moves upward to force the exhaust gases and water vapor from the cylinder. At the end of this stroke, the engine is ready to repeat the four stroke cycle.

External Combustion Engines

Most **external combustion engines** are steam engines. You explored the development and operation of these engines in Chapter 7. The steam engine uses the principle that steam (hot water vapor) occupies more space than the water from which it came. In fact, one cubic centimeter (cc) of water produces 1700 cc of steam.

The operation of the steam engine is quite simple, Fig. 27-15. Water is heated in a boiler until it changes into steam. This high pressure steam is introduced into a closed cylinder that has a free-moving piston in it. The steam forces the piston down. Next, cold water is introduced into the cylinder, condensing the steam. The resulting water takes up only 1/1700th as much space as the steam did, so a vacuum is formed in the cylinder. This causes the piston to be drawn up. At the top of the piston's stroke, a fresh supply of high pressure steam is introduced into the cylinder. The engine repeats its cycle.

Fig. 27-14. This internal combustion engine is used in Indy-class race cars.

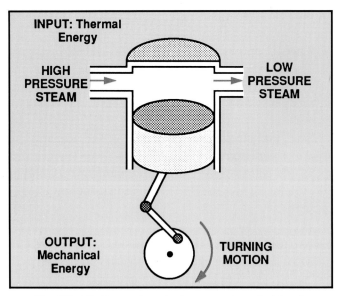

Fig. 27-15. Drawn here is the operation of a simple steam engine.

The reciprocating (up-and-down) motion of the engine is changed to rotary motion by a flywheel. This rotary motion can be used to power any number of technological devices. In past times, steam engines powered ships, locomotives, cars, and many machines in factories.

SPACE HEATING

An important use of thermal energy is heating buildings and other enclosed spaces. Three basic types of heat transfer are used to heat space. These are, as shown in Fig. 27-16:
- Conduction.
- Convection.
- Radiation.

Conduction is the movement of heat along a solid material or between two solid materials that touch each other. Conduction takes place without any flow of matter. The movement of energy is from the area with a higher temperature to the area with a lower temperature. A pan on an electric heating plate is heated by conduction.

Convection is transferring heat between or within fluids (liquids or gases). Convection involves the actual movement of the substance. This process uses currents between colder areas and warmer areas within the material. Convection can occur through natural action or through the use of technological devices. The wind is an example of natural convection action. Forced hot-air heating systems are technological devices that warm homes with convection currents.

Radiation is heat transfer using electromagnetic waves. The strength of the radiation is directly related

to the temperature of the radiating medium. Hot objects radiate more heat than cooler objects. The heat people feel on a bright, sunny day is from solar radiation. Also, if you bring your hand close to a hot metal bar you can feel the heat radiate from it.

Radiation heats only the solid objects it strikes, but it does not heat the air it travels through. Radiant heaters in warehouses keep workers warm in a building, while the air still feels cold.

PRODUCING HEAT

A number of methods are used to produce thermal energy to heat materials and buildings. These include burning fuels, capturing heat from the surroundings, and converting electrical energy.

Fuel Conversion

Typical **fuel converters** include fossil fuel furnaces, wood burning stoves, and fireplaces. A furnace has a firebox, a heat exchanger, and a means of heat distribution, Fig. 27-17. The fuel is burned in the firebox to generate thermal energy. Convection currents pass through the cells of the heat exchanger and raise its temperature. This thermal energy is transferred in the heat distribution chamber to a heating medium (water or air), which is then passed over or through the heat exchanger.

In some systems, water is heated or turned to steam. The fluid is then piped to radiators in various locations. These radiators use convection and radiation currents

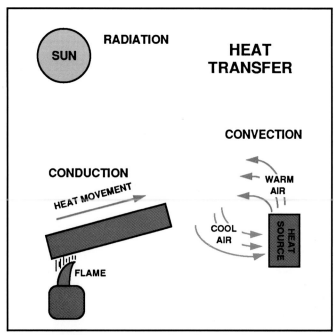

Fig. 27-16. Shown are examples of conduction, convection, and radiation, the three important heating processes.

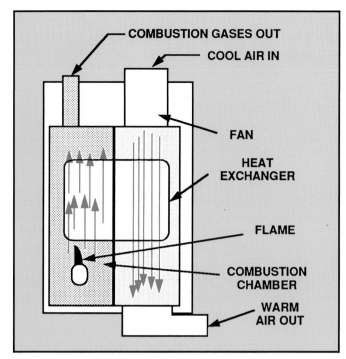

Fig. 27-17. This drawing shows the airflow through a gas fired hot-air furnace.

to heat the room. Other systems blow air through ducts to areas needing heat. Convection currents circulate the warm air within the enclosure.

Atmospheric Heat

The atmosphere has heat available no matter how cold the day seems. The standard device used to capture this heat is called a **heat pump.** It is, actually, a refrigeration unit that can be run in two directions. In one direction, the pump removes heat from the room and releases it into the atmosphere. This is part of what an air conditioning (cooling) unit does. When a heat pump is operated in the opposite direction, the pump takes heat from the outside air and releases the heat inside a building.

Heat pumps work on a simple principle: when a liquid vaporizes, it absorbs heat, and when a liquid is compressed, it releases heat. The system consists of a compressor, cooling or condenser coils, evaporator coils, and a refrigerant (volatile liquid), Fig. 27-18.

In a heat pump, a heat transfer medium such as ammonia is allowed to vaporize in the evaporator coils. The heat needed to complete this task is drawn from the material around the coils. This may be air (atmospheric heat pump) or water from a well (groundwater heat pump).

The refrigerant gas is then compressed. This action causes the material to give off heat through the condenser coils. This heat may be used to warm air or water, which is then transferred to the rooms needing heat.

The system may be reversed to produce cooling for air-conditioning. The heat for the evaporation is drawn from within the building. The heat from compressing the gas is expelled into the outside atmosphere.

Electric Heat

There are a number of ways in which electricity is used to heat a building. The heat pump described above is one way. It is electricity that powers the compressor that draws heat from the air. Another method that uses electricity is a furnace that uses an electric resistance heater.

One very common method of heating uses electric resistance heaters in each room. These heaters have special wires that have a high resistance to electrical current. The wires become very hot when electricity passes through them. The hot wires warm the air around them, and convection currents transfer the heat to all parts of the room.

Another type of electric heating heats with radiation (radiant heating). This system uses high-resistance wires installed in the ceiling. When electricity passes through them, they become warm. This warmth radiates into the room much like heat radiated from a hot bar of steel. Objects in the room are warmed by the electromagnetic waves emitted from the system.

ELECTRICAL ENERGY CONVERTERS

Life in America is closely linked with our electrical generation and distribution systems. Electricity is very important in our society. Without electricity, stores would close, food would spoil, and many homes would become dark and cold. Let's examine the typical electrical generation and distribution system.

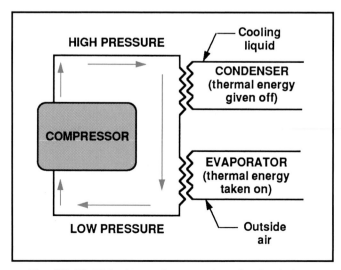

Fig. 27-18. This shows the operation of a simple heat pump.

Technology: How it Works

TURBINE: A device that transforms energy from a flowing stream of fluid into rotating mechanical energy.

Humans have harnessed the energy of nature to do work for centuries. Early efforts used animal power. Later, the forces of the wind and flowing water were harnessed. The first efforts in this arena involved the windmill and the waterwheel, Fig. 1. Both of these devices are effective but provided limited amounts of power.

Fig. 1. This wind turbine was derived from a early windmill.

The device that has served best in capturing the energy of flowing fluids is the turbine. Turbines can be used to convert energy from wind, steam, or running water into rotating motion.

One common use of turbines is in hydroelectric power plants. These turbines work at about 90 percent efficiency. This means that they convert 90 percent of the energy from the rushing water into rotating mechanical energy.

The first water turbine was developed in France in 1827. It was invented by Benoit Fourneyron. A modern water turbine is the Francis turbine, Fig. 2. In this device, water is piped into the turbine. The water swirls through the area between the casing and blade unit. The casing of the turbine includes a series of vanes that catch the rushing water and direct it onto the blades. These vanes are set so that the force of the water is tangent to the rotational plane of the blades. This angle captures the maximum amount of energy from the water. The turbine shaft is usually connected directly to a generator.

Another common type of turbine is the steam turbine, Fig. 3. The steam turbine uses a series of blade units similar to those in a jet engine. The hot steam enters at one end of the turbine. As the steam moves through the unit, it forces the blades to turn. The steam cools and expands as it travels across the blade units. To account for this, each succeeding set of blades is slightly larger in diameter. The steam finally condenses into water at the back of the turbine and is returned to the boiler for reheating.

Steam turbines are used in coal-fired, natural gas, and nuclear power plants. They also power many ocean going ships.

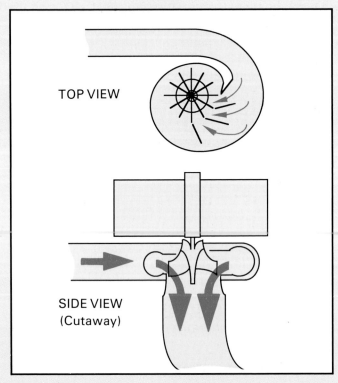

TOP VIEW

SIDE VIEW
(Cutaway)

Fig. 2. Shown are two views of the Francis water turbine.

Fig. 3. This is a cutaway model showing the inside of a steam turbine.

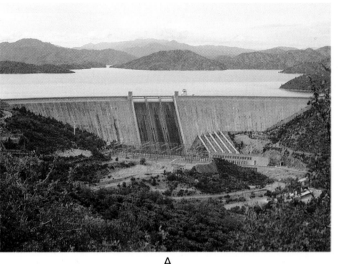
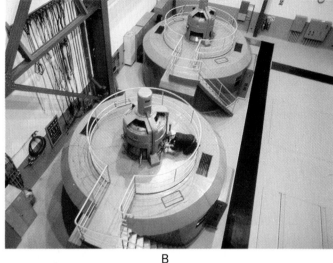

Fig. 27-19. Hydroelectric power. A–This is Shasta Dam in Northern California. Note the water delivery pipes that lead to the power plant in the lower right portion of the picture. B–This is inside a hydroelectric plant. These are the generators above the water turbines. (U.S. Department of Energy)

GENERATION

Electricity generation uses the principles of electro-mechanical energy conversion. In most commercial systems, water or steam is used to turn a turbine. A water-powered plant is called a **hydroelectric** generating plant, Fig. 27-19. This plant uses a dam to develop a water reservoir. This water is channeled through large pipes into the turbines in the generating plant.

A steam-powered electrical plant uses fossil fuels or nuclear energy to produce steam to drive the generator's turbines. Fossil fuel electrical plants burn coal, natural gas, or fuel oils to produce thermal energy.

A nuclear plant uses atomic reactions to heat water in a primary system. The heated water is used to produce steam in a secondary system. The steam then drives the generator's turbine, Fig. 27-20. Keeping the loops separate prevents the water in the reactor from entering the steam turbines. This reduces the hazards for workers in the plant and for people living near the plant.

Steam and water turbines have a series of blades attached to a shaft. As the water or steam strikes the turbine blades, the shaft turns. This shaft is attached to a generator, which changes the mechanical energy (motion) into electrical energy.

The electrical generator is the opposite of the electric motor described in Chapter 7. Recalling Chapter 7, two laws of physics are directly applied. The first law states that like poles of a magnet repel one another, and unlike

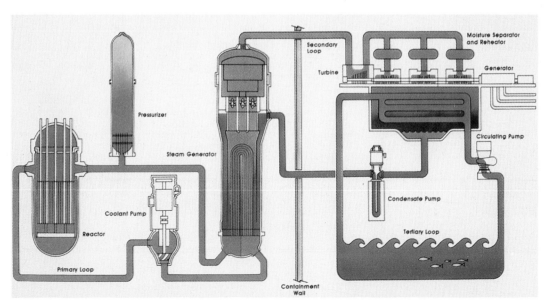

Fig. 27-20. This is a drawing of a nuclear steam supply system. Notice how the water in the primary loop is used to heat the water in the secondary system. The contents of the two loops remain separate. (Westinghouse Electric Corp.)

poles attract one another. The second law states that current flowing in a wire creates an electromagnetic field around the conductor.

Look at Fig. 27-21. You will note that there are two magnets. Like a motor, the outside magnet of the generator is a stationary electromagnet called the field magnet. The inside magnet is a series of wires wound on a core. This part is called the armature and is able to rotate on its axis.

An electrical current is allowed to flow in the coils of the field magnet. This action produces an electromagnetic field around the field magnet that cuts through the armature. When the armature is spun by the water or steam turbine, the wires on the armature cut through the magnetic lines of force around the field magnet. This induces a current in the armature. This current is drawn off through commutators and fed into the distribution system.

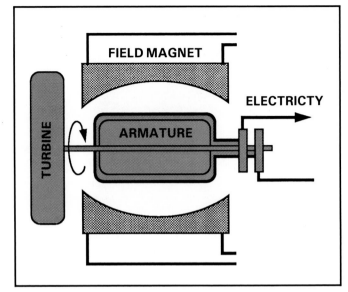

Fig. 27-21. Shown is a basic diagram of an electric generator.

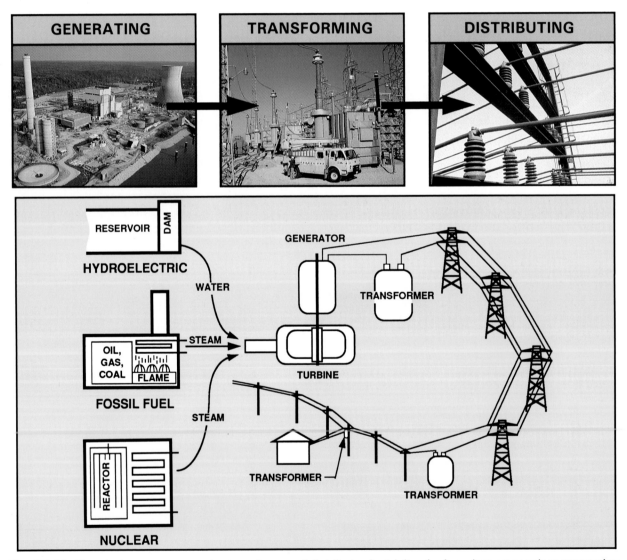

Fig. 27-22. Electrical systems generate, transform, and distribute electricity. A—A nuclear power plant generating the electricity. B—A giant transformer station stepping up and distributing electricity. C—High voltage power lines for taking electric power over long distances.

DISTRIBUTION

The electricity produced in the generating plant is passed through a step-up transformer. This transformer is called a step-up transformer because it steps up, or increases, the output voltage of the electrical current. This very high voltage reduces power losses in transmission.

The high-voltage electrical current is carried to distant locations, generally, by large transmission lines supported on tall steel towers, Fig. 27-22. When the current reaches the area in which it will be used, the electricity flows through another transformer, which reduces, or steps down, the voltage. This lower voltage electrical current moves along the distribution lines. Just before it reaches its final destination, the electricity enters another step-down transformer. This transformer generally reduces the current to 110 and 220 volts for residential use. Some industrial applications use 440 or 880 volts.

APPLYING ENERGY TO DO WORK

Work involves moving a load. Therefore, motion is always present while work is done. Three important types of motion need to be discussed.
- **Rotary motion** (spinning around an axis).
- **Linear motion** (moving in a straight line).
- **Reciprocating motion** (moving back and forth).

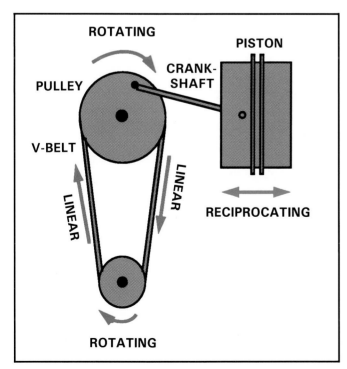

Fig. 27-23. An example of mechanisms to change the type and direction of motion. This system can change rotating motion into reciprocating motion, or vice-versa.

A main activity in energy conversion involves changing the type or direction of a load's motion. This action is called power transmission. Power transmission takes the energy generated by a converter and changes it into motion. Look at the example shown in Fig. 27-23. The figure shows the reciprocating motion of a piston in the cylinder of an internal combustion engine. This motion is changed to rotary motion by the crankshaft. The end of the crankshaft is attached to a pulley that drives the V-belt. The belt travels in a linear motion around a second pulley. The second pulley changes the linear movement back into rotary motion. Also, note that the two pulleys rotate in opposite directions. Applying motion to perform work often requires both changing the type of motion and its direction.

MOTION CHANGE MECHANISMS

Two basic types of systems are used to change the type, direction, or speed of a force. These are mechanical power and fluid power (or fluidic) systems.

Mechanical Power Systems

Mechanical systems use moving parts to transfer the motion, Fig. 27-24. This is the oldest method of

Fig. 27-24. This crane uses a mechanical means to lift a load. (American Electric Power)

transferring energy. There are a number of different mechanical methods used in technological devices. Fig. 27-25 shows six common techniques:

- **Lever:** A device that *changes the direction of a linear force.* A downward force may be applied to one end of a lever. This causes the lever arm to pivot on its fulcrum. The opposite end moves in an upward direction. You will remember from Chapter 7 that the location of the fulcrum determines if the device multiplies the amount of the output force or the distance it moves. Many door handle mechanisms in automobiles transfer motion with levers.

- **Crank:** A pivot pin near the outside edge of a wheel or disk that *changes reciprocating motion into rotating motion.* The diameter of the swing of the crank will determine if the amount of the force or distance of the force is multiplied. An internal combustion engine transfers power from the piston to the transmission using this type of drive.

- **Gears:** Two or more wheels with teeth on their circumference that *change the direction of a rotating force.* The relative diameters of the input and output gears determine if the system is a force multiplier (the output gear rotates faster) or a distance multiplier (the output gear turns over a greater area). If a smaller input gear is used, the unit will increase the output force and reduce its speed. If a larger input gear is used, the unit will decrease the output force and increase its speed. Some automobile transmissions use this type of power transmission system.

- **Cam:** A pear-shaped disk with an off-center pivot point that is used to *change rotating motion into reciprocating motion.* A cam with a large lobe (extended portion) creates longer strokes for the reciprocating member. However, the force is reduced. The valves in an internal combustion engine are opened using a cam system.

- **Gear and rack:** A rotating gear meshes with a bar that has gear teeth along its length (rack) that *changes rotating motion into linear motion.* As the gear turns, it slides the rack forward or backward. Rack and pinion (gear) steering for automobiles uses this system.

- **Pulleys and V-belt:** Two pulleys with a V-belt stretched between them that *change the speed or power of a motion.* As one pulley turns, the V-belt moves, which in turn rotates the second pulley. The two pulleys rotate in the opposite direction. If the force is applied to a larger pulley, the smaller pulley turns faster but has less power. If the power is applied to the smaller diameter pulley, the larger diameter pulley turns slower, but it turns over a greater area. Many machines are driven by pulleys and V-belts.

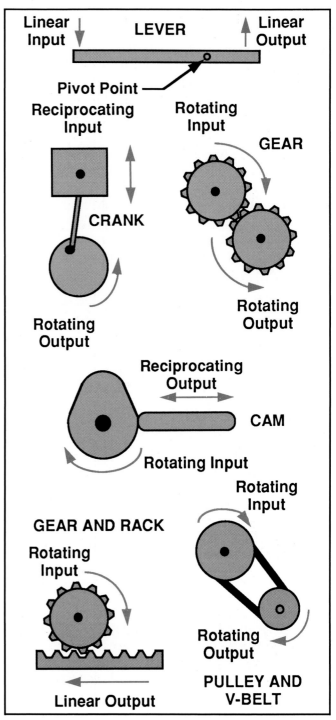

Fig. 27-25. These are six important mechanical techniques that change the type, direction, or speed of a moving force.

Under heavy loads the V-belt can slip as it drives the pulleys. The problem is overcome by using gears and a chain drive similar to the one used on bicycles.

Fluid Power Systems

Fluid power systems use either liquids or gases to transfer power form one place to another. Systems that use air as the transfer medium are called **pneumatic** systems. Liquids (usually oil) are used in **hydraulic** systems.

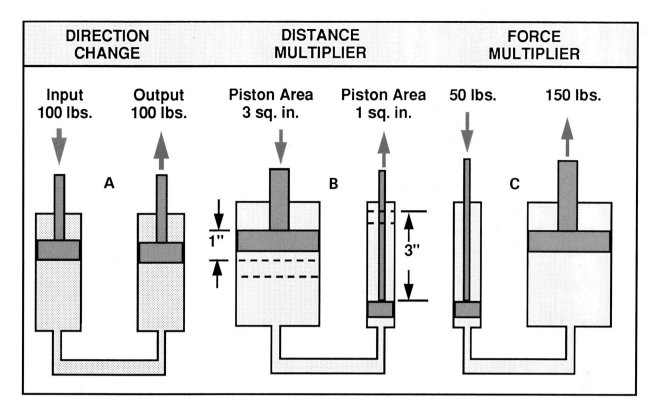

| DIRECTION CHANGE | DISTANCE MULTIPLIER | FORCE MULTIPLIER |

Fig. 27-26. Hydraulic systems can be used in three ways. A–Change the direction of a force. B–Increase the distance of the force. C–Increase the strength of the force.

Generally, these systems contain two cylinders with movable pistons, a pump, valves to control the flow, and piping to connect the components. Fig. 27-26 shows the three basic uses for hydraulic systems.

The view on the left shows a typical power transfer system. It neither increases the distance nor the intensity of the force. The system simply moves the force from one location to another, and it changes the force's direction. The fluid is forced downward by the piston in the left cylinder. This action causes the fluid to flow into the right cylinder. The fluid's movement causes the right piston to move upward.

The center drawing shows a distance multiplier. The left piston is forced downward one inch. Since the piston has an area of three square inches, the motion displaces three cubic inches of fluid. This causes three cubic inches of fluid to flow into the right cylinder. The right piston's area is only one square inch. Therefore, the piston must move upward three inches to accommodate the three cubic inches of fluid.

The right drawing shows a force multiplying system. A force of 50 pounds is applied to the left piston which has an area of one square inch. Thus, the original force is 50 pounds per square inch (PSI) on the fluid beneath the piston. This force is transferred to the right piston, which has a three square inch area. The 50 pounds per square inch force from the piston on the left exerts a 150

pound per square inch upward force to the piston on the right.

Pneumatic systems operate in a similar way. However, the force and distance movement calculations are more difficult. Liquids do not compress, so nearly all the force is transferred from one cylinder to the other. A slight amount of the force is used to overcome the friction of the piston and the fluid in the pipes.

However, air can be compressed. Therefore, some of the force in pneumatic systems is used in reducing the volume of the air in the system. The remainder of the force is applied to moving a load.

SUMMARY

Modern life and technology depend on energy that can be traced back to the sun. Inexhaustible energy is available to us through wind, water, solar, geothermal, and ocean power. Wind and running water were our earliest sources of energy. Biomass resources and biofeuls supply energy from renewable sources. Fossil fuels, and nuclear power are the most used exhaustible resources.

Often, this energy is used by converting it from one form to another. Wind energy, water energy, and fossil fuels are commonly converted into electrical energy. The electrical energy is then turned into thermal energy and radiant energy in homes and business. Through a string of technological processes, the earth's energy sources are tapped and made to work for society.

WORDS TO KNOW

All of the following words have been used in this chapter. Do you know their meanings?

Active collector
Anaerobic digestion
Biochemical process
Biomass resource
Cam
Conduction
Convection
Crank
Direct active solar system
Direct gain solar system
External combustion engine
Fermentation
Fuel converter
Gasification
Gear and rack
Gears
Heat pump
Hydraulic
Hydroelectric
Indirect active solar system
Indirect gain solar system
Insolation
Internal combustion engine
Isolated solar system
Lever
Linear motion
Liquidification
Liquefaction
Ocean mechanical energy conversion
Ocean thermal energy conversion
Passive collector
Photovoltaic cell
Pneumatic
Prime mover
Pulleys and V-belt
Pyrolysis
Radiation
Reciprocating motion
Rotary motion
Solar converter
Thermochemical conversion
Tidal energy conversion
Waterwheel
Wave energy conversion
Windmill

TEST YOUR KNOWLEDGE

Write your answers on a separate piece of paper. Please do not write in this book.

1. What is the primary source of the earth's energy?
2. True or false. Photovoltaic cells convert voltage into light.
3. True or false. A forced hot-air heating system warms a room using conduction.
4. True or false. The steam engine is an external combustion engine.
5. Match the process on the left with its category on the right:

_____ Liquidification. A. Thermochemical.
_____ Fermentation. B. Biochemical.
_____ Direct combustion.
_____ Gasification.
_____ Anaerobic digestion.

6. List the two types of heat engines.
7. Wood chips, solid municipal waste, and manure are examples of _____.
8. List the two types of ocean energy.
9. A device that uses the heat in the outside air or ground water to warm a house is called a _____.

APPLYING YOUR KNOWLEDGE

1. Design and construct a passive solar device that will heat the air in a shoe box. Prepare a sketch of the device, and prepare an explanation of how it works.
2. Select a major energy conversion system such as a coal-powered electric generating plant. Explain the advantages and disadvantages of the conversion system. Next, list at least two energy converters that could be used to replace the system you selected. List their advantages and disadvantages. Present your work on a chart similar to the one shown below.

Energy conversion system:	
ADVANTAGES	DISADVANTAGES
Alternate system #1:	
ADVANTAGES	DISADVANTAGES
Alternate system #2:	
ADVANTAGES	DISADVANTAGES

Solar powered cars use photovoltaic cells to power their motors. These cars compete in races, as shown in this photo. The cars become better every year. (General Motors)

Section Eight - Activities

ACTIVITY 8A - DESIGN PROBLEM

Background:

All technological devices use energy. Often, this energy is converted from one form to another. For example, electrical energy is changed into mechanical energy by a motor. In addition, the motion produced by an energy converter is often transformed. For example, the reciprocating motion of an engine's piston is changed to rotating motion by a crankshaft.

Situation:

You are employed as a designer for Children's Technology Museum. The curator want a series of models that show how humans change types of motion. She wants models that demonstrate how changing the rotating motion produced by a vertical waterwheel is changed into (1) reciprocating motion for a colonial sawmill or (2) pounding motion for a forge, Fig. 8-A.

Challenge:

Select one of the two models listed above. Design and build a working model using wood, cardboard, or other easily worked materials. Fig. 8-B shows a sample model. This model converts vertical rotating motion into horizontal rotating motion.

Fig. 8-B.

Optional:

Prepare a display, using your model, that explains its function and history.

ACTIVITY 8B - FABRICATION PROBLEM

Background:

Using inexhaustible energy sources is a major challenge for the future. An important source is wind.

Challenge:

Build and test a working model of a wind-powered electric generation system.

Equipment:

One piece 3/4 in. x 5 1/2 in. x 8 in. wood base (pine, fir, spruce)
One piece 3/4 in. x 18 in. strip of thin gauge sheet metal
Two 1/2 in. #6 sheet metal screws
One small electric motor (available from a hobby or electronics store)
One multimeter

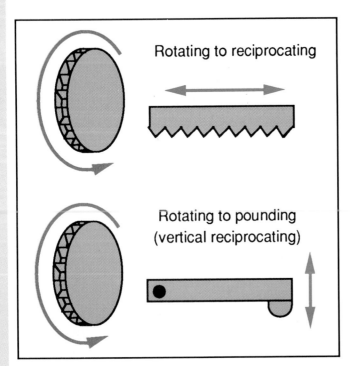

Rotating to reciprocating

Rotating to pounding (vertical reciprocating)

Fig. 8-A.

24 in. electrical hook-up wire
One 3-blade or 4-blade propeller (available from a hobby store)

Procedure:
Building the Prototype
1. Obtain a piece of wood with dimensions of 3/4 in. x 5 1/2 in. x 8 in.
2. Cut a strip of sheet metal that is 3/4 in. wide and 18 in. long.

Caution: Follow all safety rules demonstrated by your teacher!

3. Drill a 1/8 in. diameter hole 1/2 in. from each end of the strip of sheet metal.
4. Form the sheet metal strip around the motor as shown in Fig. 3. This forms the motor mount.
5. Bend the sheet metal strip to form 1 in. long tabs that can be mounted to the wood base.
6. Use the two sheet metal screws to attach the motor mount to the base.
7. Attach the propeller to the motor.
8. Press the motor into the loop of the motor mount.
9. Attach a piece of hook-up wire to each wire coming out of the motor.

10. Solder your joints and wrap them with electrical tape.
11. Set the multimeter to read current.
12. Attach the wires to the multimeter as shown in Fig. 4.

Testing the Device
1. Place the wind-powered generator in front of a window fan.
2. Turn the fan on to its "low" setting.
3. Observe the rotation of the propeller.
4. Observe and record the meter reading on the multimeter.
5. Repeat steps #3 and #4 using the fan's "medium" setting.
6. Repeat steps #3 and #4 for the fan's "high" setting.
7. Turn off the fan.

Optional:
5. Change the propeller to a model with a different number of blades.
6. Repeat steps 1 through 4.
7. Write a report that explains your observations.

Optional:
5. Change the angle at which the wind is hitting the propeller to 30° off center.
6. Repeat steps 2 through 4.
7. Write a report that explains your observations.

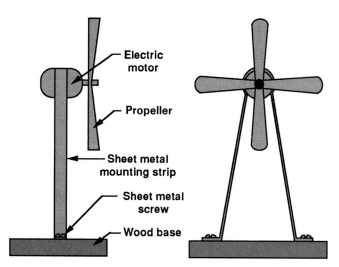

Fig. 8-C.

- Electric motor
- Propeller
- Sheet metal mounting strip
- Sheet metal screw
- Wood base

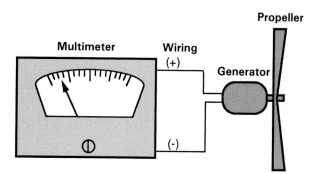

Fig. 8-D.

Multimeter   Wiring (+)   Propeller   Generator   (-)

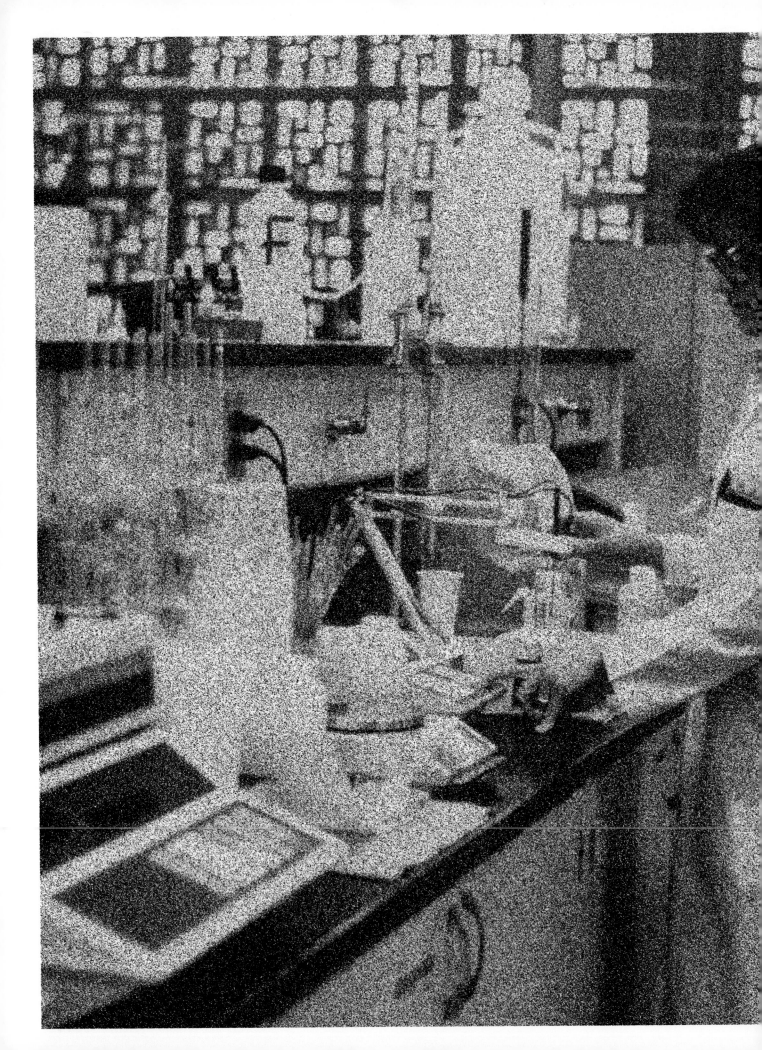

Managing Technological Systems

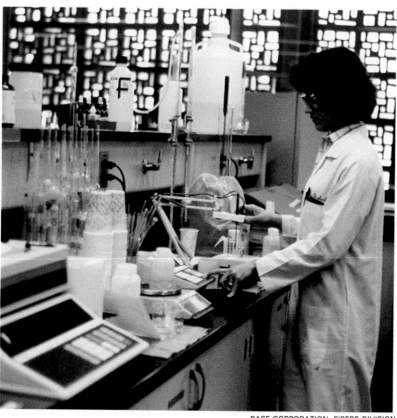

BASF CORPORATION, FIBERS DIVISION

The key to any technological enterprise is people. (D.P. Technology)

28
CHAPTER

ORGANIZING A TECHNOLOGICAL ENTERPRISE

After studying this chapter you will be able to:
- *Describe entrepreneurship and intrapreneurship.*
- *Define management.*
- *List and describe the functions of management.*
- *Describe the responsibilities of managers to owners, to workers, and to the public.*
- *List and describe the three main forms of ownership.*
- *Describe how a business is legally formed.*
- *List and describe the main levels of management in a corporation.*
- *Describe how a company is financed.*
- *Explain the differences between equity financing and debt financing.*
- *Explain who stockholders are and the rights they possess.*

You have learned a great deal about technology as you studied this book. You have learned that technology consists of human-made devices and apparatus designed to control and modify the natural environment. Technology is a system with goals, inputs, processes, and outputs. Also, you learned that technology occurs in communication, construction, manufacturing, and transportation contexts.

TECHNOLOGY AND THE ENTREPRENEUR

Have you ever thought about the source of technology? Where does it come from? Technology is developed by people to serve people, Fig. 28-1. It is the product of the human mind. However, that is not enough in the modern world. Organization is necessary

for technology to be developed, produced, and used for the benefit of people. The days when most technology was developed, produced, and used by one person are long gone. People now use very complex systems to create our technology. At the base of many of these systems are **entrepreneurs**, Fig. 28-2. These are people with very special talents. Entrepreneurs see beyond present practices and products. They see new ways to meet human needs and wants. Entrepreneurs focus on what the customers value, and they develop systems and products to meet desires and expectations. They might change the entire way something is being done.

A good example of **entrepreneurship** is the chain of McDonald's restaurants. The chain grew out of a small hamburger stand in southern California. The original owners had developed some innovative ways to make and sell their product. An outsider, Ray Kroc, saw greater possibilities. Under his leadership, managers developed something new–the national fast-food business. They carefully studied the various jobs, developed

Fig. 28-1. This airplane was designed to transport people quickly, safely, and comfortably. (United Airlines)

Fig. 28-2. Entrepreneurs have the vision to recognize consumer wants and to devise ways to meet them. (Goodyear Tire and Rubber Co.)

Fig. 28-3. What purpose do you think these boats will serve?

special management techniques, standardized the product, and developed effective training programs. This is some of the work of the entrepreneur, to improve the use of the resources and create new products or markets.

The dictionary defines the entrepreneur as any person taking the financial risks of starting a small business. But, this definition leaves out the aspect of entrepreneurial spirit, the spirit of innovation. People starting another beauty shop, deli, or bakery are taking financial risks and may become successful business operators, but they are not innovators. They do not deal with change as an opportunity to produce a new product or service.

Entrepreneurship is involved in starting a new company, but there can be entrepreneurship within an existing company. It can exist in small or large companies. Entrepreneurship involves an attitude and an approach. It is searching for opportunities for change and responding to them. Many large companies encourage entrepreneurship within their organization. In fact, a new term has evolved to describe this action. The term is **intrapreneurship**, which is *the entrepreneurship spirit and action within an existing company structure*.

MANAGEMENT AND TECHNOLOGY

Technology is purposeful, Fig. 28-3. Technology is developed to meet a problem or opportunity. But, identifying and responding to the need for change is only one part of developing technology. The production and use of technology must be managed. Therefore, *technology is a product of managed human activity*. Actually, all of your actions are managed. You manage your own activities, or they are managed by other people.

This is not bad. **Management** is simply the act of planning, directing, and evaluating any activity. It can be as simple as managing personal expenditures, or as complicated as managing an industrial complex, Fig. 28-4.

Management involves **authority** (the right to direct actions) and **responsibility** (accountability for actions). Managers have the responsibility to make decisions that ensure the business is successful. Their authority may include hiring personnel, purchasing materials, developing products, setting pay rates, etc. Likewise, managers have the responsibility of protecting the

Fig. 28-4. People manage their personal affairs as well as industrial activities. These managers are examining a model of a boiler system. (American Electric Power)

rights of a company's owners, workers, and customers. This might include securing product patents, investing company funds wisely, providing a safe work environment, and producing a quality product.

FUNCTIONS OF MANAGEMENT

To carry out their duties, company managers perform four important functions, Fig. 28-5. These are:

- **Planning:** This includes setting long-term and short-term goals for the company or segments of it. Planning also involves selecting a course of action to meet these goals. Planning activities often result in an action plan, or plan of work. An action plan lists what needs to be done, who will do it, and when it is to be done.

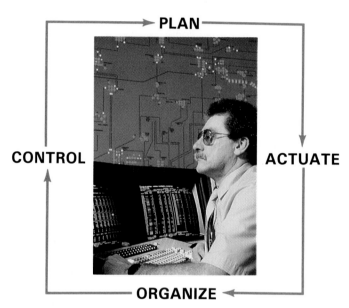

Fig. 28-5. Managers plan, organize, actuate, and control company activities. Here a manager is gathering information. He is performing the *control* function.

- **Organizing:** This involves structuring the company or work force to address company goals. Organizing could include establishing organizational charts, establishing chains of command, and determining the company's operating procedures. Organizing makes sure that people, materials, and equipment are secured and scheduled to meet the action plan.
- **Actuating:** This involves initiating the work related to the action plan. Actuating could include training employees, issuing work orders, providing a motivational work environment, or solving production problems. Actuating causes plans to take form. With this function, products and structures are built, or services are provided.
- **Controlling:** This involves comparing results against the plan. Controlling could include evaluating purchasing activities, product quality, work performance, inventory levels, etc. Typical terms applied to this area are inventory control, production control, quality control, and process control. Control actions ensure that resources are used properly and outputs meet stated standards.

AUTHORITY AND RESPONSIBILITY

Managerial functions are carried out through an organizational structure, Fig. 28-6. This structure, generally, begins with the **owners,** the people with the most authority. The owners have the ultimate control, or *final authority,* over company activities. It is their company. They can hire or fire management personnel, set policies, or close the business.

However, in most larger companies, the owners are not the managers. They frequently have other jobs or interests. Also, they often do not have the skills needed to manage a large complex business. The owners

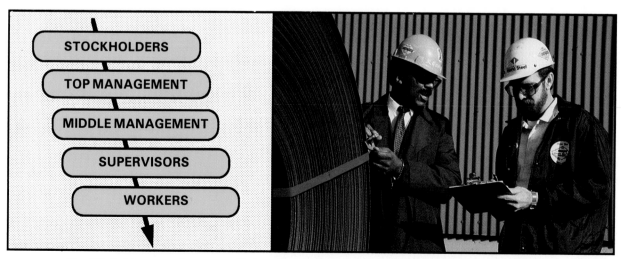

Fig. 28-6. Managers have specific levels of authority and responsibility within a company.

delegate their responsibility to full-time managers. In many companies, the top manager is the **president** or the **chief executive officer (CEO).** In small companies, the two titles may be held by the same person. In very large companies, these titles are held by two different people. In either case, this level of management is responsible for the entire company's operation. People at this level have *day-to-day control of the company.*

Few people can manage a company by themselves. Therefore, the top managers employ other managers to assist them. The number of managers and level of the managers employed will vary with size and type of company. Larger companies generally have **vice presidents** that report to the president or CEO. Each vice president will have responsibility for some segment of the company. The segment might be a functional area like marketing, production, or engineering. In other cases, the segment is a regional responsibility, such as foreign sales, or West Coast operations.

Most vice presidents have a number of managers reporting to them. This level of management is responsible for a part of the vice president's *scope of responsibilities.* Regional sales managers may report to a vice president for sales. Plant managers often report to vice presidents in charge of production. This level of management is often called **middle management.** Middle management is below **top management** (presidents and vice presidents) but above operating management.

The lowest levels of management directly oversee specific operations in the company. Managers at this level may be supervisors on the production floor, district sales managers, or employment directors. They are the managers who are closest to the people that produce the products and services that the company sells. They are often called **supervisors,** or **operating management.**

RISKS AND REWARDS

As was said earlier, management is everywhere. You manage yourself each day. You determine what you will do, when you will do it, and how well you did it. However, most technology is developed and produced through the planning of industrial companies, Fig. 28-7. The company employees design, engineer, and produce the products, structures, transportation services, and communication media you depend on daily.

Everyone involved with a company has some basic elements in common. These elements are **risks** and **rewards.** Owners risk their money to finance the company. Banks also accept a level of risk. Banks and other lending institutions make loans to finance company growth. Employees risk missing other employment opportunities by working for the company. Consumers risk their money when they buy a product.

In return, the risk takers expect a reward. The owners want their investment to grow. They also expect periodic financial returns for the use of their money. Banks expect interest to be paid on their loans. Employees expect job promotions, pay raises, safe working conditions, and job security. Consumers expect a return for the money they spend on products and service. They expect performance and value.

Fig. 28-7. Most technology is developed and produced by industrial enterprises with their greater financial resources. (The Coastal Corp.)

Technology: How it Works

INSTRUMENT LANDING SYSTEM (ILS): A navigational aid used to help pilots guide aircraft onto a runway for a safe landing.

Weather often does not allow a pilot clear visibility while approaching an airport. A modern navigational aid is the instrument landing system (ILS). It is found at almost all commercial airports, Fig. 1.

An ILS system produces two radio beams. One beam is in the aircraft's horizontal plane, and the other beam is in the aircraft's vertical plane. The horizontal beam is a rather narrow beam that gives the pilot the proper glide slope (angle of descent) for a safe landing. The other,

Fig. 1. Modern aircraft use many landing aids. (Delta Air Lines, Inc.)

wider beam radiates at a right angle to the glide slope beam. This beam allows the pilot to align the airplane with the center of the runway.

The pilot has a display screen that indicates the aircraft's alignment with the two beams. Two lines are shown on the display that tell the pilot how the plane is aligned with the runway. When the plane is making a proper approach, these lines will be in alignment with vertical and horizontal index lines on the screen. An improper approach causes the lines produced by the ILS to be displayed away from the index lines, Fig. 2 (upper left and lower right).

Arrayed along the center beam are three additional vertically broadcast beams. The first beam is known as the outer marker beacon. The beam closest to the runway is called the boundary beacon. These markers tell the pilot the distance he or she is from the runway as the plane approaches for a touchdown.

The instrument landing system guides the pilot to about 200 feet above the runway. The actual landing is completed visually.

Used with the lights that mark the boundaries of a runway and verbal help from air traffic controllers, ILS provides an effective aid in landing.

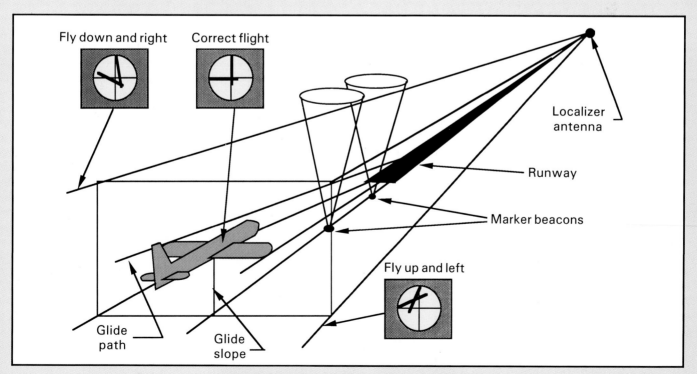

Fig. 2. Shown are the beam paths of an ILS.

FORMING A COMPANY

Each of these risk-taking enterprises is a legal business unit. Companies are organized and operated under the laws and mores (accepted traditions and practices) of our society.

There are several important features involved in the formation of a company. These features include:

- Selecting a type of ownership.
- Establishing the enterprise.
- Securing financing.

SELECTING A TYPE OF OWNERSHIP

Business enterprises can be divided into two different sectors: public and private. **Public enterprises** are those enterprises that are controlled by the government. They are generally those enterprises that meet two criteria. First, they are for the general welfare of the society, and second, they cannot or should not make a profit. For example, consider a police department, which is run like a business. Police are commissioned to protect people and their property. They have managers (police chiefs and captains) and workers (patrol officers, traffic officers, narcotics officers, etc.). However, some aspects of law enforcement might get cut back if the police department had to show a profit and attract private investment. This could limit the department's market to the people who could pay for the service. The police would solve the crimes that showed the most profit.

Another example of public ownership is road construction. We pay for it through taxes. However, each segment of road does not have to show a profit. If it did, we would not have many of our rural roads.

Private enterprises are owned by individuals or groups of people. This ownership can be through a direct means of investment or through an indirect means, such as a pension or investment fund. Private enterprises can be either publicly or privately *held*. A publicly held enterprise is a business in which the public has the opportunity to purchase an interest in the company. Privately held enterprises do not present members of the public this opportunity.

Private owners invest their money, take their risks, and hope to reap a profit. The owners, within legal limits, are free to select business activities, produce the products and services they choose, and divide the profits as they see fit.

There are three main types of private business ownership, Fig. 28-8. The first type is a **proprietorship**. A proprietorship is a business owned by a single owner. The owner has complete control of the company. He or

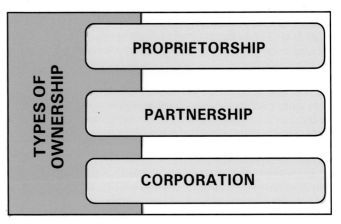

Fig. 28-8. Businesses are either owned by the government (public) or by individuals (private).

she sets goals, manages activities, and has the right to all business profits. A proprietorship is a fairly easy type of ownership to form.

However, proprietors might have difficulty raising money. The company's finances are often limited to the owner's wealth. Banks are hesitant to loan a large quantity of money on unproven businesses or individuals.

An additional problem is that many individuals do not have the knowledge and ability needed to run all aspects of the business. This causes inefficiency in operating the company. Finally, the proprietor is responsible for all the debts the business incurs. Proprietors cannot separate their business debts from their personal wealth. This is a major disadvantage. It is called **unlimited liability.**

Proprietorships are the most common type of ownership in the United States. However, proprietorships are generally limited to small retail, service, and farming businesses, Fig. 28-9. Thus, the dollar impact of this form of ownership on the economy is considerably less than the impact made by the large corporation.

Fig. 28-9. Small businesses and farms are generally proprietorships.

A second form of private ownership is the **partnership.** Partnerships are businesses owned and operated by two or more people (partners). Partnerships have more sources of money to finance the company. Also, the partners interests and abilities may complement each other. One partner may be strong in production, while another may have sales skills.

However, having more than one owner active in a business may cause confusion. Employees might receive conflicting directions. They may not know who really is their boss. Also, partnerships, like proprietorships, have unlimited liability. This is a particularly touchy problem, since one partner can commit the entire partnership to financial risk.

The last form of ownership is a **corporation.** A corporation is a business owned by individuals who have purchased a portion, or share, of the company. Legally, the corporation is like a "person." A corporation can own property, sue or be sued, enter into contracts, and contribute to worthy causes.

Generally, the owners of corporations do not manage the company. The owners employ professional managers for this task. The owners invest their money and expect to receive a **dividend** (periodic payment from the company's profits).

Since a corporation is a legal person, the company, not its owners, is responsible for the debts. This feature is called **limited liability.** This means that an owner's loss, if the company fails, is limited to the amount of money that was invested in the corporation.

ESTABLISHING THE ENTERPRISE

Once the type of ownership is selected, the company must be established. There are few legal requirements for proprietorships and partnerships. In many cases, a license from the city is all that is needed. However, corporations are a different story. They are, generally, large and can have a serious financial impact on people and communities. Therefore, their formation is placed under state control. Each state establishes its own rules for forming a corporation. However, most states require the following steps, Fig. 28-10.

Filing Articles of Incorporation

A corporation is an artificial being. A corporation, like a person, must be born. This process is started by the filing of **articles of incorporation** with one of the states. If the company obtains recognition from that state, the other 49 will also recognize it. The states will then allow the company to conduct its business within their borders.

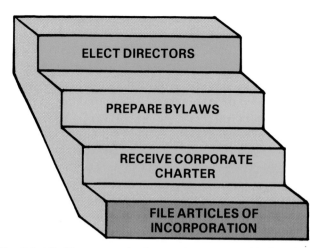

Fig. 28-10. Shown are the four main steps used in forming a corporation.

The articles of incorporation serve as an application for a corporate charter (a birth certificate for the corporation). The form usually asks for the company name, type of business the company plans to enter, the location of the company offices, and the type and value of any stock that will be issued.

RECEIVING A CORPORATE CHARTER

The articles of incorporation are filed with the appropriate state official. The state officers then review the articles. The officers try to determine if the business will operate legal activities and provide customers with appropriate products or services. If they believe that the business meets all state laws, a **corporate charter** (operating permit) is issued. This allows the company to conduct the specified business in the state.

Approving Bylaws

All incorporated business must have a set of **bylaws.** These are the general rules by which the company will operate. A set of bylaws includes the information contained in the charter: name of the company, purpose of the business, and location of the corporate offices. In addition, the bylaws will list:
- Date and location of annual stockholders' meeting.
- Date, location, and frequency of the board of directors' meetings.
- Corporate officers.
- Duties, term of office, and method of selecting corporate officers.
- Number of directors, as well their duties and terms of office.
- Types of proposals that can be presented at the annual stockholders' meeting.
- Procedure for changing the bylaws.

Electing Directors

The charter and bylaws allow the company to operate. However, the stockholders want their investment to be wisely managed. This requires oversight and management. Many companies have hundreds or thousands of stockholders. Few can, or want, to be involved in the managing of the company. Therefore, a **board of directors** is elected to represent the interests of the stockholders. This group of people is charged with forming company policy and providing the overall direction for the company.

A typical board of directors includes two groups of people. One group is made up of the top managers of the company. These individuals are known as **inside directors.** They come from within the company's managerial structure. Other directors are not involved with the day-to-day operation of the company. They are selected to provide a different view of the company's operation. Since these people are outside the managerial structure, they are called **outside directors.**

The directors are elected using a voting system similar to our political system. The main difference is that companies do not use a one-person-one-vote rule. Instead, they use a one-share-one-vote procedure. Each **share of stock** (equal portion of the total company), Fig. 28-11, has a vote assigned to it. The stockholders have as many votes as they have shares of stock. Therefore, people who own a large share of the company and accept a large risk have a greater say in forming company policy.

FINANCING THE COMPANY

So far you have learned how a company is formed. However, there is more to starting a company. The company needs money to operate. There are two basic methods of raising operating funds. These are equity financing and debt financing, Fig. 28-12.

Equity Financing

Equity financing involves selling ownership in the company. This is an important way in which corporations are financed. The company is authorized by its charter to sell a specific number of shares of stock. Individuals may buy these and, by this action, become an owner of part of the company. Owners receive certain rights with the shares they own. These include:

- The right to attend and to vote at the annual stockholders' meeting.
- The right to sell their stock to another individual.
- The right to receive the same dividend (portion of the company's profits) per share as other stockholders.

Fig. 28-11. This is an example of a stock certificate (share of stock).

- The right to a portion of the company's assets (property, money, etc.) if the company is liquidated.

Debt Financing

Debt financing involves borrowing money from a financial institution or private investors. Banks and insurance companies loan corporations money to finance new buildings, equipment, and the company's daily operations. The banks and insurance companies charge interest for the use of their money. They charge their best (safest) customers lower interest rates. This rate is called the **prime interest rate** or **prime.** Other borrowers

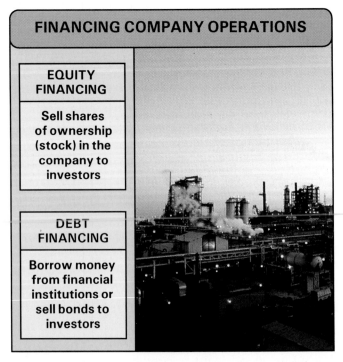

FINANCING COMPANY OPERATIONS

EQUITY FINANCING

Sell shares of ownership (stock) in the company to investors

DEBT FINANCING

Borrow money from financial institutions or sell bonds to investors

Fig. 28-12. To build and expand a company, some method of financing is used.

pay a rate higher than the prime rate. This rate is often quoted in terms of the prime, such as prime plus two percent.

Corporations can also sell debt securities called **bonds.** These instruments are usually long term (10- to 20-year) securities. They are often sold in fairly large denominations, usually $5,000 or more. The company pays quarterly or yearly interest on the face value (original value) of the bonds. At maturity (end of the bond's term) the company pays back the original investment.

SUMMARY

Technology is developed by people for people. At the very foundation of this action are entrepreneurs. They are people that can see possibilities and are willing to take risks. Often, the results of an entrepreneur's work is a company. A company is a business enterprise that is organized and managed to produce goods and services. The company is operated by a group of people that are called managers. They plan, organize, actuate, and control company activities. The managers set goals, structure systems, direct operations, and measure results. Without management, technological activities would become less efficient. With proper management, technology promises to serve our needs.

WORDS TO KNOW

All of the following words have been used in this chapter. Do you know their meanings?

Actuating
Articles of incorporation
Authority
Board of directors
Bond
Bylaws
Chief executive officer
Controlling
Corporate charter
Corporation
Debt financing
Dividend
Entrepreneur
Entrepreneurship
Equity financing
Inside directors
Intrapreneurship
Limited liability
Management
Middle management
Operating management
Organizing
Outside directors

Owner
Partnership
Planning
President
Prime
Prime interest rate
Private enterprise
Proprietorship
Public enterprise
Responsibility
Reward
Risk
Stock
Supervisor
Top management
Unlimited liability
Vice president

TEST YOUR KNOWLEDGE

Write your answers on a separate piece of paper. Please do not write in this book.

1. List the three major types of ownership.
2. True or false. Owners of proprietorships are totally responsible for the company's debts. This is called unlimited liability.
3. Forms filed with officials in a state that lead to a corporate charter are called the _____.
4. Place the steps in forming a company, listed on the right, in their proper order:

   _____ First step.   A. Elect corporate
   _____ Second step.   directors.
   _____ Third step.   B. Prepare bylaws.
   _____ Fourth step.   C. Receive corporate
                                charter.
                             D. File articles of
                                incorporation.

5. True or false. Managers of a company cannot become directors of the same corporation.
6. A certificate that indicates a person owns a portion of a corporation is called _____.
7. List the two types of financing.

APPLYING YOUR KNOWLEDGE

1. Assume your class is going to produce and sell popcorn at basketball games. Develop an organization chart for the enterprise. Indicate who has the most authority and the chain of command from that person to all other members of the enterprise.
2. Make an appointment with a stockbroker. Ask him or her to tell you about stocks and bonds and how they are traded. You may want the person to speak to your class.

Organization is needed to make a technological enterprise function. (Northern Telecom)

29
CHAPTER

OPERATING TECHNOLOGICAL ENTERPRISES

After studying this chapter you will be able to:
* *Define an economic enterprise.*
* *List and describe the five main managed areas of activities found in technological enterprises.*
* *Diagram the relationships among the managed areas of activity within an enterprise.*
* *List and describe the functions of research and development.*
* *List and describe the functions of production.*
* *Describe the common systems of manufacturing.*
* *Explain the processes involved in maintaining quality.*
* *List and describe the functions of marketing.*
* *Explain how producers distribute products to consumers.*
* *List and describe the functions of industrial relations.*
* *Describe the major types of programs that are included in industrial relations.*
* *List and describe the functions of financial affairs.*
* *Explain the relationship between expenses, income, and profit.*
* *Explain the steps a product or service goes through from creation to implementation.*

SOCIETAL INSTITUTIONS

Over time, humans have developed a complex society to meet their wants and needs. Within this society are a series of institutions. Five basic institutions, as shown in Fig. 29-1, are:
* **Family:** The basic unit within the society that provides the foundation for social and economic actions.

* **Religion:** An institution that develops and communicates values and beliefs about life and appropriate living.
* **Education:** An institution that communicates information, ideas, and skills from one person to another and from one generation to another.
* **Political/Legal:** An institution that establishes and enforces the societal rules of behavior and conduct.
* **Economic:** An institution that designs, produces, and delivers the basic goods and services required by the society.

All of these institutions use technology because they are concerned with efficient and appropriate action. They apply resources to meet human wants and needs. People in each institution use technical means to make their job more efficient.

However, almost all technology originates in the economic institution. For example: teachers work in the educational institution. They may use computers to make their teaching more efficient. Yet, the computer is not a product of the educational institution. The computer is a product of the economic institution. Likewise,

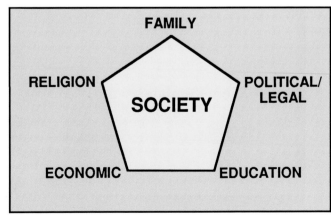

Fig. 29-1. These are five important institutions in society.

politicians are part of the political/legal institution. They may use television to help win an election. Again, television and all associated communication devices are not developed in the political system. They are outputs of economic activity.

ECONOMIC ENTERPRISES

Technology is directly associated with economic enterprises. **Economic enterprises** are enterprises that engage in economic activity. **Economic activity** can be described as *business efforts directed toward making a profit.* Thus, economic activity includes all trade in goods and services paid for with money, Fig. 29-2.

Not all economic action *develops* technology. One type of economic activity is called commercial trade. Commercial trade includes all the wholesale and retail merchants who move products from the producers to the consumers. These traders do not change the prod-

Fig. 29-2. Economic activity includes four major types of activity.

ucts they distribute. They simply make them easily available for people to buy.

Banks, insurance companies, and stockbrokerage firms provide financial services to large numbers of people. They protect our wealth, buy and sell stocks and bonds, or insure our lives and possessions. But again, these companies use technology. They do not develop it.

Similarly, any number of service businesses repair products and structures for a fee. They service and maintain technological devices. Their work extends the life of devices, but it does not develop them.

INDUSTRY

Most technological design, development, and production takes place within a special type of economic enterprise. We call these businesses **industry**, Fig. 29-3. The term, industry, can have several meanings. One definition groups all businesses that make similar products together. This definition leads people to talk about the steel industry, the forest products industry, and the electronics industry.

However, in this book, a more restricted definition is used. In this definition, industry is *the area of economic activity that uses resources and systems to produce products, structures, and services with intent to make a profit.*

This definition includes the four major human productive activities. Therefore, we have communication industries, construction industries, manufacturing industries, and transportation industries. These industries apply the four basic types of technologies to meet human needs. They provide us with products, structures, and information. In addition, they move us and our goods.

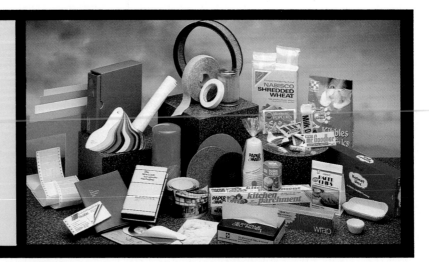

Fig. 29-3. All of these household products were made through industry.

AREAS OF ACTIVITY

In each industry, a number of efficient and appropriate actions take place. These actions are designed to capture, develop, produce, and market creative ideas. These technological activities form the link from the inventor/innovator to the customer.

There are thousands of individual actions that cause a new product or service to take shape. However, they can be gathered into one of five different areas of activity, Fig. 29-4. These are:

- Research and development: The managed activities that may result in new or improved products and processes.
- Production: The managed activities that develop methods for producing products or services, and the activities that produce the desired outputs.
- Marketing: The managed activities that encourage the flow of goods and services from the producer to the consumer.
- Industrial Relations: The managed activities that develop an efficient work force, and the activities that maintain positive relations with the workers and the public.
- Financial Affairs: The managed activities that obtain, account for, and disburse funds.

The first three activity areas are product- or service-centered. They directly contribute to the design,

production, and delivery of the planned outputs. The other two areas are support areas. Financial affairs provides financial support, while industrial relations contributes to human or personnel support.

RESEARCH AND DEVELOPMENT

Research and development can be viewed as the "idea mill" of the enterprise. In this area, employees work with the true raw material of technology, human ideas. They convert what the human mind envisions into physical products and services. These actions can be divided into three steps: research, development, and engineering.

RESEARCH

Research is the process of *scientifically seeking and discovering knowledge,* Fig. 29-5. Research explores the universe systematically and with purpose. Research also determines, to a large extent, the technology we will have in the future. It determines the type of human-built world we will live in.

Most technology is a result of two types of research. The first is **basic research.** Basic research seeks knowledge for its own sake. We conduct basic research to enlarge the scope and depth of human understanding. People working in basic research are not concerned

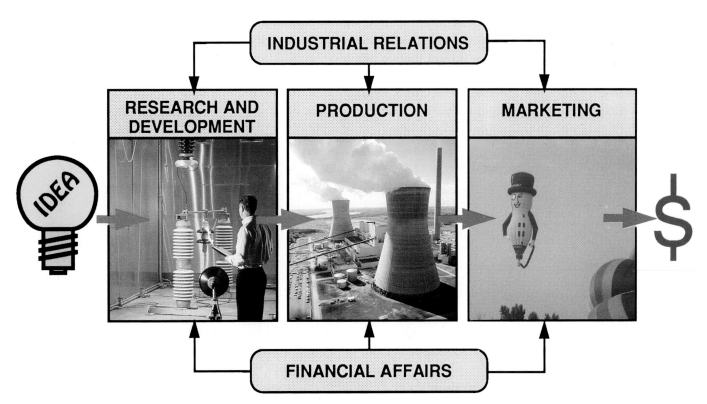

Fig. 29-4. Shown are five major areas of managed activities that change ideas into products, structures, and services. Notice how each area relates with the others.

Fig. 29-5. Research seeks to discover knowledge. Computers have proven a powerful aid for researchers. (AT&T)

about creating new products. Their focus is on generating knowledge.

The second type of research is **applied research.** This activity seeks to reach a commercial goal by selecting, applying, and adapting knowledge gathered during basic research. The focus of applied research is on *tangible* results such as products, structures, and technological systems.

Basic research and applied research complement one another. The former finds knowledge, while the latter finds a use for it. For example, basic research may develop knowledge about the reaction of different materials to high temperatures. Applied research might then determine which material is appropriate for a missile nose cone.

DEVELOPMENT

Development uses knowledge gained from research to derive specific answers to problems. Development converts knowledge into a physical form. The inputs for development are two-dimensional information such as sketches, drawings, or reports. The outputs are models of three-dimensional artifacts such as products or structures.

Development takes place in two areas, Fig. 29-6. The first is **product or structure development.** Product or structure development applies knowledge to design new or improved products, structures, and services. Development may derive a totally new product or structure. It may, also, improve on one that already exists. For example, the bicycle was originally a product from the

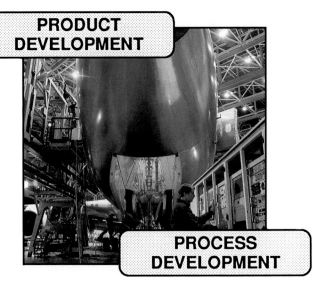

Fig. 29-6. The two types of development are product and process development. Product and process development were used in designing this aircraft. (Boeing)

1800s. A ten-speed bicycle was later developed from the standard bicycle. Likewise, the laser printer uses many processes found in photocopiers.

The second type of development is **process development.** This activity devises new or improved ways of completing manufacturing, construction, communication, or transportation tasks. Process development may result in something totally new, such as fiber-optic communication. It may also improve on old processes, such as inert gas welding, which was derived from standard arc welding.

ENGINEERING

Developed products and structures must be built, and developed processes must be implemented. To do this, people need information. Product and structure

Fig. 29-7. These employees are using a computer system to develop a product drawing.

engineering are responsible for this activity. **Engineering** develops the specifications for products, structures, processes, and services. This is done through two basic activities–design interpretation and engineering testing.

Design interpretation conveys the information needed to produce the artifact (product or structure). This includes three main types of documents: engineering and architectural drawings, bills of materials, and specification sheets.

Engineering drawings convey the characteristics of manufactured products, Fig. 29-7. A set of engineering drawings includes:

- Detail drawings that convey the size, shape, and surface finish of individual parts.
- Assembly drawings that show how parts go together to produce assemblies and finished products.
- Systems drawings that show the relationship of components in mechanical, fluidic (hydraulic and pneumatic), and electrical/electronic systems.

Architectural drawings are used to specify characteristics of buildings and other structures, Fig. 29-8. They include floor, plumbing, and electrical plans for a structure. The drawings also include elevations showing interior and exterior walls.

However, drawings do not convey all the information needed to build products and structures. The people who implement designs need to know the quantities, types, and sizes of the materials and hardware needed. This information is included on a **bill of materials.**

Finally, data about material characteristics are contained on **specification sheets.** You may look back in Chapter 12 for a more extensive presentation of these three types of documents.

PRODUCTION

Research and development develops and specifies ideas for products and structures. **Production,** then, must produce (manufacture or construct) the physical item. There are a number of different systems used to produce products and structures. The four common manufacturing systems are:

- **Custom manufacturing:** This is producing a limited quantity of products to a customer's specifications. Generally, the products are produced only once. This system requires highly skilled workers and has a low production rate. Custom manufacturing, therefore, is an expensive system to operate. Two examples of custom manufactured products are tailor-made clothing and the space shuttle.
- **Intermittent (job lot) manufacturing:** With intermittent manufacturing, a group of products is man-

Fig. 29-8. The engineers shown here are reviewing architectural drawings for a oil refinery. (Exxon Corp.)

ufactured to the company's or a customer's specification. The parts move through the manufacturing sequence in a single batch. All parts are processed at each workstation before the batch moves to the next station, Fig. 29-9. Often, repeat orders for the product are expected. This manufacturing activity is relatively inexpensive, but considerable setup time is required between batches of new products.

- **Continuous manufacturing:** In continuous manufacturing, a production line manufactures or assembles products continuously. The materials flow down a manufacturing line that is specifically de-

Fig. 29-9. These kitchen cabinets are being produced with an intermittent manufacturing system. (American Woodmark)

Fig. 29-10. The continuous manufacturing process brought about the rapid expansion in automobile production. This is the end of a continuous automobile body line. (Daimler-Benz)

signed to produce that product, Fig. 29-10. The parts flow from station to station in a steady stream. This manufacturing line handles a high volume and has relatively low production costs. But, continuous manufacturing lines are fairly inflexible and can be used for very few different products (often only one).

- **Flexible manufacturing:** This is a relatively new, computer-based manufacturing system. Flexible manufacturing combines the advantages of intermittent manufacturing (short runs) with the advantages of continuous manufacturing (low unit production cost). Machine setup and adjustment is computer-controlled. This allows for quick and relatively inexpensive product changeovers.

Similar production systems are used in the other technologies. Housing can be built to an owner's specifications (custom-built), Fig. 29-11. Other dwellings are built in tracts with a common plan used to make a large number of buildings.

Fig. 29-11. Some homeowners choose to have their houses custom-built.

An analogy could be made in transportation. Driving an automobile is much like custom manufacturing. It is flexible but relatively expensive. Rapid transit buses and trains move people on set lines but at lower costs. This is more like continuous manufacturing.

The actual production of products and structures can be divided into three important tasks, planning to produce, producing, and maintaining quality of the product or structure, Fig. 29-12.

PLANNING

Planning determines the sequence of operations needed to complete a particular task. Planning is the backbone of most production systems. It determines the needs for human, machine, and material resources. Planning also assigns people and work to various workstations.

Closely associated with planning and scheduling is production engineering. Production engineers design

Fig. 29-12. These are the three tasks of producing. Pictured are the planning and producing stages in the production of communication messages. (Hewlett-Packard, Ball State University)

and install the system that is used to build the product or structure. They are concerned with the physical arrangement of the machines and workstations needed to produce the product.

PRODUCING

Producing is the actual fabrication of the artifact. In manufacturing, it involves changing the form of materials to add to their worth or value. These activities include locating and securing material resources, producing standard stock, and manufacturing the products.

Product **manufacturing processes** are used to change the size, shape, combination, and composition of materials. These processes include using casting and molding, forming, and separating processes to size or shape materials. Conditioning processes change the internal properties of the material. Assembling processes put products together, and finishing processes protect their surfaces.

Construction processes are used to produce buildings and heavy engineering structures. Typical construction processes include preparing the building site, setting foundations, erecting superstructures, enclosing and finishing structures, and installing utility systems.

Communication processes are used to produce graphic and electronic media. Generally, communication messages are designed, prepared for production, produced, and delivered. This is done through the processes of encoding, storing, transmitting, receiving, and decoding operations.

Transportation processes are used to move people and cargo. They are used in land, water, air, and space systems. Typical transportation production processes include loading, moving, and unloading vehicles.

MAINTAINING QUALITY

Throughout these processes, a standard of perfection is maintained. Customers want products, structures, and systems to meet their needs and desires. For this to happen, a process called **quality control** is used. Quality control includes all systems and programs that *ensure the outputs of technological systems meet engineering standards and customer expectations.*

Often, people think "quality" means smooth, shiny, and exact size. This is not always true. A smooth, shiny road makes a poor driving surface. Cars would have controlling and braking difficulty on such a surface. Likewise, holding the length of a nail to +/- .001 in. tolerance is inappropriate. The cost of manufacturing to that tolerance is too high for the product. The important quality consideration in a nail is holding power, not exact length. Quality can be measured only when a person knows how the product or part is to be used. The product's function dictates quality standards.

Inspection

An important part of a quality control program is **inspection.** Inspection compares materials and products with set quality standards. There are three phases to an inspection program, Fig. 29-13. The first phase inspects materials and purchased parts as they enter production operations. The second phase inspects work during production. The final phase inspects the end product or structure.

Fig. 29-13. Quality control inspects materials entering the plant, work-in-progress, and, finally, finished products. (Goodyear Tire and Rubber Co.)

Inspection can be done on every product or on a representative sample of the products. Expensive, complex, or critical components and products are subjected to 100 percent inspection. This means that every part is inspected at least once. Products such as aircraft components and space satellites are examples of outputs that receive 100 percent inspection.

Less expensive and less critical parts receive random inspections. A sample of the product is selected that represents a typical production run. The sample size and the frequency of inspections is determined using statistics (mathematically based predictions). Sample-type inspection is part of a program called statistical quality control.

The selected sample is inspected. If the sample passes, the entire run is accepted. If the sample fails to meet the quality standards, the entire run is rejected. Rejected production lots can be sorted to remove rejects (parts that fail to meet standards), the run can be discarded, or the run can be reworked.

Random inspection is used whenever it is cost-effective. Often, the cost of 100 percent inspection outweighs its value. For example, it would be expensive to use 100 percent inspection on roofing nails. Also, the occasional defect that slips past random inspection can be discarded by the user without endangering the product or customers.

Price and Value

In addition to quality, customers expect value. Price and value are two different things. **Price** is what someone must pay to buy or use the product or service. Initial prices are established by businesses and reflect market conditions. **Value** is determined by the customer. Value is a measure of the functional worth that the customer sees in the product. The customer expects the product or structure to deliver service and satisfaction equal to or greater than its cost. Answering the question, "Was the product worth what I paid for it?" can establish its value.

Cost-cutting Systems

A number of new production systems have been developed to reduce product cost and, in turn, increase product value, Fig. 29-14. These systems include **computer-aided design (CAD)** systems, which reduce product design and engineering costs. **Just-in-time (JIT)** inventory control systems schedule materials to arrive at manufacturing when they are needed. This reduces warehousing costs. Flexible manufacturing, which was discussed earlier, reduces machine setup time.

Similar systems are used in the other technologies. Computer scheduling is used in construction to ensure

Fig. 29-14. Computer systems, like this transportation vehicle tracking system, reduce costs and increase value. (Norfolk-Southern)

that human and material resources are effectively used. Computer ticketing reduces transportation costs. Computer systems make layout and color photograph preparation for printed products more economical.

MARKETING

Producing artifacts is of little value to companies unless they can sell those artifacts to customers. The products and structures must be exchanged for money.

Fig. 29-15. These people are participating in a quality audit program. This program examines customers' reactions to new recipes. (National Live Stock and Meat Board)

This is the challenge for marketing personnel. **Marketing** efforts *promote, sell, and distribute products, structures, and services.* Specifically, marketing involves four important activities:

- **Market research** gathers information about the product's market, Fig. 29-15. This could include data about who will buy the product and where these people are located, in addition to their age, gender, and marital status. Also, market research can measure the effectiveness of advertising campaigns, sales channels, or other marketing activities.

- **Advertising** includes the print and electronic messages that promote a company or its products. Advertising can also present ideas such as "Say no to drugs!" Advertising is designed to cause people to take action (buy a certain product) or think differently (buckle your seat belt while riding in an automobile).

  Closely related to advertising is packaging, Fig. 29-16. This activity deals with designing, producing, and filling containers. The packages are designed to promote the product through colorful or interesting designs. Packaging also protects the product during shipment and display. Finally, the packages must include information that helps the customer select and use the product wisely.

- **Sales** is the activity that involves the physical exchange of products for money. Sales includes sales planning, which develops selling methods and selects and trains sales personnel for their efforts. Sales, also, includes the act of selling. This involves approaching customers, presenting the product, and closing the sale. This series of steps is all part of sales operations. The end result of sales operations is an order from the customer and some income for the company.

- **Distribution** is physically moving the product from the producer to the consumer. This consumer may be another company or a retail customer. Consumer

Fig. 29-16. Packages are designed and produced to promote and protect the product as well as to inform the customer. (James River Corp.)

products follow at least three common channels, Fig. 29-17. The product may move from the producer directly to the consumer. This channel is called direct sales. Sales of homes, encyclopedias, cosmetics, vacuum cleaners, and transportation services often use this channel.

In another channel, the producer sells the products or service directly to a retailer. The retailer then makes the items available to the customer. Franchised dealers, such as new automobile dealers, are an example of this distribution channel. This channel allows the producer to regulate the number of sales outlets and the quality of service that those outlets provide.

A third channel has the producer selling products to wholesalers. These businesses buy and take possession of the products. They, in turn, sell their commodities to retailers. The retailers then sell the commodities to the customers. In this channel, producers have little control over the retailers who are selling their products.

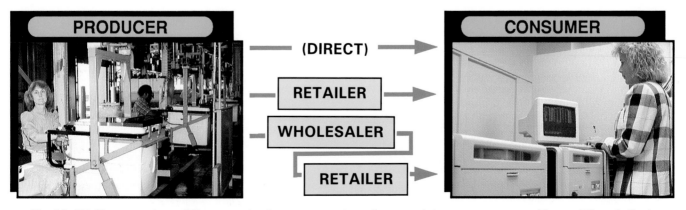

Fig. 29-17. There are three different paths that move products from producers to consumers. (General Electric Co.)

Technology: How it Works

INFORMATION SUPERHIGHWAY: The vast information and services available through global computer networks.

Rapid advances in computer technology have placed large amounts of information and services at the fingertips of anybody with a computer and a modem. A *modem* is a device that allows a computer to communicate with other computers over phone lines. As computer technology has advanced, the price of the machines has declined. This has allowed more people than ever before the opportunity to own a computer.

There are many different on-line service networks that a computer user can connect to. A *network* is several computers connected together so that they can use the same information and programs. The phrase *on-line* means that a computer is connected to a network. In other words, if you use your computer and modem to connect to one of these networks, you are "on-line."

The Internet is a worldwide network of thousands of computers connected through the phone system. The Internet began between 30 and 40 years ago as a system connecting researchers and universities together. The system has grown and spread so that now anyone with computer access can use the system. Nearly any type of information you need can be found on the Internet, from recreational to educational. In addition, many private companies offer a home page.

A *home page* is an electronic address that can be accessed through a part of the Internet called the World Wide Web, Figure 1. Companies and individual users can set up home pages so that people can get access to information they have to share. Many companies use a home page to sell their products or services. Orders can usually be placed directly with the company while on-line.

As computer technology continues to grow, so do on-line services. As the networks grow, so will the amount of information and services available.

Fig. 1. The World Wide Web is just one feature of on-line services. (Reprinted with the permission of America Online, Inc.)

Research and development develops the product, structure, or service. Production produces the item, while marketing promotes and sells it. But, these three activities cannot stand alone. They require money and people to make everything work. These requirements are the responsibility of two managed support areas–industrial relations and financial affairs.

INDUSTRIAL RELATIONS

You have learned that humans are the foundation for all technology. They create it, develop it, produce it, and use it. Thus, people are fundamental to all company operations. They work for the company, buy the company's products, and pass laws that regulate company operations.

Companies, therefore, are very concerned about their relationships with people. They nurture positive relationships with people by engaging in industrial relations activities. These activities can be grouped under three main programs. These programs are, as shown in Fig. 29-18:

- **Employee relations:** Programs that recruit, select, develop, and reward the company's employees.
- **Labor relations:** Programs that deal with the employees' labor unions.
- **Public relations:** Programs that communicate the company's policies and practices to governmental officials, community leaders, as well as to the general population.

EMPLOYEE RELATIONS

All companies have employees. These employees did not magically appear, fully qualified to work. They are the result of managed employee relations activities. These activities select, train, and reward the people that produce products and manage operations.

Selecting Employees

The first action in the employee relations process is called **employment.** Employment makes sure that the company is staffed with qualified workers. This task involves determining the company's employment needs. Then, qualified applicants are acquired through a searching process called **recruiting.** Recruiting can be done through newspaper advertisements, school recruiting visits, or employment agencies. Other applicants may come to the company seeking employment on their own.

Next, job applicants go through a **screening** process. Screening allows qualified people to be selected from the applicant pool. Generally, this selection process starts with an application form to gather personal and work experience data. The promising applicants are interviewed to gather additional information, Fig. 29-19. Finally, some jobs require special abilities and knowledge. In these cases, tests may be given to determine which applicants possess the necessary characteristics. In all cases, the questions asked on application blanks, during interviews, and on tests must relate to job performance. It is illegal to ask questions that contain an age, racial, or gender bias.

Training Employees

Successful applicants gain employment. However, few people are ready to begin work at this point. Most new employees need some basic training, and some may need special instruction. Basic information about the

Fig. 29-18. Industrial relations develops and administers people-centered programs. (Harris Corp)

Fig. 29-19. This job applicant is being interviewed to gain information not gathered on the application form. (Inland Steel Co.)

company and its rules and policies is provided to all workers. This is called induction training.

Special job skills can be provided through one of three programs. Simple skills are generally taught through **on-the-job training.** With this program, the new worker is trained at his or her workstation by an experienced worker or manager. More specialized skills may be developed in **classroom training** sessions. With classroom training, qualified instructors provide information and demonstrate practices that each employee needs.

Highly skilled workers are developed through **apprenticeship training.** Apprentices receive a combination of on-the-job training and classroom training over an extended period of time. Apprenticeships last about four years.

In all three programs, workplace safety is stressed. New employees are informed about company safety, shown how to work safely, and in many cases, tested on safe work practices. Safety training is closely coupled with providing a safe work place. State and federal agencies provide companies with rules and regulations dealing with safety. A principal source of these regulations is OSHA (Occupational Safety and Health Administration).

Executives and professional employees (engineers, salespeople, etc.) also receive training. This training may be called executive development, sales training, or managerial training. Since most of the work done by these employees is not associated with the making of products, these training sessions are often given in classroom settings, Fig. 29-20. The entire training activity is called human resource development (HRD) in many companies.

Rewarding Employees

People want to be recognized and rewarded for their work. Companies do this in two ways. First, companies pay employees a wage, salary, or commission as a direct reward for work accomplished. A **wage** is a set rate that is paid for each hour worked. A **salary** is a payment rate based on a longer period of time, such as a week, month, or year. Wage earners are often called hourly workers. Hourly workers are usually the production workers who build products, erect building, print products, or provide transportations services.

Salaried employees tend to be the technical and managerial workers. They develop products, engineer facilities, maintain financial records, and direct the work of other people. These employees often have more education than hourly workers. Salaried employees are held accountable more in terms of the amount of work that they do than the hours that they work.

Fig. 29-20. This group of managers is receiving training to improve their communication skills. (Clorox Co.)

Some sales people are paid in a different way. They receive a **commission** for each sale they make. Usually, the commission is a percentage of the total dollar value of the goods sold.

The second type of reward is called **benefits.** These are the insurance plans, vacations, holidays, and other programs provided by the company. These items cost the company money and, therefore, are a part of the total pay package for an employee.

LABOR RELATIONS

Many larger companies work with labor unions that represent their employees. These companies require a labor relations program. This program works on two levels. First, **labor agreements** (contracts) are negotiated with the union. Labor agreements establish pay rates, hours, and working conditions for all employees covered by the contract. The agreements cover a specific period of time, which generally ranges from one to three years.

During the contract period disputes often arise over its interpretation. These disputes are called **grievances.** This is the second level of work for labor relations. Labor relations officials work with union representatives to settle the grievances.

PUBLIC RELATIONS

Companies are operating entities in society. They hire people, pay taxes, and have direct impacts on communities. Company managers form policies and have practices that they feel benefit the company. These practices often are subject to government regulation and may be affected by community pressures.

A company's public relations program is designed to gain acceptance for company operations and policies. The program informs governmental officials about the need for, and impact of, laws and regulations. Public relations also communicates with the community leaders so that local actions do not hamper the company's legitimate interests. Finally, public relations communicates with the general public. This communication presents the company as a positive force in the community, Fig. 29-21. This image improves the company's ability to hire qualified workers and to sell its products.

FINANCIAL AFFAIRS

Just as a company needs people, it also needs money. Companies must buy materials and equipment, pay wages and salaries, and rent or buy buildings. Taxes and insurance premiums must be paid. This action can be shown in a simple cash flowchart, Fig. 29-22. If you start at the upper left corner and read down you will see that:

Management employs people to use machines to change the form of materials, which is paid for by money and produces products, which are converted in the marketplace into money.

You will note that the word *money* appears twice. Money pays for resources: people's time, materials, and machines. This money is called **expenses.** Money is also the end result–sales **income.** The difference between income and expenses is profit or loss. The goal of a company is to have more income than expenses. This makes a company profitable.

Profits must also be managed. **Profits** serve two important purposes. First, they can become **retained earnings.** These are profits that are held by the company and used to enlarge its operations. Profits are an important source of money for financing new

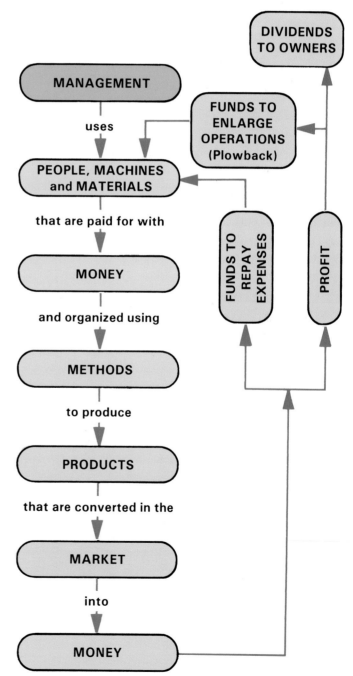

Fig. 29-22. A simple cash flowchart that shows how money cycles within company activities.

Fig. 29-21. This float from the Rose Bowl Parade is part of a company's public relations program. The float is intended to help create a positive image for the company.

products, plant expansions, and mergers. Loans, bonds, and stock sales are other sources of financing.

Another portion of the profits of many companies is paid out as **dividends.** These are quarterly or annual payments to the stockholders. Dividends are the rewards for investing in the company and sharing the risks of owning a business.

Fig. 29-23. One responsibility of financial affairs personnel is purchasing materials for the company. (Brush Wellman)

Managing the use of this money is the responsibility of financial affairs employees. They raise money, pay for insurance, collect from customers, and pay taxes.

They also keep records of the financial transactions of the company. This area is called **accounting.** Each financial action is recorded as either an income or expense item.

Finally, financial affairs purchases the materials, machines, and other items needed to operate the company, Fig. 29-23. Purchasing officers seek the "best" items in relation to company needs. The term "best" takes into account price, quality, and delivery date.

INDUSTRY-CONSUMER CYCLE

We have been looking at company activities as a linear action. The activities were presented as starting with research and development. Production was the next step and, finally, marketing. This path is correct if you look at a single model or version of one product or structure. However, our system is much more complex. The economy is actually described as dynamic (always changing), Fig. 29-24. Products are developed, produced, and sold. Consumers select, use, maintain, and discard the products. They communicate their satisfaction or dissatisfaction with current products. This causes companies to redesign existing products or, in some cases, to develop new ones. These new entries are

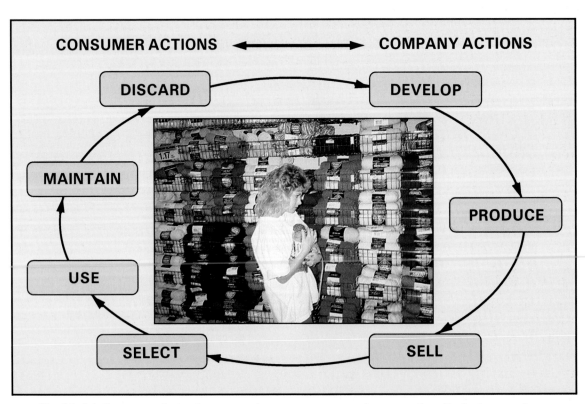

Fig. 29-24. Customers' reactions to products cause companies to continually design, produce, and sell new items.

sold. They, too, are selected, used, maintained, and discarded. This cycle continues with a constant array of new products being developed and obsolete ones disappearing.

SUMMARY

Most technology is developed and produced by industrial enterprises. These enterprises create a steady flow of products, structures, and services to meet human wants and needs. This is done through five main areas of activity. Research and development develops and specifies the item; production produces it; marketing promotes, sells, and distributes it. These product- and structure-centered functions are supported by industrial relations and financial affairs activities. Industrial relations recruits and develops the workforce. It also deals with labor unions and promotes the company's image and policies. Financial affairs maintains financial records, pays the company's bills, and purchases material resources.

WORDS TO KNOW

All of the following words have been used in this chapter. Do you know their meanings?

Accounting
Advertising
Applied research
Apprenticeship training
Architectural drawings
Basic research
Benefit
Bill of materials
Classroom training
Commission
Communication processes
Computer-aided design (CAD)
Construction processes
Continuous manufacturing
Custom manufacturing
Development
Distribution
Dividend
Economic activity
Economic enterprise
Employee relations
Employment
Engineering
Engineering drawings
Expense
Financial affairs
Flexible manufacturing
Grievance

Income
Industrial relations
Industry
Inspection
Intermittent manufacturing
Just-in-time (JIT)
Labor agreement
Labor relations
Manufacturing processes
Market research
Marketing
On-the-job training
Price
Process development
Product development
Production
Profit
Public relations
Quality control
Recruiting
Research
Retained earning
Salary
Sales
Screening
Specification sheets
Transportation processes
Value
Wage

TEST YOUR KNOWLEDGE

Write your answers on a separate piece of paper. Please do not write in this book.

1. Define industry.
2. True or false. All businesses that intend to make a profit are called industry.
3. True or false. Price and value are the same thing.
4. True or false. Many skilled workers are trained through apprenticeship programs.
5. Match the statements on the left with managed area of activity on the right that has that responsibility:

_____ Accounts for income and expenditures.   A. Research and development.

_____ Makes structures.   B. Production.

_____ Designs services.   C. Marketing.

_____ Specifies characteristics of products.   D. Industrial relations.

_____ Promotes products.   E. Financial affairs.

_____ Responsible for quality control.

_____ Includes public relations.

_____ Purchases materials.

6. List the four types of manufacturing systems.
7. True or false. Promoting products is called sales operations.
8. Making sure that products meet specified standards is called _____.
9. List the three important programs included in a total industrial relations program.
10. A dispute over an interpretation of the labor agreement is called a _____.
11. True or false. Profit is the money paid to stockholders as a reward for investing in the company.

APPLYING YOUR KNOWLEDGE

1. Select a simple product, such as a kite. Apply the principles of research and development, production, and marketing to design, produce, and advertise it.
2. Set up a production line for chocolate suckers, cookies, or another food product. Describe how you will (1) plan for the product, (2) produce the product, and (3) maintain quality. Use a form similar to the one to the right for your planning.

Product:	
PLANNING FOR PRODUCTION	
Steps in making the product: 1. 2. etc.	Resources needed: Human: Materials: Equipment:
CONTROLLING QUALITY	
Standards to be met: 1. 2. etc.	Method of inspecting: Human: Materials: Equipment:

People must work together if they are to make a quality product.

Section Nine - Activities

ACTIVITY 9A - FORMING THE COMPANY

Background:

Many technological devices and products are produced by companies. These enterprises are the product of human actions. They are formed and structured to efficiently use resources to produce artifacts that meet our wants and needs.

Situation:

Students have mentioned that they need a way to be informed about important sporting and social events at school. They also want a way to publicize happenings that they feel are important. From these comments, you have concluded that an inexpensive personalized calendar would meet their needs and earn you a profit. This type of calendar can be produced with limited finances by using new computer software, Fig. 9-A.

Challenge:

Organize a company to produce a 9-month calendar. The calendar should span the school year and should have selected days personalized. Consider the tasks to be completed, and consider the managerial structure needed to complete them. Be sure to recognize that there are production tasks and marketing tasks. Also, there are two distinct phases of the company operations, which may require two different organizations. One operation could finance the company, design the calendar, and sell calendar entries. The other operation would then maintain financial records, produce the calendars, and sell the finished products, Fig. 9-B.

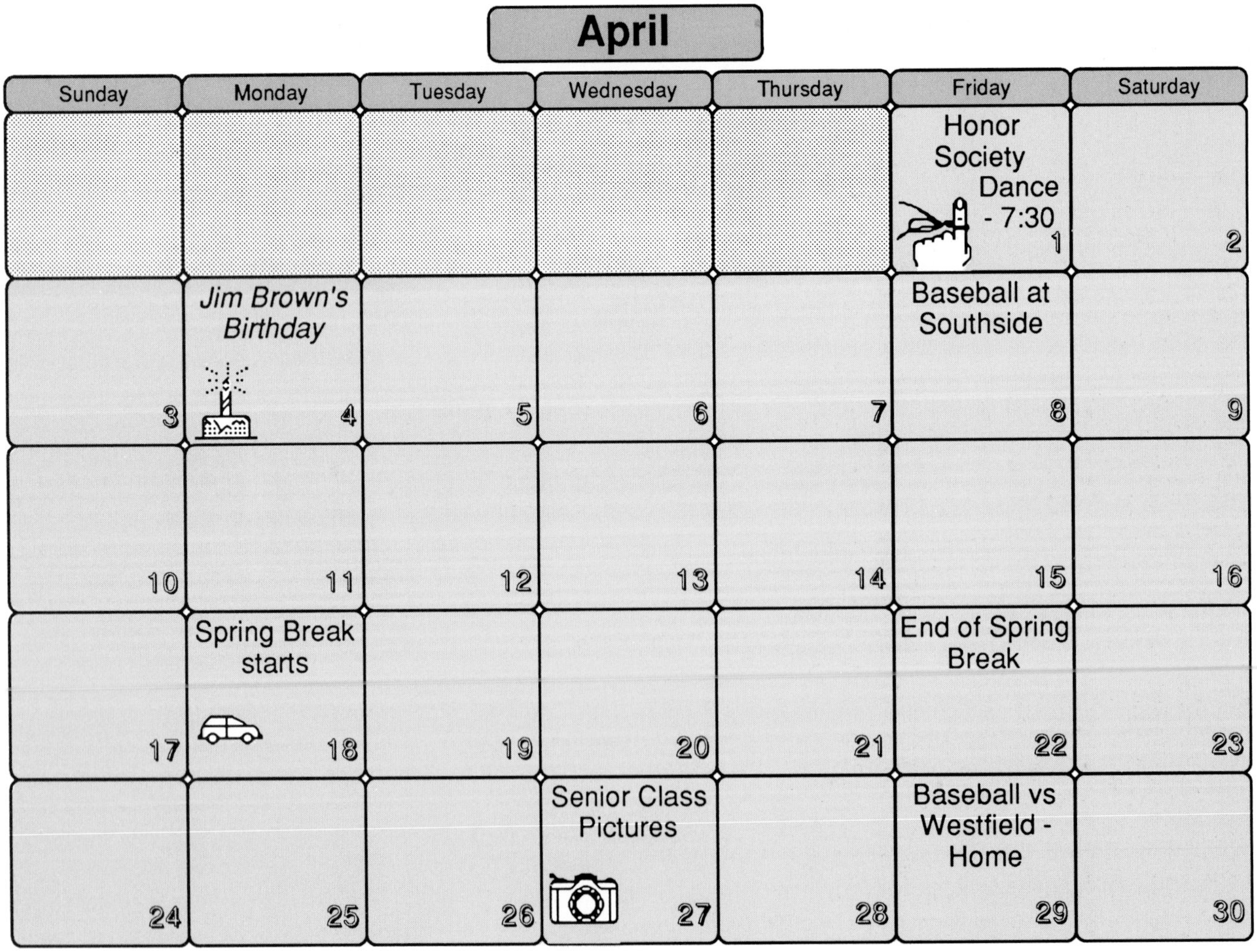

April

Sunday	Monday	Tuesday	Wednesday	Thursday	Friday	Saturday
					Honor Society Dance - 7:30 1	2
3	Jim Brown's Birthday 4	5	6	7	Baseball at Southside 8	9
10	11	12	13	14	15	16
17	Spring Break starts 18	19	20	21	End of Spring Break 22	23
24	25	26	Senior Class Pictures 27	28	Baseball vs Westfield - Home 29	30

Fig. 9-A.

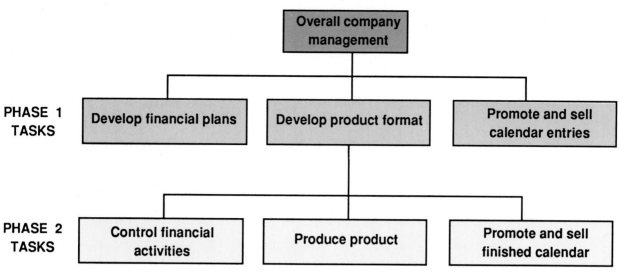

Fig. 9-B.

Optional:
- Write a one-page job description for each job in the organization chart.
- Develop a set of goals for each major department in the company.

ACTIVITY 9B - OPERATING THE COMPANY

Background:

Companies are a series of independent tasks that have been integrated into a functioning enterprise. Each task must be planned for and carried out with efficiency.

Challenge:

Identify, schedule, and complete the several tasks required to produce and market a personalized calendar for your school.

Procedure:

Design and Development Department
1. Obtain software that can be used to produce a personalized calendar. An example is CalendarMaker™ by CE Software Inc.
2. Follow the instructions to produce a calendar for a single month. This will acquaint the department members with the operation of the software.
3. Establish the layout for the calendar.
4. Produce a common layout sheet for the marketing group to use in selling calendar entries.

Calendar Entry Marketing Department
1. Determine the selling price for a calendar entry.
2. Develop a calendar entry order form.
3. Develop posters to promote the sale of calendar entries.
4. Sell calendar entries.

Production Department
1. Receive calendar entry forms from the marketing department.
2. Enter data on calendar layouts.
3. Print proof of the calendar.
4. Submit proof to marketing for approval.
5. Correct calendar entries.
6. Print master calendar.
7. Reproduce calendar.

Calendar Marketing Department
1. Produce and distribute advertisements for the sale of calendars.
2. Select and train calendar sales persons.
3. Sell calendars.
4. Maintain sales records.

Finance Department
1. Set budgets for company operations.
2. Sell stock and maintain stockholder records.
3. Purchase materials and supplies.
4. Maintain all financial records.

Executive Committee (president and vice presidents)
1. Set deadlines for important activities.
2. Monitor progress in completing tasks.
3. Set budgets.
4. Set selling prices for calendar entries and finished calendars.
5. After the calendar copies have been sold, close the company and liquidate the assets.

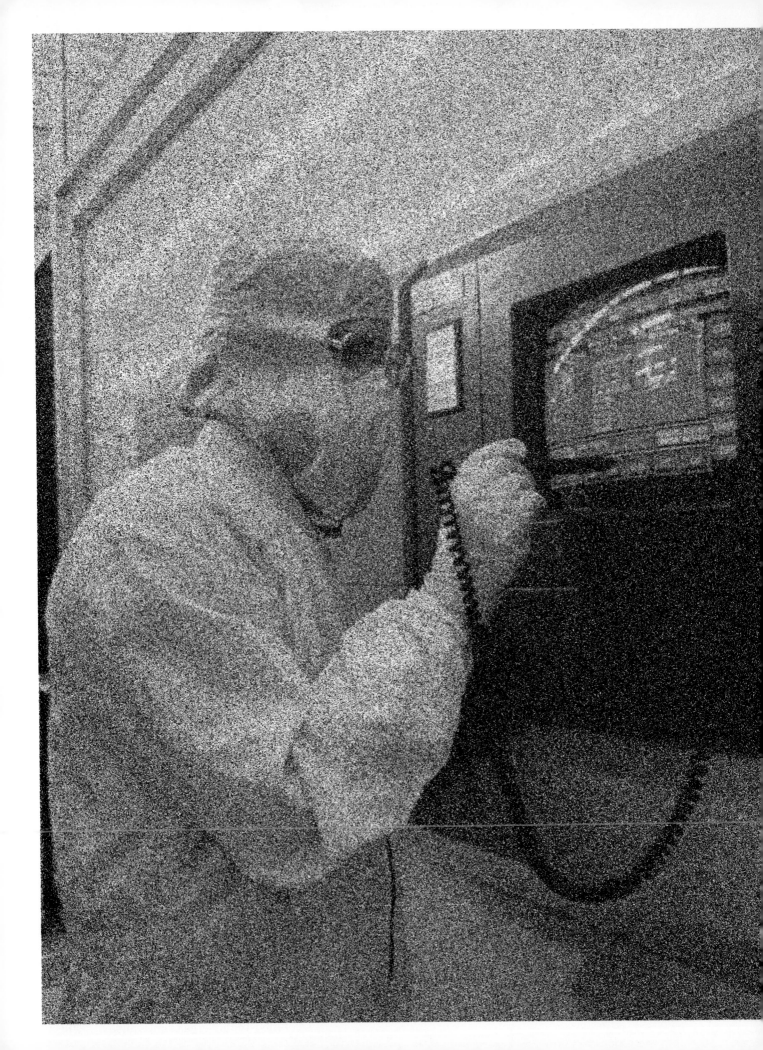

Technological Systems in Modern Society

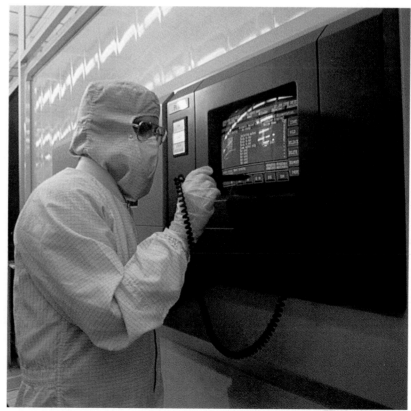

TEXAS INSTRUMENTS

Society makes choices between many alternatives. Each person
makes decisions that influence the choices a society makes.
(U.S. Department of Energy, Bud Smith)

30

CHAPTER

TECHNOLOGY: A SOCIETAL VIEW

After studying this chapter you will be able to:
- *Describe some destructive natural forces that affect human life.*
- *Explain how technology can help to control destructive natural forces.*
- *Discuss how humans have a responsibility to care for the planet on which they live.*
- *Describe technology in a global context.*
- *Explain futuring as a method of determining the value of technological actions.*
- *List and describe the important features of futures research.*
- *Describe the four major types of futures that are considered in futures research.*
- *List three factors that are causing a condition called an environmental crisis.*
- *Describe some actions that each of us can take to reduce our use of energy and other resources.*
- *List and describe actions that companies can take to remain competitive in a world economy.*

You have been learning about technology for almost 30 chapters in this book. You've been introduced to many concepts and a great deal of information. Key to all this learning is one principle: *Technology is developed for only one reason: to serve people.* You've learned that technology is the product of human knowledge and ability. It is developed by people and for people. It is designed to help people modify and control the natural world. Also, technology is the sum total of all human-built systems and products. *Technology is the human-built world*, Fig. 30-1.

How we use technology determines if it is good or bad, helpful or harmful. Technology, itself, is neutral. Technology, by itself, does not affect people or the environment. *How* people use technology can *help* or *harm* the world around them.

TECHNOLOGY CONTROLS AND HARNESSES NATURAL FORCES

Natural forces affect, and sometimes disrupt, human life, Fig. 30-2. Hurricanes wreck ships and destroy coastal settlements. Tornadoes can level entire sections of towns and cities. Floods wash away farmland and homes. Floods carry vital topsoil away, reducing the productivity of crop lands. Fires burn buildings, crops, and forests. Earthquakes shake structures until they collapse. In all these natural events, people can be killed and their possessions destroyed.

One of the earliest uses of technology was to harness natural forces. Humans started to design a human-built world that could reduce natural destructive forces, Fig. 30-3. Today, dams hold back floodwater and produce energy in hydroelectric plants. Fire is tamed to heat our homes and process industrial materials. The inner heat of the earth is captured by geothermal power plants. Wind is used to generate electricity. Solar energy provides heat and electricity for businesses and homes.

However, controlling and using natural forces is not enough. It is not a battle of people against nature. Humans are starting to realize that they must live in harmony with the natural world. Many people understand that humans have control over the earth's future. However, this control carries with it serious responsibilities. We must protect the environment and the plants and animals that live with us on this planet.

from SLOW, ROUGH TRANSPORTATION

to RAPID, COMFORTABLE TRANSPORTATION

from HARD, HUMAN LABOR

to EASIER, MECHANIZED LABOR

from RUSTIC, ESSENTIAL HOUSING

to MODERN, COMFORTABLE HOMES

Fig. 30-1. Technology has greatly changed our world and our way of life. (Daimler Benz, GE Plastics)

TECHNOLOGY HAS A GLOBAL IMPACT

This responsibility requires that people realize we live in a very different world than our grandparents did. The technology that we develop and use today has impacts beyond our homes, cities, states, and countries. We must look at the impacts of our decisions on the entire world.

For example, the attachment that North Americans have to the automobile as the primary transportation vehicle has *global impacts*. The citizens of the United States and Canada represent a small percentage of the world's population. However, they use a large percentage of the world's petroleum. Also, automobiles discharge great quantities of pollutants into the atmosphere.

Likewise, many North Americans are concerned about high birth rates in Third World countries. The impact of each new North American child on the world's resources, however, is many times greater than a child from the Third World. We must be concerned about population *and* resource use.

Fig. 30-2. Nature's destructive forces can destroy forests with fire (left) or cause rivers to flood (right). Technology is used to prevent damage from these forces. (U.S. Department of Agriculture).

TECHNOLOGY AND THE FUTURE

Since technology is a product of human activity, it can be controlled by humans. To do this, we must have an idea of the kind of future we want. This can be developed by a new research technique called **futuring** or **futures research**.

This process helps people select the best of many possible courses of action, Fig. 30-4. Futuring emphasizes five distinct features:

- Alternate avenues. The futurist looks for *many* possible answers rather than *the* answer.
- Different futures. Traditional planning sees the future as a refinement of the present. A futurist looks for an entirely new future.
- Rational decision-making. Traditional planners rely heavily on statistical projections. They use mathematical formulas to help them predict events and impacts. A futurist uses logical thinking and considers the consequences when making decisions.

- Designing the future. Futurists are not concerned with improving present or past practices. They do not see the future as a variation of the present. They focus on predicting a possible future that can be created.
- Interrelationships. Traditional planners use linear models that suggest one step leads to the next step. Futurists see alternatives, cross-impacts, and leaps forward.

Using a futures approach, exciting new technologies can be developed. The futurist must have a dual view. One view must be of a present challenge or problem. The other view must be of the world of the future, where his or her grandchildren will live. Looking at both of these views requires a combination of short-term and long-term goals. The most important tool to develop these futures is the human mind. It is the only tool capable of reasoning and making value judgments.

Futuring will require the two types of thinking introduced in Chapter 10. The first is divergent thinking.

Fig. 30-3. People use technology to harness and control the forces of nature. Fire can destroy a forest (left) or can be harnessed to cook pastries (right).

Fig. 30-4. People can use futuring to determine the type of future they want.

This type of thinking lets the mind soar. People are encouraged to explore all possible and, in many cases, impossible solutions. They look for interrelationships and connections. Out of this activity emerges a number of futures.

One possible future, however, must be selected as the one that will be "created." This requires convergent thinking. The final solution must receive focus and attention. Its positive and negative impacts must be carefully analyzed.

This analysis must be done with four types of futures in mind. The first is the **social future**. It suggests the type of relationships people want with each other. The second is the **technological future**. This looks at the type of human-built world we desire. The third is the **biological future**. It deals with the type of plant and animal life that we want. The fourth is the **human-psyche future.** This future deals with the mental condition of people. It stresses the spirit rather than the mind, attitude instead of physical condition. It is concerned about how people will feel about life and themselves.

These futures are listed separately. Actually, they are interconnected, Fig. 30-5. For example, new technologies directly impact how people relate with one another. Television dramatically changed family life, recreation, and the entertainment industry. Technology also changes the natural environment. Acid-rain caused by automobile emissions and coal-burning electric plants has destroyed forests in Canada and the eastern United States. Likewise, technology has changed how we view ourselves. Some people feel trapped by technology while other people feel empowered by it.

TECHNOLOGY: CHALLENGES AND PROMISES

It is impossible to explore how each new technology has impacted our lives and how it will impact the future. However, we can explore some examples to provide a foundation for personal study. Later you can use the examples as you assess other technologies.

Three issues that are constantly being discussed are energy use, the environmental protection, and global economic competition. Let's look at these issues in terms of their challenges and promises.

Fig. 30-5. Technology impacts the environment, people, and the society as a whole. (FMC Corp)

Technology: How it Works

MECHANIZATION: Replacing human labor with machines that are operated or controlled by people or other machines.

Increasing productivity is key to a rising standard of living. For increased productivity, more goods and services must be produced with the same consumption of materials or labor. One method of increasing productivity is mechanizing work. Mechanization is moving tasks from human operation and control to a mechanical control. This action reduces the labor component in each job.

Early mechanization was done with simple procedures. Recent mechanization often uses complex machinery. Agriculture provides an example of this situation, Fig. 1. An animal-drawn plow replaced the digging stick. Later, that plow was replaced by the horse-drawn moldboard plow. Currently, we have large tractors that pull complex tilling machines. But, even these machines are being replaced by no-till or minimum-till farming practices, which reduce the amount of work it takes to prepare the soil for planting.

In the industrial setting, robots, automatic machines, and other computer-controlled machines are reducing the amount of human labor needed for each job. These mechanized changes increase the level of education and skill required for remaining jobs.

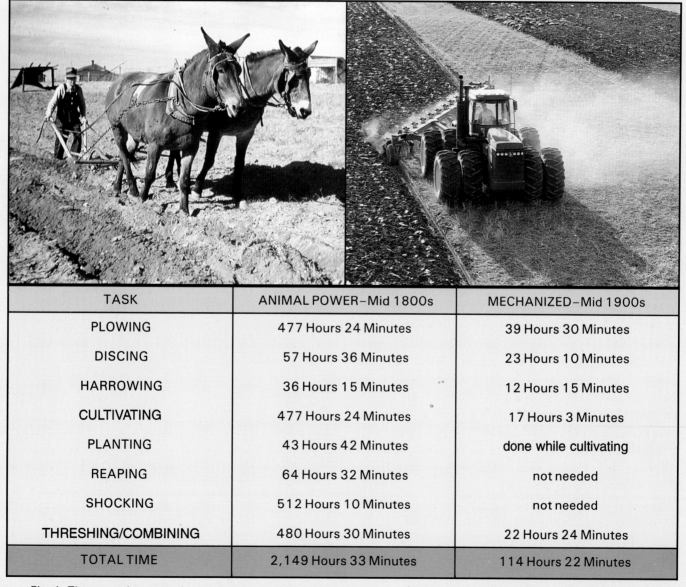

TASK	ANIMAL POWER–Mid 1800s	MECHANIZED–Mid 1900s
PLOWING	477 Hours 24 Minutes	39 Hours 30 Minutes
DISCING	57 Hours 36 Minutes	23 Hours 10 Minutes
HARROWING	36 Hours 15 Minutes	12 Hours 15 Minutes
CULTIVATING	477 Hours 24 Minutes	17 Hours 3 Minutes
PLANTING	43 Hours 42 Minutes	done while cultivating
REAPING	64 Hours 32 Minutes	not needed
SHOCKING	512 Hours 10 Minutes	not needed
THRESHING/COMBINING	480 Hours 30 Minutes	22 Hours 24 Minutes
TOTAL TIME	2,149 Hours 33 Minutes	114 Hours 22 Minutes

Fig. 1. These are the approximate times devoted to the important tasks in planting and harvesting 80 acres of grain.

ENERGY USE

The world as we know it would come to a grinding halt without energy. Almost everything in the human-built environment depends on energy. Therefore, the supply and use of energy resources is very important. This issue hit the headlines during the Arab oil embargo in the early 1970s. At that time many people realized their dependence on petroleum as an energy source. Long lines formed at gas stations and heating oil was in short supply in many areas.

Every person uses energy resources. Many of these resources are exhaustible, or *nonrenewable*. When we burn them all, there will be no more. The supply is said to be **finite** - there is a limited quantity of the resource available. One of these resources, already mentioned, is petroleum. It is still the fuel that powers most transportation vehicles. Other exhaustible energy resources are coal and natural gas.

The shrinking supply of these resources is a major concern, particularly in the case of petroleum. We are challenged to reduce our dependence on these resources. One alternative to using finite resources is to shift to renewable resources. These are the resources that have a life cycle. They are the products of farming, forestry, and fishing. However, these resources are in limited supply at any one time. For example, in many poor countries, wood is the primary fuel for cooking. The growing population causes villagers to roam a greater distance to find the firewood they need. Thus, large regions are being stripped bare of trees. Also, using wood for fuel removes it as a source of building material.

Likewise, corn can be used to make ethanol. Ethanol can be used as a fuel. However, if we shift large quantities of corn from food production to fuel production, world hunger could be worsened.

Shifting from exhaustible resources to renewable resources may be a partial solution. Another solution is capturing inexhaustible resources, Fig. 30-6. The most common of these are solar, wind, and water energy. We can generate electricity using all three of these energy sources. However, this requires a large expenditure of money and human energy. It will also cover large tracts of land with solar and wind generators. Additionally, it will take time to develop the technology to fully use these resources. For example, solar-powered automobiles are now an interesting experiment, Fig. 30-7. Commercial vehicles of this type are years away.

An immediate solution to our energy problems is to *use energy more efficiently*. This will require that all people think about and change their lifestyle. The members of society will have to ask themselves a

SHIFT FROM EXHAUSTIBLE RESOURCES

TO RENEWABLE RESOURCES

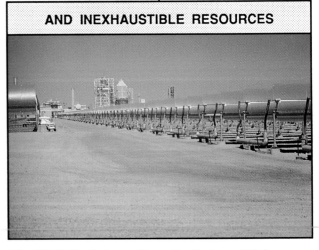

AND INEXHAUSTIBLE RESOURCES

Fig. 30-6. The future will require that we shift from exhaustible energy sources to renewable and inexhaustible sources. These photos show (top to bottom) petroleum storage, trees, and a solar power station.

number of very difficult questions. Should people drive to work alone in a personal car? Should people heat their homes to 75° (24°C) in the winter and cooled to 65° (18°C) in the summer? Should we make buildings

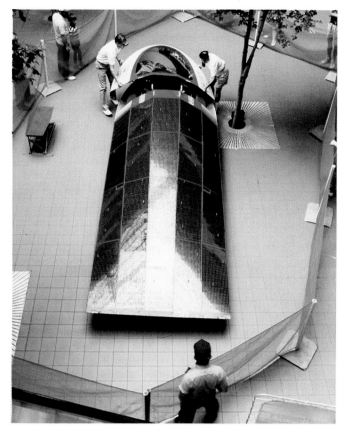

Fig. 30-7. Solar-powered cars like this one are a transportation vehicle of the future. (AC-Rochester)

more costly by using more insulation and installing double- and triple-glazed windows? Should people be strongly encouraged, through taxes or fees, to use public transportation instead of their cars? Should gas-guzzling cars be removed from the market?

ENVIRONMENTAL PROTECTION

Open space, clean air, land for a home or farm, and safe drinking water were once viewed as a birthright for people in North America. We simply expected them to be available. Today, we know better. We have discovered that our unwise use of technology can threaten our quality life.

This understanding was slow to come. These issues came into focus in the late 1960s. There took place what was called an environmental revolution. It brought forward three basic principles:

• The natural environment has a direct affect on the safety and health of people.
• The long-term survival of any civilization is based on wisely managing natural resources.
• A healthy natural environment is essential for human life.

Protecting the environment involves studying the relationship between human population and the use of technology. These two factors directly impact a number of environmental conditions including the climate, the supply of food, water, energy, and material resources.

Many scientists say that we are creating an environmental crisis. This means that we must take action or the environment will be permanently damaged. Three important forces that are contributing to this crisis are overpopulation, resource depletion, and pollution.

Overpopulation

The world's population reached four billion around 1975. This was twice the population of the early 1900s and four times that of 1800. The rate of increase of the global population is about 2 percent per year. This means that in approximately 40 years the number of people on the globe will double again. It is predicted that by about 2030 over 10 billion people will live on earth.

The world's population is not evenly distributed around the world. The number of people per square mile of cultivated land varies greatly. In Europe, this density is three times that of North America. In south Asia it is four times, and in China is eight times that of North America. Likewise, the population growth rate also varies. North America and western Europe have about a 1 percent annual growth rate while Latin America has a 3 percent rate.

This is a "good news-bad news" situation. The good news is that people are living longer. Technology has given us better health care, more food, better disease control, and better sanitation. The bad news is that this is only true in developed countries. "Third world" countries are experiencing most of the population growth. The economies of these poor countries cannot support rapidly growing populations. This leads to tremendous hardships. One of every three persons in poor countries go to sleep hungry each night. More than twelve million people starve to death each year. More that 30 million people die of disease.

As we continue to control diseases, use diplomacy instead of war to solve national conflicts, and improve health care, the population will grow even faster. The increased population will contribute to the other two problems: pollution and resource depletion. Concerted human effort will be needed to limit the world's population.

Resource Depletion

Each person places demands on the resources of the planet. Many material resources, like energy resources, have a finite supply. Once used, they are gone *forever*. These resources include metal ores, petroleum, natural gas, sulfur, and gypsum.

As with energy resources, we can shift from using nonrenewable resources to employing renewable resources. These are primarily the food and fiber produced by farming, forestry, and fishing, Fig. 30-8. However, their supply is limited by the fertility and availability of land and the productivity of the oceans, rivers, and lakes. If we take too much from the water and land, they will be damaged.

An example of using too much is the intensive forestry practices used in many parts of the world. The logs, limbs, and bark of the tree are processed into products. This may seem wise, but it is not. The forest floor is left without the limbs and the bark. These parts of the tree would normally break down and provide nutrients for other plants and trees. Without the limbs and bark, the soil becomes less fertile. This, coupled with the acid rain that is falling on many forests, produces trees that are not as healthy as before. Therefore, their production of wood fiber is reduced.

Likewise, intensive farming practices will deplete the soil. This makes heavy applications of commercial fertilizers necessary. Also, using prime farmland for housing forces us to cultivate less productive areas. We also strain precious water resources to irrigate crops growing in the desert. We draw the water from rivers and wells. This, in turn, reduces the amount of water downstream and depletes aquifers. (Aquifers are underground water-bearing layers of rock, sand, or gravel.)

A problem related to these issues is land use. Throughout the country conflicts are raging between environmentalists (people who are most concerned about leaving the environment unspoiled) and commercial interests. An example of one controversy is the spotted owl habitat in the forests of Oregon and Washington. Commercial interests want to cut virgin timber and manage the forest for maximum fiber production. Environmentalists want large tracts of these forests left untouched as a habitat for the owls.

Debate continues over how much wilderness is enough. Some people want large tracts of range and forest land to remain untouched. They want only hiking trails in these areas.

Few people dispute the need to save some unspoiled areas. The debate is over the size and number of these areas. The difficulty with these debates is that there are no clearcut answers. There are only opinions of what is right or wrong, good or bad.

As with energy, short-term resource depletion solutions lie in using our material, land, and water resources better. We must use our land, but not abuse it. Some of it should be set aside as nature preserves. Other areas should be considered multi-use land. It would combine recreational and commercial uses. Hunters, campers, ranchers, farmers, and loggers would all use the land. Finally, some tracts of land would be devoted to commercial, residential, and transportation uses.

We must reconsider how we use the earth's resources. We must learn to use all materials more efficiently. This will mean buying fewer items that we really don't need. Buying products that will last becomes very important, as does maintaining and repairing them instead of throwing things away. Finally, when a product can no longer be used the materials must be recycled.

Pollution

People living in our major cities rarely see clear skies. Their vision is often reduced by haze and their health is affected by smog. Water in many parts of the world is unsafe to drink. In some areas the land is contaminated with hazardous waste. People cannot live on this land or travel over it. All of these things are called **pollution**, Fig. 30-9. Pollution is most often a product of human activity.

In recent years, pollution has been clearly brought to our attention. In the 1960s an accident on the Cuyahoga River in northern Ohio dramatized the extent of the problem. The river, polluted with flammable liquids, caught fire. No one believed a river could burn, but it did. Later, astronauts traveled above the atmosphere. They looked out of their spacecraft and were saddened at the sight of the dirty, brown blanket of air covering the earth.

Pollution affects more than the air we breath, the land we walk upon, and the water we drink. Many sci-

Fig. 30-8. Farming is one source of renewable material resources and food.

Fig. 30-9. Pollution can affect the land, water, and air that we depend on for life.

entists think that it is changing the climate of our planet. In the spring of 1983, scientists observed a brown layer of air pollution over the Arctic region. Later, other scientists discovered a hole in the ozone layer over the Antarctic. Since that time a controversy has raged. Some scientists say that the hole will let ultraviolet light into the atmosphere. They believe these rays, combined with increased levels of carbon dioxide and other gases, will cause the earth to retain more heat. This will cause higher land and water temperatures. Scientists are concerned that a warmer climate will melt the polar ice caps and cause the oceans to rise. In turn, large coastal areas, including some major cities, will be flooded. Scientists call this problem the "greenhouse effect."

Other scientists feel that global warming is part of a natural cycle. They say that warming and cooling has occurred a number of times over history. No one is sure what will happen. However, it is an area of great concern. Changes have been called for in our technological actions. Certain chemicals in air-conditioners, refrigerators, foam containers, and aerosol sprays are blamed for part of the problem. These chemicals are being removed from the market as replacements are developed.

The greenhouse effect is worsened by the cutting down of rain forests. Finally, gases from burning coal and petroleum products also contribute to global warming.

A plan is needed to protect the atmosphere and the ozone layer. Inaction on our part could cause a significant change in the climate of our world. The greenhouse effect could turn lush farmland into desert, cause the extinction of some species of plants and animals, and create widespread human suffering.

GLOBAL ECONOMIC COMPETITION

A third issue directly related to people and technology is mainly economic. It deals with the distribution of wealth and industrial power. Some historians suggest that technical knowledge and power has moved steadily around the world from east to west. At one time China was a global power and the source of many innovations. This center of power moved across what is now India into the area we now call the Middle East. The area around the Mediterranean was the dominant economic center about 2000 years ago. Later, northern Europe become the economic leader with the industrial revolution of the 1800s. This was followed by the dominance of the United States in the 1900s. Now we are seeing the area called the Pacific Rim become more important. The industries of North America and Europe are being challenged by Japan, Korea, and other Far East countries. This represents nearly a full cycle of industrial development around the globe.

Today, we understand economics better than at any time in history. This allows countries on the backside of the economic wave to resist losing economic power. They can take a number of actions, including:

• Changing management styles. Companies are redefining the role of workers and managers. They are creating teams that design and produce products, Fig. 30-10. These teams include designers, engineers, production workers, quality control specialists, marketing people, and managers. Their goal is to create products that are "designed for manufacture." This means that products can be easily produced and assembled. Managers are not seen as bosses. Managers help others do their work better and with greater ease.

Fig. 30-10. People work together to design and produce world-class products. (Texaco)

- Many purchasing and warehousing operations are also computer controlled. These systems include computer controlled warehousing in which robot vehicles store and retrieve parts. For example, when new parts are received, the computer selects an empty storage bay. The system directs a vehicle to place the parts in that bay and notes the location. The system retrieves the parts on command when they are needed.

- Producing world-class products. Successful companies now make products that meet the needs of customers around the world. This means that the products must function well, be fairly priced, and deliver excellent value to the customer. No longer can a local or national area provide a safe market for a company. Political forces are causing the world to

- Increasing their use of computers. Competitive companies are rapidly assigning routine work to computers. This means there is less manual labor and more computer-aided or controlled work, Fig. 30-11. We now hear about an "alphabet" of computer actions.
  - CAD (computer-aided design) systems make drawings easy to produce, correct, and store.
  - CIM (computer-integrated manufacturing) ties many manufacturing actions to the computer. Machine control, quality control, parts movement, and an array of other operations are monitored by computer systems.
  - Just-in-time (JIT) inventory control. This computer system monitors material orders so that supplies arrive at the plant just before they are needed. Also, finished products are made only when they are needed. In neither case are they stored in a warehouse. Costs are thus reduced.

Fig. 30-12. Quality circles are teams of workers and managers that work together to improve the quality of product. (Ohio Art Co.)

Fig. 30-11. Modern industrial practices replace manual operations (left) with computer systems (right) when possible.

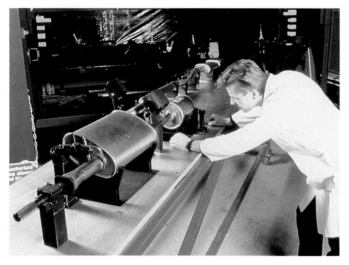

Fig. 30-13. Modern products are designed with quality in mind (left) and are often produced on flexible, automated production lines. (Arvin Industries)

become one large market for all countries. To meet the demand for quality products, *quality circles* have been developed. These are groups of managers and employees who meet on a regular basis. The group identifies quality problems and recommends ways to correct them. This interaction builds concern for and pride in making quality products, Fig. 30-12.

- Using flexible, automatic manufacturing systems. Traditional manufacturing is based on long production runs. Semi-skilled workers on a manufacturing line made a single product. This type of production is being replaced by automated production lines called flexible manufacturing systems, Fig. 30-13. Computer controlled, such systems can produce a number of different products with simple tooling changes. Flexible manufacturing is cost effective for small quantities of products.

SUMMARY

Technology and human life cannot be separated. We use technology, depend on technology, and feel we must have technology. It lets us travel easily, communicate quickly, and live in comfort. Technology has also caused us concern. Poor application of technology pollutes the air, water, and land around us. It can threaten our very lives.

The challenge facing all people is to determine the type of future we want. Then, we must develop the technology that will promote that future, Fig. 30-14. Likewise, we will have to change our personal habits and preferences. Some technologies that we now depend upon may need to be modified or abandoned. Only then can future generations enjoy a good quality of life.

Fig. 30-14. We will need to use new technologies to develop the type of future we want. (U.S. Department of Energy)

WORDS TO KNOW

All of the following words have been used in this chapter. Do you know their meanings?

Biological future
Finite
Futures research
Futuring
Human-psyche future
Pollution
Social future
Technological future

TEST YOUR KNOWLEDGE

Write your answers on a separate piece of paper. Please do not write in this book.

1. Determining the type of future we want uses a new process called _____.
2. True or false. Technology is the human-built world.
3. List the four types of futures that can be suggested.
4. Human-made chemicals in the air, water, or land are called _____.
5. True or false. A product that can be competitive anywhere in the world is called a world-class product.

APPLYING YOUR KNOWLEDGE

1. Choose an environmental problem and gather information about it. Summarize your knowledge on a chart like the one to the right.
2. Prepare a drawing or a photo montage that shows an "ideal" future that you would like to live in.

Problem:
Factors that cause or contribute to the problem:
Possible solution #1:
Problems that solution #1 might cause:
Possible solution #2:
Problems that solution #2 might cause:

31
C H A P T E R

TECHNOLOGY: A PERSONAL VIEW

After studying this chapter you will be able to:
- *Describe how technology impacts your personal life.*
- *Describe how jobs have evolved during the agricultural (colonial), industrial, and information periods.*
- *Explain the employment opportunities related to technological activities.*
- *Describe the common levels of technology-based jobs and the types of education required for each.*
- *List the factors that should be considered in selecting a career.*
- *Describe the demands of technology-based jobs in the future.*
- *Explain how working with people, data, and machines changes with different occupations.*
- *Describe how technological activities we now think of as fictional may become modern processes.*

Our personal life is highly dependent on the technology that people have developed. In a brief 50 years, life in the western world has changed dramatically. We live in different housing, travel on different systems, have different products to purchase, and communicate in ways far different from the past. To help investigate the changes that will affect everyday life in the 21st century, let's look at four areas:
- Technology and lifestyle.
- Technology and employment.
- Technology and individual control.
- Technology and new horizons.

TECHNOLOGY AND LIFESTYLE

Each person lives in a specific way or has a lifestyle. A **lifestyle** is what a person does with their business and family life–their work, social, and recreational activities, Fig. 31-1.

COLONIAL LIFE AND TECHNOLOGY

The lifestyle during our colonial period was a harsh contrast to that of today. Housing was simple and modest. Most products were designed and produced to meet the basic human needs. There were few decorative items available, and those items were mostly owned by the wealthy. Transportation systems included simple boats and animal-drawn wagons. The communication systems available were crude.

Most people lived on farms. A few people practiced basic crafts. These people were the carpenters,

Fig. 31-1. What kind of lifestyle do you think these two people enjoy?

blacksmiths, and other tradesmen needed to produce basic products required by the community.

The men and boys did most of the field work on the farm and practiced the trades. Women and girls tended gardens and did household work, Fig. 31-2.

Everyone had to work long hours, six days a week, to raise small amounts of food. The seventh day was set aside by most colonial families for church activities. People did not take vacations, and they celebrated few holidays.

However, with the exception of slaves, most people were their own bosses. They owned their farms and stores, or they practiced their crafts as independent workers. Each was an owner, manager, and worker rolled into one.

Fig. 31-2. This woman is practicing the old craft of candlemaking, a common craft in colonial life.

THE INDUSTRIAL REVOLUTION

The industrial revolution of the mid-1800s changed this lifestyle. Several events took place. First, advanced technology was developed for the farm. This technology included the moldboard plow, the reaper, and the steam tractor. These and other devices made farmers more efficient. Fewer people could farm more land and produce more food to sustain the society. Second, the opening of the land west of the Appalachians allowed for larger more efficient farms. During this period, the percentage of the work force engaged in farming began dropping rapidly. At the start of the industrial revolution, more than 90 percent of the work force was engaged in farming. Today, less than three percent of workers are employed on farms.

Also, a large number of people from Europe immigrated to the United States during this time. These immigrants, plus the farmers who were no longer needed to till the soil, provided a vast labor supply.

This labor supply was a basic resource for the factory system that was then being developed. The demand for goods could no longer be met by the local tradesmen working in their shops. These activities were being replaced by centralized manufacturing operations. These operations included several important factors, Fig. 31-3:

- Professional managers, who established procedures, employed resources, and supervised work.
- Division of labor, which assigned portions of the total job to individual workers. Each worker did only part of a total job, allowing them to quickly develop the specialized skill needed to do the assigned task.

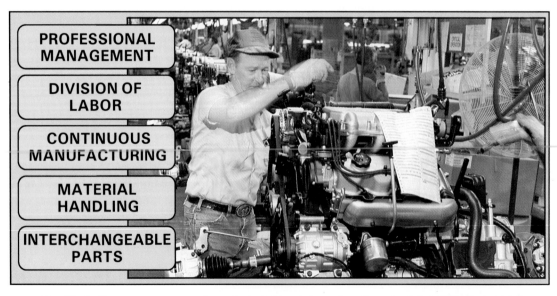

Fig. 31-3. Many current manufacturing plants are based on the principles of the industrial revolution. (Chrysler Corp.)

- Continuous manufacturing techniques, which increased production speed. Raw materials generally entered a production line at one end, and finished products left the line at the other end.
- Material handling devices, which were used to move the products from work station to work station. Workers remained at their work stations, and the products moved to them.
- Interchangeable parts, which allowed production of large quantities of uniform products.

Men and women, as well as boys and girls, worked long hours in the factories for low pay. In addition, they had little control over their working conditions. They provided manual labor without any managerial voice. For the first time in American history, the vast majority of people worked for someone other than themselves.

The low wages and poor working conditions caused widespread worker unrest. Labor unions were formed to give the workers a voice in determining working conditions and pay rates. Bloody battles erupted between the workers and management. The government usually supported managerial positions, resisting attempts of the unions to deal with the issues.

These conflicts were finally settled with changes in governmental attitudes, new laws, and different management stances. This led to a strong industrial period for the country. The period was characterized by broad employment opportunities. The workers enjoyed a high standard of living. The 40-hour five-day workweek with a number of holidays and paid vacation time became fairly common.

THE INFORMATION AGE

The invention of the computer, and later the microchip, changed the industrial revolution. During the industrial period, the company that could efficiently process the greatest amount of materials succeeded. This took large investments in big continuous manufacturing plants. These were characterized by the huge automobile and steel manufacturing plants. They employed thousands of people and used millions of tons of materials.

The computer allowed development of a new type of manufacturing, flexible manufacturing. This type of manufacturing can quickly and inexpensively respond to change. People with few or narrow skills are replaced with computer-controlled machines, Fig. 31-4. The workers who are left have more training and motivation to work. They accept change and responsibility more readily than the worker of the industrial revolution.

Also, management has changed greatly. Management is less distant from the worker. The entire work

Fig. 31-4. This robot is placing cartons on a pallet. This work used to be done by semi-skilled workers. (Cincinnati Milicron)

force is seen as a team, with each person having an area of responsibility, Fig. 31-5. Managers may be responsible for setting goals and controlling money. Workers are responsible for producing products. However, everyone is responsible for work procedures and product quality.

The information age has given us opportunities and challenges. The economically advanced areas of the world have created a high standard of living.

Fig. 31-5. In modern industry, managers and workers make up a team that cooperate together to reach company goals. (Air Products Co.)

Communication systems now link us with every point on the globe. We can travel around the world quickly, safely, and in great comfort. There are high quality products and highly functional structures available everywhere.

These changes give people a new lifestyle. The people who change and adjust to the demands of the new age can live very differently. They are better informed, work more with their brains than their brawn, travel greater distances, and have more control over their work, Fig. 31-6.

However, not all people reap these rewards. People need a good education, a desire to continue to learn, an ability to deal with change, and a willingness to take personal risks. The days of living well with a marginal education and limited skills are rapidly disappearing.

Also, people in the lesser-developed countries are failing to receive the benefits from the information-age revolution. Their economies do not deliver the basic essentials of life. Their lifestyle has seen little improvement over the recent decades. Their hunger and poverty persist.

TECHNOLOGY AND EMPLOYMENT

Lifestyle and employment are closely connected. Most people need the money they earn through working to afford the type of life they want. Some general requirements for most jobs in the future can be identified.

Fig. 31-6. Modern workers often work more with their brains than their brawn.

Fig. 31-7. Many technical jobs require specialized training beyond high school. (General Electric Co.)

Fundamental to almost all jobs will be a high school education. Also, quite often, specialized technical training beyond that level will be needed, Fig. 31-7. This may be technical training in career centers and community colleges or a university education.

In addition, the worker in the information age must be willing to:

- Pursue additional education and training throughout their work life.
- Accept job and career changes several time during their work life.
- Work in teams, and place team goals above personal ambition.
- Exercise leadership, and accept responsibility for one's work.

TYPES OF TECHNICAL JOBS

There will be a wide variety of jobs that require technical knowledge. These jobs can meet the interests and abilities of all different types of people, Fig. 31-8. They include five levels, as shown in Fig. 31-9. The first is the technically trained **production worker.** These job opportunities are available in all four technological areas. They include the people that process materials and make products in manufacturing companies, erect structures, operate transportation vehicles, service products and structures, and produce and deliver communication products.

Most production workers operate technological equipment. These positions require a high school education for job entry. Many of these workers will have

Fig. 31-8. There are technology-based jobs for people with a variety of interests and abilities. (Inland Steel Co.)

attended a vocational education program at a secondary school. The programs cover a specific area such as automotive, building trades, or drafting. Often, the company will provide any specific training or retraining needed do assigned work.

A large group of more highly trained people will also be required in the future. These people are called technicians. **Technicians** work closely with production workers but do more specialized jobs. Generally, they have more training and experience than production

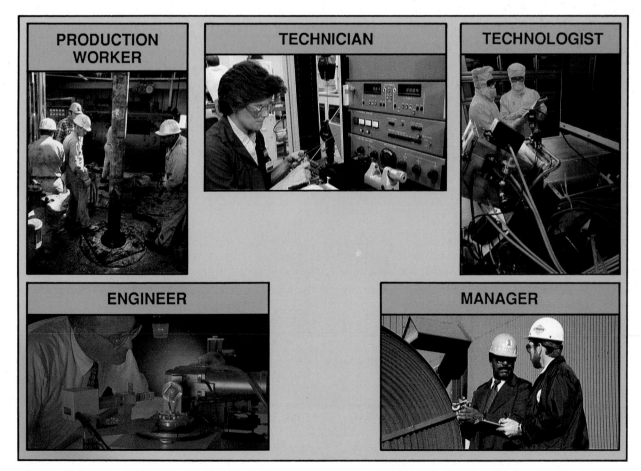

Fig. 31-9. People with backgrounds in technology can become production workers, technicians, technologists, engineers, or managers. Many people will work in more than one of these areas during their careers. (American Petroleum Institute, Goodyear Tire and Rubber Co., AT&T, Inland Steel Co.)

workers. Typically, technicians set up and repair equipment, service machinery, conduct product tests and laboratory experiments, and work in quality control. Quite often, these people will have an associate degree (two-year college level) in a specific area such as electronics, computer repair, or heating and air-conditioning.

Many companies have a need for people with advanced technological knowledge. These needs are handled by engineers. **Engineers** design products and structures, conduct research, and develop production processes and systems. Engineers have a university degree(s) in engineering. Depending on the company and the work they do, engineers may have bachelor's (B.S.), master's (M.S.), or doctorate (Ph.D.) degrees. These degrees may be in the traditional areas of chemical, electrical, civil, or mechanical engineering. Other emerging areas are industrial, manufacturing, and construction engineering.

A fairly new job category is that of the **technologist.** Technologists are highly trained technical employees. They form the bridge between the engineers who design systems and the workers and technicians who must implement them. The technologist, usually, has a bachelor's degree in industrial technology or engineering technology.

Finally, companies will need technically trained **managers.** These managers would have the background of technicians and engineers. Additionally, they possess managerial skills and training. They can set goals, plot courses of action, and motivate people to work together. They are people-oriented leaders, who also have technical knowledge.

SELECTING A JOB

People should not just take the first job that they can find. Whenever possible, a job should match the person's interests and abilities. When deciding on employment opportunities, a person should consider at least three factors: lifestyle, job requirements, and job satisfaction.

Lifestyle

Each of us wants to live comfortably. A person's job has a direct impact on both life at work and life away from it.

Some people like to travel and meet new people. An industrial sales job could provide these aspects. A factory job that pays well and has good holiday and vacation benefits can also meet this need.

Some people enjoy being at home every night and do not want to move away from their hometown. For these people, cross-country truck driving would be a poor career choice. Working for a local production, communication, or transportation company would be a better idea.

Many people enjoy facing new challenges and being in a position of responsibility. Management positions provide these opportunities. However, managers must accept a great deal of accountability for their actions and the stress that goes along with it.

Requirements

There are a variety of job requirements that affect each employee. In selecting a job, a person should consider the job requirements from three important points.

Authority and Responsibility. First, is the job's requirement to accept authority and responsibility. This view looks at your freedom to organize the work assigned to the job. This also considers the level of accountability that goes with the job. Generally, management, scientific, and engineering positions offer more freedom and responsibility, Fig. 31-10. People are usually paid more as they gain higher levels of authority and responsibility.

Data, Machines, and People. The second consideration is the balance of three items. These items are working with data, machines, and people, Fig. 31-11.

Fig. 31-10. People interested in high levels of authority and responsibility could look into managerial or engineering positions. (AT&T)

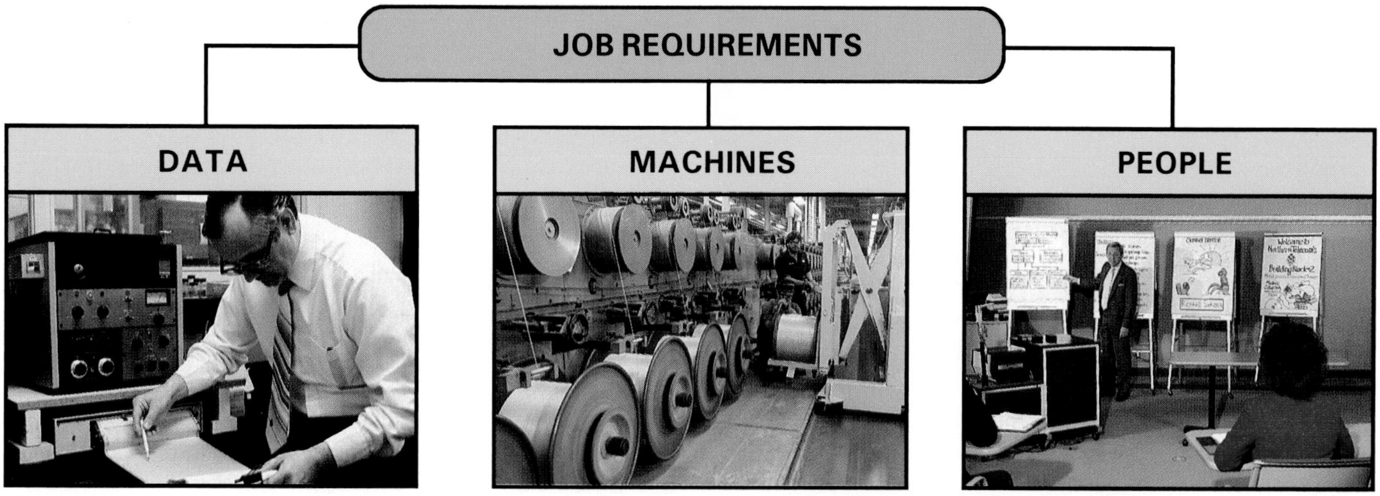

Fig. 31-11. Each job deals, in varying degrees, with machines, people, and data. Which areas do you prefer? (Goodyear Tire and Rubber Co.)

Actually, no job emphasizes only one of these factors. However, many jobs will place an emphasis on one over the other two. For example, accounting is a job that places a strong emphasis on working with data. A welder should like to work with machines and materials. Sales representatives should enjoy working with people. Managers have to work with both people and data. They have to organize work and make reports. They must also motivate workers and solve disputes.

Education. The last factor is the level of education a job requires. Not all people have the same ability to learn or retain information. Each person will eventually reach some maximum level of education. For some people, completing high school is a challenge. Others will move on through university systems until they have a Ph.D. People should select jobs that have educational requirements that are matched with their ability to learn.

Job Satisfaction

Job satisfaction is how happy a worker is with his or her job. Three factors that strongly affect job satisfaction are values, recognition, and pay.

Values. A job should match the values of a person. Each person wants to think that what he or she does is important. Custodians should take pride in a clean building. Likewise, engineers should be proud of the products, structures, or systems they design.

People should match their job with their belief system. Individuals who are against war should not work for military ordnance manufacturers. Similarly, a person who is concerned about human health should not sell alcohol or tobacco products.

Recognition. A second factor that impacts job satisfaction is recognition. Many people want to be recognized for what they do. Some jobs allow for more visible recognition. Can you name some important musicians, politicians, or sports figures? Most likely you can. However, can you name any important medical researchers, industrial designers, or architects? This may prove more difficult.

Even in specific fields, recognition varies widely. Name ten famous musicians. How many of them are classical musicians or jazz musicians? This little exercise should help you see that some occupations provide more public recognition than others.

Public recognition is not the only type of recognition that you can earn. Professional recognition is important to many people. Professional recognition involves being recognized by those people who do the same type of work as you. Many professions award a "woman-of-the-year" award or a "man-of-the-year" award for their group.

Pay. The third factor that relates to job satisfaction is pay. The money received for work is important. However, money is not the most important thing in life for many people. Several studies have been conducted to determine what factors contribute to job satisfaction. These studies have found that doing a job that you feel is important ranks very high. Recognition for your work is equally important. Money often comes out third, fourth, or fifth in importance. Seldom does a high salary make a distasteful job satisfying.

TECHNOLOGY AND INDIVIDUAL CONTROL

Technology holds great promise and hidden dangers. People will make the difference, Fig. 31-12. We often say that *they* should control this, or *they* should

Fig. 31-12. This quiet fishing lake was once a strip mine. After the coal was removed, people restored the area. (American Electric Power Co.)

do that, or *they* should stop doing something else. The harsh reality is that "they" will never accomplish anything. Only when someone says "I" instead of "they," does anything meaningful get done.

The future lies in the hands of people who believe they can make a difference. Examples like Thomas Edison, the Wright brothers, and Albert Einstein may come to mind. These individuals did not wait for a group to do something. They pursued their own vision of what was important and needed by society.

This type of action requires that people understand technology. People must also comprehend the political and economic systems that direct its development and implementation.

Individuals control technology in a number of ways. The first way is through the role of the **consumer.** This involves:

- Selecting proper products, structures, and services.
- Using products, structures, and services properly.
- Maintaining and servicing the products and structures owned.
- Properly disposing worn or obsolete artifacts.

Consumer action causes appropriate technology to be developed, Fig. 31-13, and it helps inappropriate technology disappear. When a product or service does not sell, its production stops.

An individual also has **political power.** All companies must operate under governmental regulations. These regulations include the Occupational Safety and Heath Act, the Pure Food and Drug Act, the Environmental Protection Act, and the Clean Water Act. Few laws are passed solely because an elected official thinks they are important. Most of them come from people asking their elected representatives to deal with a problem or concern.

Finally, an individual can make a difference as an activist. **Activists** use public opinion to shape practices and societal values. Some people see this as a bad thing. They worry about environmental activists, right-to-life activists, and women's rights activists. Actually, an activist is simply a person practicing his or her constitutional right. Each person has the right to freedom of speech and assembly. You can meet, discuss, and promote your points of view.

Fig. 31-13. The types of products and structures you demand as consumers make a difference. Solar heating is becoming popular. This solar heated home will save energy without harm to the environment. (U.S. Department of Energy)

Technology: How it Works

EARTH-SHELTERED BUILDINGS: Structures that are built into the earth to take advantage of the insulation value and thermal properties of soil.

Energy efficient buildings are important to many people. A number of different building techniques can be used to make buildings more efficient. One way is to use *earth-sheltered construction*. This method of construction covers part or all of the dwelling with soil. The purpose is to use the soil's thermal qualities to keep the house at an even temperature. The temperature a few feet down in soil does not vary much during the year.

Fig. 1. This cliff dwelling in the Walnut Creek National Monument in Arizona used earth sheltering.

Using the earth to shelter a dwelling is not a new idea. Native Americans in Arizona used earth-sheltering in 1000 A.D. This was *400 years* before Columbus discovered America. Using protective cliff faces to shelter communities was common throughout the southwestern United States. Today, a number of these sites have been uncovered and restored, Fig. 1.

Earth-sheltered construction requires careful planning. Special problems exist, such as: moisture, loads from soil, and orientation for solar heat gain. Buildings can be partially earth-sheltered or totally covered, Fig. 2. In the northern hemisphere the building is normally sited so it faces south. This is so the house can absorb solar heat through large windows on the south face. The wall of windows helps make the front rooms light and airy. These are the daytime living rooms. Bathrooms and bedrooms can be placed toward the back of the structure. These rooms are normally used at night, when artificial light is needed.

Earth-sheltered construction can be used for more than housing. It is also appropriate for theaters, shopping malls, convention centers, and warehouses. The function of these structures does not require windows or exterior views.

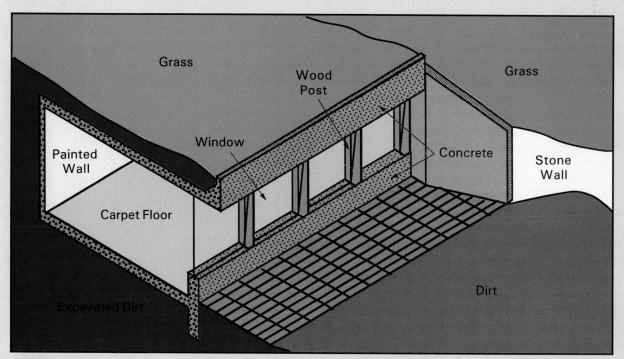

Fig. 2. This drawing shows the design for an earth sheltered house that has the roof covered with earth. This house would be built using reinforced concrete.

TECHNOLOGY AND MAJOR CONCERNS

Technology concerns many people. People are worried about pollution, energy use, and unemployment. However, technology is designed, produced, and used by people. These issues can only be met through human action. A few areas of concern are at the forefront of our attention. These concerns include:

- Nuclear power and waste disposal: Is nuclear power generation an appropriate activity? How do we design and build nuclear power plants that are safe and reliable? How can the radioactive waste produced by nuclear power plants be disposed of safely?
- Technological unemployment: Should technology that causes unemployment be applied? Does a company have a responsibility to provide training and other benefits to workers who are unemployed through new technologies? Should foreign products that cause technological unemployment be barred from domestic markets? What are the individual worker's responsibilities in seeking training to deal with changing job requirements?
- Genetic engineering: Is it right to change the genetic structure of living organisms? Who will decide what genetic engineering activities are appropriate? Is it acceptable to alter the genetic structure of humans? How are religious and technological conflicts dealt with?
- Energy use: How can we reduce our dependence on petroleum? Can society's dependence on the private automobile be changed, Fig. 31-14? What alternate energy sources and converters should be developed? How can we control the environmental damage caused by burning fossil fuels? Should mass transit be financed with tax money? Should we place high taxes on gasoline?
- Land use: What are the rights of landowners? Should the desires of the majority overrule individual landowner rights? What responsibilities do governmental officials have for public lands? How do you balance environmental protection issues with economic issues, Fig. 31-15?
- Pollution: Should products that pollute the environment be banned from manufacture and use? How much, and what type of, evidence is needed before a product can be banned? How do you handle the economic and social impact of banning products? Should strict pollution controls apply equally to individuals and companies? Should there be a pollution tax on fuels that damage the environment? Should we limit the use of wood as a fuel? Should the solid waste (garbage) from one state be allowed to enter another state?

Fig. 31-14. The future will require that we make greater use of mass transit to prevent fuel and congestion crises.

This book does not attempt to answer these or other technological problems. Remember, the *right answer* to one person or group is often rejected by other people and other groups. The best we can hope for is *an answer* that most people will support.

TECHNOLOGY AND NEW HORIZONS

Technology continues to be developed at a very rapid pace, Fig. 31-16. Many ideas that seem impossible today will be commonplace tomorrow. This is not new. In the 1800s, Jules Verne wrote a fictional story called 20,000 Leagues Under the Sea. It dealt with the then

Fig. 31-15. The future will demand that we balance concern for the environment with our commercial interests.

Fig. 31-16. In less than 30 years, the Apple computer moved from crude prototype to a polished, powerful microcomputer. (Apple Computer)

Fig. 31-17. In the future, we may build complex manufacturing systems in space. (NASA)

Some experiments have dealt with growing zeolite crystals. These crystals may be used in portable kidney dialysis machines and in cleanup efforts with radioactive waste. Some manufacturing processes are adversely affected by earth's atmosphere and gravitational pull. Will these processes be moved into space? Will we have space stations where manufacturing is routinely done?

Fig. 31-18. People born during the era of the chuck wagon saw a man walk on the moon during their lifetime.

impossible feat of traveling under the water in a submarine. People in the 1940s and 50s read a fictional comic strip called Buck Rogers. Buck and his colleagues did the then ridiculous feat of traveling in space. What is fiction today, your children and grandchildren may use regularly.

Many futuristic ideas have been proposed. These include a number of "impossible" activities. One is mining the resources of outer space. Will we have colonies on the moon that extract and process precious mineral resources? How about mining the riches of the sea? Our oceans are the last of the vast resource beds on earth. Will they be mined, or will concerns regarding pollution keep them off-limits?

Another "fictional" prospect is manufacturing in space, Fig. 31-17. Experiments have shown that protein crystals can be grown in the microgravity (very low gravity) of space. Earth-grown crystals are often small and flawed. Those grown in space can be larger and more complex.

A third future technological activity is commercial space travel. To date, space travel is government financed. Most of it is restricted to military and scientific missions or communication satellite launching. Will we routinely be traveling into space and back? Will later generations take vacation trips to the moon?

These are only a few examples of possible technological advancements. Society has moved from the covered wagon to space travel in one lifetime, Fig. 31-18. What could be next?

SUMMARY

Technology is as old as humanity. Technology helps us lead better lives by enhancing our abilities. Technology gives us better jobs and helps us evolve new opportunities. Technology also threatens us with pollution, a potential loss of personal worth, and possibly even a nuclear winter. How we design, produce, and use technology makes the difference. The choices are yours to make. You now know more about technology. Using technological knowledge, you can select products wisely, pursue appropriate employment, and better understand societal issues. You can be a better consumer, worker, and citizen.

WORDS TO KNOW

All of the following words have been used in this chapter. Do you know their meanings?
Activist
Consumer
Engineer
Lifestyle
Political power
Production worker
Technical manager
Technician
Technologist

TEST YOUR KNOWLEDGE

Write your answers on a separate piece of paper. Please do not write in this book.
1. List the five factors that characterize the companies of the industrial revolution.
2. True or false. Most jobs in the future will require at least a high school diploma.
3. Match the job on the left with the factor(s) it emphasizes.

_____ Sales person.	A. Data.
_____ Machinist.	B. People.
_____ Accountant.	C. Machines.
_____ Quality control inspector.	
_____ Computer analyst.	
_____ Manager.	

4. List three roles a person can play that impact the type of technology that is developed.
5. True or false. For most people, money is the most important factor in job satisfaction.

APPLYING YOUR KNOWLEDGE

1. Select a technological idea that is now considered fiction. Write a description of how it will be used as an everyday item or process in the future.
2. Read a book about a person(s), like Thomas Edison or the Wright brothers, who developed new technology. Write a report on their attitude toward change and criticism as well as their inventions or innovations.

Section Ten - Activities

ACTIVITY 10A - DESIGN PROBLEM

Background:

Technology, by itself, is neither good nor bad. How a technological object is used can create benefits or drawbacks. Each person should strive to use technology wisely to make the future better, protect the environment, and help people live in harmony with nature.

Challenge:

Identify an important issue to people today that is impacted by technology. List the positive and negative factors related to the issue. Develop a plan to maximize the positive features and reduce the negative impacts.

ACTIVITY 10B - PRODUCTION PROBLEM

Background:

Humans can control the use of technology. However, many people do not know about the impacts that technology has on individuals and society.

Challenge:

You are the communication director of a citizens group. Your group is concerned about public policy issues. Your group has determined that people in your community are not participating in the local recycling program. Design a 60-second public-service commercial for television. The commercial needs to explain the importance of recycling to your community. Identify the materials that can be recycled and the benefits of recycling.

Equipment:

Storyboard forms
Pencils, magic markers
Video camera, recorder, and monitor

Procedure:
Designing the Commercial
1. Select the theme of the commercial.
2. List the major points that will be emphasized in the commercial.
3. Develop a storyboard for the commercial. (Use photocopies of Fig. 10-A.)
4. Develop a script for the actors.
5. Develop a shot chart for the director and camera person to follow.

Producing the Commercial
1. Recruit and select actors for the commercial.
2. Present the script to the actors and have them rehearse their parts.
3. Walk through the commercial with the director. Have the actors and camera operators block (plan) their movements.
4. Record the commercial.
5. Edit in any titles you need.
6. Present the commercial to an audience. Ask for their reactions.

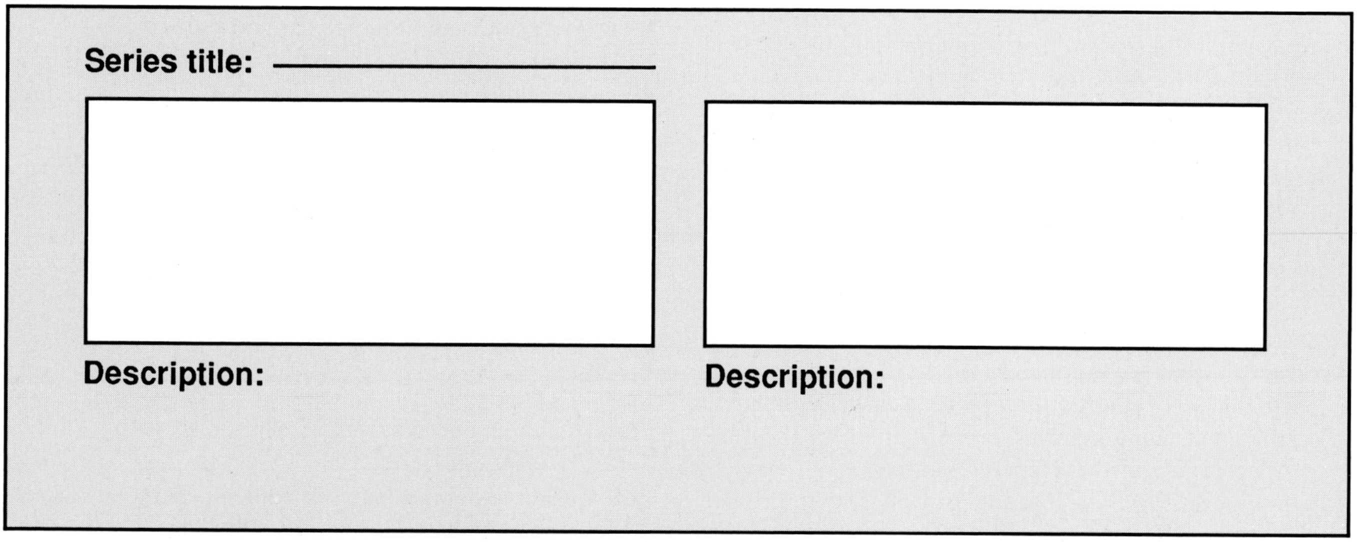

Series title: ————————————

Description:

Description:

Fig. 10-A.

TECHNICAL TERMS

A

Activist: a person who speaks out upon, or works to change, some element in society.

Actuating: starting a system in operation by assigning and supervising work.

Adhesive bonding: joining materials by making use of the stickiness or tackiness of the bonding substance.

Adjusting devices: devices that adjust the inputs to a system.

Air transportation: systems that use airplanes and helicopters to lift passengers and cargo into the air so that they can be moved from place to place.

Airfoil: an object designed to produce some directional motion when in movement relative to the air, such as the wings on a plane.

Altering: the process of making changes to an object or action.

Alternating current: electron flow in a conductor that reverses at regular intervals.

Alumina: aluminum oxide.

Amplitude: the height of a wave.

Amplitude modulation: system in which a message is encoded onto a carrier wave by changing its amplitude.

Anaerobic digestion: a controlled decaying process that takes place without oxygen. Changes waste products into methane.

Annealing: a process that softens or removes stress from a material.

Apogee: the farthest point from the earth that a satellite will reach in an elliptical orbit.

Applied research: research that seeks a commercial goal through applying knowledge gained in basic research.

Apprenticeship training: combination of on-the-job training and classroom training over an extended period of time.

Arch bridge: bridge that uses curved members to support the deck.

Architectural drawings: drawings which specify characteristics of buildings and other structures.

Articles of incorporation: papers filed with a state government that serve as an application for a corporate charter.

Assembling: a manufacturing process in which the final product is put together from separate parts.

Assembly drawings: engineering drawings that provide information to show assemblers how parts of a product fit together.

Audience assessment: tests which determine who the audience is, their likes and dislikes, plus additional information that could be useful for marketing.

Audio: the sound portion of a message.

Automatic control: devices are used to monitor, compare, and adjust a system without human intervention. A thermostat on a home heating system is an example.

Avionics: the instruments that help pilots monitor and control aircraft, or the development of such devices.

B

Balance: an equal visual weight on both sides of a center line in printed communication.

Basic research: research that seeks knowledge for its own sake.

Bauxite: aluminum ore.

Beam bridge: bridge that uses concrete or steel beams to support the deck. Commonly used on the interstate highway system.

Benefit: reward, other than monetary, that companies give to their employees.

Bill of materials: sheet which contains the information regarding the materials and hardware needed to complete a project.

Billet: a long, square piece of steel.

Biofuel: organic matter that can be burned directly for energy.

Biogas: organic matter that is converted into a flammable gas to be used as an energy source.

Biomass: organic waste that is used as an energy source.

Biotechnology: the activities that improve the types and the quantities of grown resources.

Blast furnace: a furnace commonly used in iron smelting. Uses convection heating with a current of air under high pressure.

Blimp: a non-rigid (frameless) aircraft that is filled with a light gas.

Blow molding: a process that uses air pressure to form hollow glass or plastic objects.

Board of directors: a group of people in charge of forming company policy and providing overall direction.

Bond: a debt security sold by governments and corporations.

Bonding: using cohesive or adhesive forces to hold parts together.

Brainstorming: the seeking of creative solutions to an identified problem.

Building: an enclosure to protect people, materials, and equipment from the elements.

Buoyancy: the upward force exerted by a fluid on an object.

Buttress dam: a non-solid dam that uses its structure to hold back the water.

Bylaws: a set of general rules by which a company will operate.

C

Cantilever bridge: bridge that uses trusses that extend out like arms to carry the load.

Carrier frequency: the frequency on which a radio communication message is imposed for transmission.

Casting and molding: the process in which a liquid material is poured into a cavity in a mold, where it solidifies.

Casting (theater): the procedure through which performers are selected.

Ceiling joists: beams that rest on the outside walls and some interior walls that support the weight of the ceiling.

Center lines: lines used to locate the center of a hole in drafting.

Channel: the carrier medium between a transmitter and a receiver.

Chart: a diagram that shows the relationship between people, actions, or operations.

Chemical conditioning: a method of changing the internal properties of a material through chemical reaction.

Chemical processing: breaking down or building up materials by changing their chemical composition.

Chief executive officer: person with final day-to-day control of a company.

Classroom training: job training using instructors in a classroom setting.

Closed-loop control: exists when feedback is used to control a system.

Coatings: a film of finishing material that is applied to a product or base material to protect or add color.

Cold bonding: a means of joining ductile metals, such as aluminum or copper, by applying very high pressure in a small area.

Cold composition: all methods of producing type except hot-metal composition. Includes transfer letters, strike-on type, and laser printer output.

Commercial buildings: buildings used for business and government purposes.

Commercial transportation: enterprises that move people and goods for money.

Commission: monetary reward paid to an employee based on the number of sales made.

Communication: the sending and receiving of information.

Comprehensive layout: the final layout for a design. Includes all information needed to produce the design.

Computer: a device consisting of input, processing, memory, and output units. Programs can be entered into memory so the computer can perform different tasks.

Computer-aided design (CAD): a computer-based system used to create, modify, and communicate a plan or product design.

Conceptual model: a model that shows a general view of the components and their relationships. Often the first step in evaluating a design solution.

Conditioning processes: processes in which the internal structure of a material is changed to alter its properties.

Conduction (electrical): the movement of electrons along a solid material.

Conduction (heat): the movement of heat through a solid material.

Construction: using technological actions to erect a structure on the site where it will be used.

Consumer: a person who utilizes economic goods.

Consumer product: product developed for the end users in the product cycle, such as home owners or students.

Contact printing: photographic printing procedure in which the negative is placed directly on top of a piece of light-sensitive paper.

Container ships: ships in which the cargo is loaded into large steel containers for faster loading and unloading.

Contrast: the degree of difference between items.

Control: the feedback loop that causes management and production activities to change.

Convection: the transfer of heat between two fluids (liquids or gases).

Convergent thinking: the refinement of ideas by narrowing and focusing until the most workable solution is found.

Copy: the words in printed communication.

Corporate charter: an operating permit that allows a company to do a specified business.

Corporation: a business owned by individuals who have purchased a portion, or share, of the company.

Creative abilities: the skills that are used by people to solve problems.

Criteria: the characteristics a product or structure must have in order to meet the expectations of the customer.

Crossbands: the layers in plywood with grain running at a 90° angle to the grain of the face layers.

Cutting motion: relative movement between the workpiece and the tool that causes material to be removed.

Cutting tool: a tool used for cutting material. The cutting tool must be harder than the material it is cutting and have the correct shape.

Cylindrical grinders: machines that use the lathe principle to machine the material. The workpiece is rotated against a grinding wheel that is rotating in the opposite direction.

D

Data: the raw facts and figures collected by people and machines.

Decode: to change a coded message into an understandable form.

Degrees of freedom: the limited number of ways or directions that a vehicle can move.

Depth of field: the range of distances in which the camera will capture objects in focus.

Desktop publishing: a simple computer system that produces type and line illustration layouts for printed messages.

Detail drawings: engineering drawings that provide information needed to produce a particular part.

Detailed sketch: a sketch that communicates the information needed to build a model of the product or structure.

Developing: the chemical treatment of light-sensitive materials in film emulsion to bring out the image.

Development: the work of technologists to build products and structures.

Diagnosis: determining a cause, often of a problem.

Diaphragm: the part that regulates the amount of light that enters a camera.

Dies: shaping devices in which material is squeezed between or formed over to achieve a new shape.

Dipping: a finishing method in which the product is dipped into a vat of molten metal or liquid coating.

Direct current: electron flow in conductor that moves in only one direction.

Dirigible: a rigid airship with a metal frame that is covered with a skin of fabric.

Divergent thinking: a type of thinking that seeks to create as many different (divergent) solutions as possible. The most promising solutions are then refined until one "best" answer is found.

Dividend: a periodic payment from a company's profits.

Domestic transportation: transportation within the geographic boundaries of one country.

Drafting standards: a set of rules used in drafting. The rules standardize the drawings so that they can communicate effectively.

Drawing machines: machines that stretch metal into desired shapes.

Drilling: a machining operation that produces or enlarges holes with a rotating cutter.

Drilling machine: a machine that produces a straight, round hole in a workpiece.

Dry cargo ships: ships that are used to haul both crated and bulk cargo.

Duplex system: a mobile communication system that uses two channels for broadcasting and receiving.

E

Economic activity: business efforts directed toward making a profit.

Economic enterprise: an enterprise that engages in economic activity.

Edutainment: communication that creates a situation in which people want to gain information.

Elastic range: the range between a material at rest and the material's yield point.

Electrical discharge machining: a process that uses a spark (electrical discharge) to erode a small chip from the workpiece. Used widely to make cavities for forging and stamping dies.

Electrochemical processing: breaking down or building up materials by changing their chemical composition using electrochemical processes.

Electronic publishing: a computer system that combines text and illustrations into one layout.

Enamel: a varnish to which a pigment has been added to produce a colored coating.

Encode: to change a message into a form that can be transmitted.

Energy: the ability to do work.

Engineering: process that develops the specifications for products, structures, processes, and services.

Engineers: people who apply scientific and technological knowledge to the design of products, structures, and systems.

Entrepreneur: a person who organizes, manages, and assumes the risks of a business enterprise.

Ergonomics: the science of designing products and structures around the people who use them. Also called human factors analysis.

Exhaustible resources: resources that, when used up, cannot be replaced.

Expendable molds: inexpensive molds that are destroyed when the casting is removed.

Expense: money that a company spends to purchase goods and services.

Exploded view: shows the parts that make up a product as if it were taken apart.

Extrusion: a process in which material is pushed through a hole in a die.

F

Fact script: script that develops a list of facts or characteristics.

Fare: the cost of a ticket on commercial transportation.

Fascia: board used to finish the ends of the rafters and an overhang.

Fasteners: permanent, semi-permanent, or temporary devices that hold parts together.

Feed motion: the action that brings new material into the cutter.

Feedback: using information about the output of a system to regulate the inputs to a system.

Fermentation: a process that uses yeast to decompose a material.

Fiberglass: strands of glass that are formed, cooled, and annealed. Used as the matrix for composite materials and as insulation.

Filmstrip: a series of transparencies designed to be viewed one at a time in fixed sequence.

Financial affairs: the area of managerial technology concerned with raising money and maintaining financial records.

Finishing: coating or modifying the surface of parts and products to protect them, or to make them more appealing to the customer.

Finishing processes: processes which give a product a protective or decorative coating or surface treatment.

Fission: the release of energy through the splitting of an atom.

Fixed wing aircraft: the most common aircraft. Uses fixed airfoils to generate its lift.

Fixing: the procedure that removes the unexposed silver halide crystals from film.

Flame cutting: a process in which a mixtures of gases is burned to melt a path between the workpiece and the excess material.

Flexography: an adaption of the letterpress printing process. Uses a plastic or rubber image carrier.

Float glass: a substance formed by floating molten glass on a bed of molten tin.

Floor joists: beams that carry the weight of the floor.

Flow bonding: a method of joining materials that uses a filler metal that melts onto a heated base metal. When cool, the filler metal acts as an adhesive.

Flow coating: a process that floods a surface with a finishing material. The excess is allowed to drip off the surface.

Fluid mining: mining technique in which hot water is pumped down a well. The water dissolves and forces mineral deposits up a second well.

Forming: a process that changes the shape and size of a material by a combination of force and a shaped form (such as a die or forming rolls).

Fossil fuels: mixtures of carbon and hydrogen produced by decaying or decayed organic material. Commonly used as energy sources.

Fractional distillation: procedure used in the separation process of petroleum refining.

Fracture point: in forming, the point at which force is strong enough to break the material.

Frequency: the number of cycles that pass some point in one second.

Frequency modulation: system in which a message is encoded on a carrier wave by changing its frequency.

f-stop: a number that indicates the size of the opening of the diaphragm in a camera.

Fuel converter: a device that converts a fuel into energy.

Full script: a script that contains all words, sound effects, visual effects, and production notes.

Fuselage: the part of the body of an aircraft that contains the crew, passenger, and cargo units.

Fusion: the release of energy through the combining of two atoms into a new, larger atom.

Fusion bonding: a type of bonding that melts the edges of the materials being joined so that they can flow together.

G

Galvanized steel: a zinc coated steel.

Gasification: a process in which materials are heated in the absence of oxygen to produce usable fuels. Also known as pyrolysis and liquidification.

General aviation: travel for pleasure or business in an aircraft owned by a person or business.

Genetic materials: organic material that is obtained during the normal life cycle of plants or animals.

Geosynchronous orbit: an orbit in which a satellite travels the same speed at which the earth is turning.

Germination: the origin or birth of genetic materials.

Glass: a substance made primarily of silica (70 percent), lime (13 percent), and soda (12 percent). Made by solidifying molten silica in an amorphous state.

Graphic communication: type of communication that uses two-dimensional visual messages.

Graphic model: a model used to explore ideas for components and systems. Examples are conceptual drawings, graphs, charts, and diagrams.

Gravity dam: a dam that uses the sheer weight of its concrete to hold water back. The water side is vertical, while the opposite side slopes outward.

Gravure printing: printing process that uses an etched or scribed image carrier.

Green sand casting: casting using an expendable mold made of sand that is bound together with oil or water.

Grievance: a formal complaint by a worker about a manager's violation of the union contract.

Guidance system: a group of instruments that provide information to the operator of a vehicle.

H

Hardening: a process used to make a material more resistant to denting, scratching, or other damage.

Hard-wired system: a system that sends its signals through a physical channel.

Harmony: the blending of the parts of a design to make a pleasing message.

Heat pump: a device that takes heat from one area or environment and places it into another area or environment.

Heavy engineering structure: a structure that supports transportation and communication systems.

Hertz: the unit measure of frequency. Equal to one cycle per second.

High technology: new technologies that are not in wide use today. These technologies may become common in time.

Horsepower: measurement of power that is equal to the force needed to move 500 pounds a distance of one foot.

Hot-metal composition: casting molten metal into type to produce a message.

Hovercraft: a boat that is suspended on a cushion of air. Often used in shallow water.

Hub-and-spoke system: a system made up of small local airports on the ends of two-way routes that radiate from a large regional airport.

Human-to-human communication: people communicating information to people.

Human-to-machine communication: people communicating information to machines.

Hydraulics: system that uses a liquid medium for power transfer.

Hydroelectric: electrical energy generated from a water source.

Hydrofoil: a boat fit with special underwater "wings," which raise the hull out of the water while the boat is traveling at high speeds.

I

Idea: a mental image of what a person thinks something should be.

Ideation: the thinking up of solutions

Illustrations: the pictures and symbols that add interest and clarity to printed communication.

Image carrier: a device that places an image onto a substrate.

Inclined plane: a sloping surface. Inclined planes are used to move an object from one level to another.

Induction: the process through which the change in a magnetic field induces current in a nearby conductor.

Industrial buildings: buildings that house and protect the machines that make products.

Industrial material: the input materials are the result of primary processing activities. They are used to manufacture products for both industrial and consumer markets.

Industrial product: items and products used by a company in conducting business.

Industrial relations: the area of managerial technology concerned with the human aspects of the manufacturing enterprise, such as personnel and labor relations.

Industry: the area of economic activity that uses resources and systems to produce products, structures, and services with intent to make a profit.

Inexhaustible energy source: a source of energy that is not depleted by use. Solar energy is an example.

Inexhaustible resources: resources that will last indefinitely.

Information: organized data.

Infotainment: a program provides information in an entertaining way.

Inland waterway: water transportation along rivers, lakes, and coastal waterways.

Inorganic material: material that does not come from living organisms. For example: metals and ceramics are inorganic.

Inputs: materials that flow into the system and are consumed or processed by the system.

Insolation: the solar energy available in a specific location at any given time.

Interference: anything that hinders the accurate communication of a message.

Intermodal shipping: the situation in which people or cargo travel on two or more modes of transport before they reach their destination.

International transportation: transportation that moves passengers and cargo between two nations.

Interstate commerce: business dealings that extend across state lines and are subject to additional regulations.

Investment casting: a process that uses plaster poured around a wax pattern. When heated, the wax melts away leaving the plaster mold.

Ionosphere: the region that extends from 50 to 62 miles (80 to 99 km) above the earth. Also called the thermosphere.

Isometric sketch: a type of sketch where the angles formed by the lines in the upper right hand corners are equal.

J

Jet engine: an engine that projects a steam of fluid rearward to produce motion.

Joule: the metric unit for measuring work, equal to one newton per meter.

Just-in-time (JIT) manufacturing: a system that reduces inventories of raw materials and product components to the absolute minimum. Materials and parts are scheduled to arrive at the production line just when they are needed for use.

K

Kinetic energy: energy in motion.

Knowledge: information applied to a task.

L

Land transportation: systems that move people and goods on the surface of the earth.

Landscaping: the activities that help prevent erosion and improve the appearance of a land area.

Laser machining: a cutting method that uses a beam of focused light to melt a path in a material and separate the excess from the workpiece.

Lathe: a machine that produces cutting motion by rotating the workpiece.

Layout: the physical act of designing a message.

Lens: a piece of translucent material that focuses light.

Letterpress: a printing process that uses metal plates or metal type as the image carrier.

Lever: a simple machine that multiplies the force that is applied to it.

Lifestyle: what a person does with their business and family life.

Light meter: a device that measures the amount of light available for a photo.

Lighter-than-air vehicles: vehicles that use either a light gas or hot air to produce lift.

Limited liability: the situation in which the owner(s) of a business are not monetarily responsible for all debts that a business acquires.

Linear motion: movement in a straight line; often used in reference to a machining operation.

Liquefaction: a process in which a biofuel is heated at moderate temperatures under high pressure.

Liquid: visible, fluid materials that will not normally hold their size and shape. Liquids cannot be easily compressed.

Liquidification: a process in which materials are heated in the absence of oxygen to produce usable fuels. Also known as pyrolysis and gasification.

Locomotive: a unit on a rail system that contains the power and operator components.

M

Machine tools: machines that are used to make other machines.

Machines: simple machines change the speed, direction, or amount of a force. Complex machines are groups of mechanisms that work together to perform a task.

Machine-to-human communication: machines communicating information to people.

Machine-to-machine communication: machines communicating information to machines.

Machining: a process that changes the size or shape of a workpiece by removing excess material as chips or shavings.

Maintenance: the process of keeping products in good condition and in good working order.

Management processes: actions that ensure production processes operate efficiently and appropriately. These processes are also used to direct the design, development, production, and marketing of the technological device, service, structure, or system.

Manual control: a system that requires human supervision and adjustment.

Manufactured home: a home in which most of building is built in a factory. The home is then transported to a foundation.

Manufacturing: the process that takes industrial materials and converts them into products for an end user.

Marketing: promoting, selling, and delivering a product, structure, or service.

Material processing: changing the form, shape, and size of materials using tools and machines.

Mathematical model: a model that shows a relationship in terms of a formula.

Measurement: the process of objectively describing the physical qualities of an object.

Mechanical conditioning: changes to the internal structure of a material as a result of manufacturing processes. The changes are not always suitable; the material may have to be conditioned further to restore it to the desired state.

Mechanical processing: using mechanical forces to change the form of natural resources.

Mechanisms: basic devices that are used to power or adjust equipment and machines.

Merchant ships: ships that carry cargo.

Mesosphere: the region that extends from 22 to 50 miles (35 to 80 km) above the earth.

Metric system: a system of measurement based on a unit of length called a meter. The metric system is used throughout the world.

Milling machine: a machine that uses a rotating cutter for the cutting motion.

Mock-up: a model designed to show people how a product or structure will look. Used to evaluate the styling, balance, color, or other aesthetic features of a technological artifact.

Modeling: the activity of simulating expected conditions to test design ideas.

Motion pictures: a series of transparencies that can produce the illusion of motion when viewed.

Multiple-point tool: a tool with more than one cutting edge, such as a saw blade.

Multiplex system: a system that allows several unrelated messages to travel down a single conductor at the same time.

N

Natural material: a material that occurs naturally on earth.

Noise: unwanted sounds or signals that become mixed in with desired information.

Non-traditional machining: machining processes that use electrical, sound, chemical, and light energy to size and shape materials.

O

Oblique sketch: a sketch that shows the front view as if a person was looking directly at it. The sides and top extend back from the front view.

Offset lithography: a printing process that uses the principle that oil and water do not mix.

One-view drawings: drawings used to show the layout of flat sheet metal parts.

On-the-job training: a method of teaching skills at the workstation, often under the guidance of an experienced worker.

Open-loop control: exists when a system runs without any feedback.

Opportunity: exists when a technological device can be sold to consumers even though an existing device serves the purpose.

Ores: earth or rock from which metal can be commercially extracted.

Organic material: a material that comes from living things, such as cotton or wool.

Organizing: a function of management that develops a structure so that goals can be reached.

Orthographic assembly drawings: drawings that use a single view to show the mating of the parts.

Orthographic projection: drawing showing a component or product from the front, the side, and (often) the top.

Outputs: the results, good and bad, of the operation of any system.

P

Pagination: complex computer system that allow the operator to accurately merge text and illustrations.

Panchromatic: the most common form of black-and-white film.

Partnership: a business owned and operated by two or more people.

Passenger ships: ships that carry people.

Pasteup: act in which all type and illustrations are assembled onto a sheet that looks like the finished message.

Pathways: the structures along which vehicles travel, such as roads, railways, etc.

Perigee: the closest point to the earth that a satellite will reach in an elliptical orbit.

Permanent molds: molds that can be used again and again to cast or mold a part.

Personal transportation: travel using a vehicle owned by one person.

Perspective sketch: a sketch that shows the object as the human eye or a camera would see it.

Petroleum: a mixture of a large number of hydrocarbons found in various parts of the earth. It can be refined into gasoline and other products.

Photographic communication: the process of using photographs to communicate a message.

Photovoltaic cell: a device that converts solar energy into electricity.

Physical model: a three-dimensional representation of reality.

Pictorial assembly drawings: drawings that show an assembly using oblique, isometric, or perspective views.

Planning: the process of developing goals and objectives.

Plastic range: the range in which a force will bend, stretch, or compress a material but not fracture or break it.

Pneumatics: system that uses air as a medium for power transfer.

Point of interest: the place where your eye is drawn in a photograph.

Political power: the ability of a person or organization to gain the attention and cooperation of elected officials.

Pollution: the release of substances into the environment due to human activity.

Polymerization: the linking together of chain-like molecules of a plastic, changing it from a liquid to a solid.

Potential energy: stored energy that has the ability or potential to do work.

Power: the rate at which work is done.

Precision measurement: a type of measurement that is used when size is critical to the function of a device. Accuracy is very important.

Price: what a person must pay to buy or use a product or service.

Primary processing: the steps in a production process that come between obtaining the resource and manufacturing the product or constructing a structure.

Prime interest rate: the rate of interest that banks and insurance companies charge their best customers.

Prime mover: a device that changes a natural source of energy into mechanical power.

Printed graphic communication: communication through a printed medium.

Problem: exists when people encounter a difficulty or obstacle.

Problem solving: the process that is used to develop solutions to a problem or opportunity.

Processes: the steps needed to complete a series of tasks. The actions undertaken to create structures and products.

Production: developing and operating systems that produce products and structures, or provide services.

Production processes: actions taken to fulfill the function of a technological system.

Production worker: a person who processes materials, erects structures, operates transportation vehicles, services products, or produces and delivers communication products.

Profit: the amount of money left over after all the expenses of a business have been paid.

Projection printing: photographic printing procedure in which light is projected through a negative onto light-sensitive paper.

Proportion: the height-width relationships of parts within a design.

Proprietorship: a business owned by a single owner.

Propulsion: the system on a vehicle that converts energy into motion.

Prototype: a working model of a new product, intended to test its operation.

Pyrolysis: a process in which materials are heated in the absence of oxygen to produce usable fuels. Also known as liquidification and gasification.

Q

Quality control: the process of setting standards, measuring the size of a product, comparing the size to the standards, and making adjustments.

R

Radiation: the transfer of heat using electromagnetic waves.

Receiver: a device that acquires and decodes a signal.

Receiving: recognizing and changing coded information into a usable form.

Reciprocating motion: a back and forth movement.

Recruiting: the process of attracting potential workers to a company.

Recycling: the process of reclaiming or reusing old materials to make new products.

Refined sketch: a sketch that brings together the best ideas from rough sketches.

Renewable energy source: a source of energy that can be replaced with the life cycle of organisms, such as trees or animal power.

Renewable resources: biological resources that can be grown and harvested.

Repair: the process of putting a product back into good working order.

Research and development: the process of designing,

developing, and specifying the characteristics of a product, structure, or service.

Residential buildings: buildings in which people live.

Retained earning: the portion of profits that a company uses to expand its operations or to develop new products.

Rhythm: the flow of a communication message.

Rotary motion: motion around a central axis.

Rotary wing aircraft: an aircraft that develops lift by spinning an airfoil, such as a helicopter or a gyrocopter.

Rough sketch: the type of sketch that a designer uses to work through ideas.

Route: the path that a vehicle follows.

S

Salary: a rate of payment based on a long period of time (month, year).

Sawing machines: machines that use a blade with teeth to cut material.

Schedule: list of the departure and arrival times for trips on commercial transportation systems.

Screening: the selection process that pulls qualified personnel out of a pool of job applicants.

Screw: a mechanism that uses an inclined plane wrapped around a shaft. It is a force multiplier.

Script: the object that identifies the characters, develops a situation, and communicates a story.

Secondary manufacturing processes: the steps in the manufacturing process in which industrial materials are converted to useful products.

Semi-skilled worker: workers who perform tasks that require a limited amount of training. These workers are usually trained on the job.

Separating: the process of using tools to remove unwanted material.

Shearing: a process that uses force applied to opposed edges to fracture excess material away from the workpiece.

Shell molding: uses a sand and resin mixture poured over a heated metal pattern. The heat melts the resin, bonding the sand into a thin shell.

Show format script: a script that lists the various film or show segments.

Shutter: the device that permits or prevents light from entering a camera.

Silkscreening: printing through a stencil mounted on a fabric screen. Also called screen printing.

Simplex systems: a mobile communication system that uses the same channel for broadcasting an receiving.

Single-point tool: a tool with only one cutting edge.

Skilled worker: highly skilled individuals that have extensive training and work experience. They receive training in special schools or through apprenticeship programs.

Slide: a single transparency designed to be viewed independently.

Smelting: using heat to extract metals from their ores.

Solar converter: an energy converter that uses the sun as its energy source.

Solar weather system: the cycle of inexhaustible energy that originates from the sun.

Solid: a material that holds its size and shape and can support loads.

Solid model: a computer representation that takes into account both the surface and the interior substance of an object.

Space transportation: systems that move people and goods outside the earth's atmosphere.

Square: when the angle between two surfaces is 90°.

Stamping: process in which both forming and cutting are done to a material.

Standard measurement: a type of measurement used when the size of a part is not critical to the function of a product.

Steel: an alloy, or mixture, of iron and carbon.

Stock: a portion, or share, of a company.

Stop bath: an acidic solution that neutralizes the photographic developing solution.

Stratosphere: the region that extends from 7 to 22 miles (11 to 35 km) above the earth.

Submersible: a vessel that can travel on the surface or underwater.

Substrate: a material on which an image is printed.

Surface grinding: a machine that moves the work under a spinning grinding wheel.

Surface model: a computer model in which an object is defined only by its surface.

Suspension bridge: bridge that uses cables to carry the load. Often used for bridges that span great distances.

Suspension system: system that keeps a vehicle in contact with a road or rail and, also, separates the passenger compartment from the drive system for passenger comfort.

Synthetic material: materials that are made by humans, such as plastics. They do not occur naturally.

T

Technical graphic communication: information communicated through engineering drawings or technical illustrations.

Technical manager: a manager who has the background of technicians and engineers.

Technician: a person who sets up and repairs equipment, services machinery, conducts product tests and laboratory experiments, or works in quality control.

Technologist: a person who helps manufacturing and construction personnel apply engineering designs and solve problems.

Technology: humans using objects (tools, machines, systems, and materials) to change the natural and human-made (built) environment. It is conscious, purposeful actions by people designed to extend human ability or potential to do work.

Telecommunication: communication over a distance.

Tempering: a process that relieves the internal strains in a material that are caused by hardening or other treatments.

Terminal: the area where passengers and cargo are loaded onto and unloaded from vehicles.

Thermal conditioning: a process that changes the internal structure of a material through controlled heating and cooling.

Thermal processing: using heat to melt and reform a natural resource.

Three-view drawings: drawings used to show the size and shape of rectangular and complex parts.

Thumbnail sketch: a basic sketch of a design that allows the designer to experiment with different ideas.

Tolerance: the amount a dimension can vary and still be acceptable.

Tool: a simple object that is used by humans to complete a task.

Transducer: a device that changes energy of one form into energy of another form.

Transformation: the process of changing resources into a more usable form.

Transmit: sending coded transmissions to a receiver.

Transmitter: a device that codes and sends a communication message.

Transportation: all acts that relocate humans or their possessions.

Troposphere: the region that includes the first six miles (9.7 km) of space above the surface of the earth.

Truss bridge: a bridge that uses small parts arranged in triangles to support the deck. Commonly used for railroad bridges.

Turning machine: a machine that rotates the work against a single-point tool to produce the cutting motion.

Twist drill: a cutting device with points on the end of a shaft.

Two-view drawings: drawings used to show the size and shape of cylindrical parts.

Typesetting: function which produces the words of a message. Includes generating, justifying, and producing the type.

U

UHF: ultra high frequency.

Unlimited liability: the situation in which the owner(s) of a business are monetarily responsible for all debts that a business acquires.

Unskilled worker: a person who performs tasks that require only a minimum of training.

Utilities: the systems of a structure that provide water, electricity, heat, cooling, and communications.

V

Value: a measure of the functional worth that a customer sees in a product.

Vehicle: a powered carrier that supports, protects, and moves cargo or people within a transportation system.

Veneer: a thin sheet of wood that is sliced, sawed, or peeled from a log.

VHF: very high frequency.

Video: the visual part of a broadcast message.

W

Wage: a set rate paid for each hour worked.

Waste water system: system that carries the used water from sinks, showers, toilets, and washing machines into the sewage system.

Water transportation: systems that use water to support the vehicles moving people and goods.

Waterwheel: a series of paddles that extend into flowing water, which produces a rotating mechanical motion.

Watt: unit measure of power, equal to one joule per second.

Wavelength: the distance from the beginning to the end of one wave cycle.

Wheel and axle: a lever mechanism that has a shaft attached to a disk. They can be either a distance or a force multiplier.

Windmill: a wind-driven wheel that produces a rotating mechanical motion.

Wing: an airfoil on a fixed wing aircraft. Develops the aircraft's lift.

Wireframe model: a series of lines, arcs, and circles that together form the outline of an object in a computer model.

Work: applying a force that moves an object a distance in the direction of the applied force.

Y

Yield point: the lowest point at which a material can be permanently deformed by and applied force.

INDEX